Edexcel GCSE Mathematics

16+

Karen Hughes

Peter Jolly

David Kent

Keith Pledger

Heinemann Educational Publishers
Halley Court, Jordan Hill, Oxford OX2 8EJ
a division of Harcourt Education Limited

Heinemann is a registered trademark of
Harcourt Education Limited

OXFORD MELBOURNE AUCKLAND
JOHANNESBURG BLANTYRE GABORONE
IBADAN PORTSMOUTH (NH) USA CHICAGO

First published 2002

07 06 05 04 03 02
10 9 8 7 6 5 4 3 2 1

British Library Cataloguing in Publication Data is available
from the British Library on request.

ISBN 0 435 53284 7

Designed by Gecko Ltd
Typeset by Tech-Set Ltd, Gateshead, Tyne and Wear
Original illustrations © Harcourt Education Limited, 2002
Illustrated by Tech-Set Ltd and Barking Dog Art
Cover design by Miller Craig and Cocking Ltd
Printed and bound in Italy by printer Trento
Cover photo: © Stone/Tom Till
Picture research by Jenny Silkstone

Acknowledgements
The publishers' and authors' thanks are due to Edexcel for permission to reproduce questions from past examination papers. The answers have been provided by the authors and are not the responsibility of the examining board.
Every effort has been made to contact copyright holders of material reproduced in this book. Any omissions will be rectified in subsequent printings if notice is given to the publishers.

Photo acknowledgements
p1 Corbis/Amos Nachoum; p22 Taxi/Juergen Mueller; p36 Corbis/Patrick Ward; p44 Source Unknown; p59 Stone/Frank Herholdt; p144 Stone/Alan Thornton; p147 Corbis/Patrick Ward; p156 Corbis/James Marshall; p167 Stone/Steve Outram; p173 Edifice/Philippa Lewis; p176 PhotoDisc/Ryan McVay; p185 PhotoDisc/Jeff Maloney

Tel: 01865 888058 email: info.he@heinemann.co.uk

Contents

About this book

This book is carefully designed and structured for 16+ students to help you through your Edexcel GCSE mathematics course in one year. The short units for each of the four key areas are:

- Number
- Algebra
- Shape, space and measure
- Data handling

The book leads you step by step through the key knowledge, skills and concepts you need to obtain a grade C in the Edexcel GCSE mathematics intermediate tier. Each unit gives a clear explanation of the key ideas and techniques, with exemplar questions and authored solutions. These are followed by practice exercises, so that you can check your understanding before moving on. The exercises are carefully graded and include past paper questions marked with an *E*. Answers are given at the end of the book.

Teaching points giving guidance on how exemplar solutions have been worked out are shown in the margin like this:

In fraction questions, 'of' means the same as multiply.

Examiners' hints are shown in boxes like this:

Hint: You need to learn the answers to questions 1–3 for the examination.

Worksheets giving you access to higher level questions (enabling you to gain a grade B) are included in the Teachers' Resource Pack. Look for the link (E). You will also see links marked (S) to indicate that a support worksheet is available for that topic.

Coursework counts for 20% of your final mark. This consists of a handling data project and an investigation. Help with coursework can be found in the Teachers' Resource Pack and on the web at www.heinemann.co.uk.

Learn the language

Angles

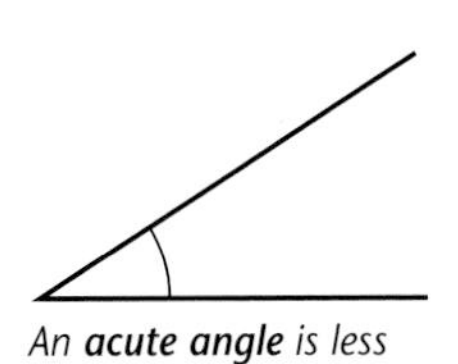

*An **acute angle** is less than 90°.*

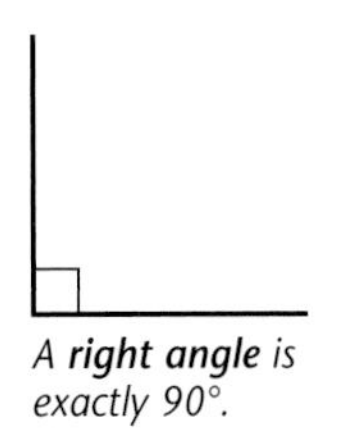

*A **right angle** is exactly 90°.*

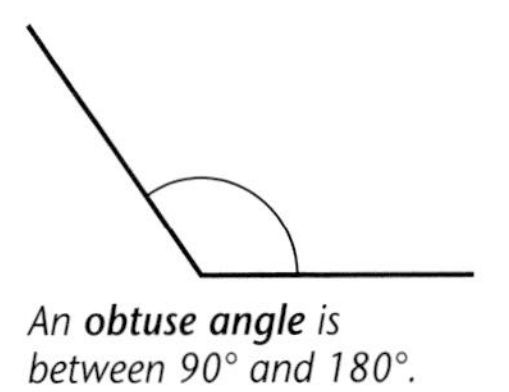

*An **obtuse angle** is between 90° and 180°.*

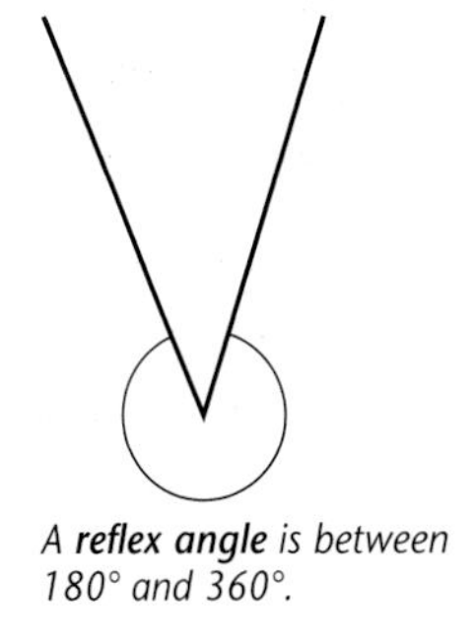

*A **reflex angle** is between 180° and 360°.*

Triangles

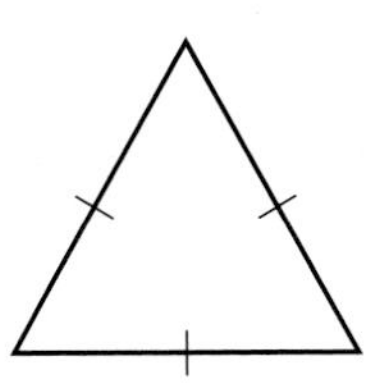

*An **equilateral** triangle has 3 equal sides and 3 equal angles.*

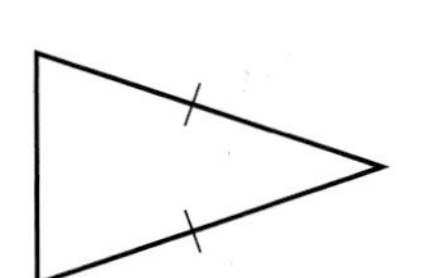

*An **isosceles** triangle has 2 equal sides and 2 equal angles.*

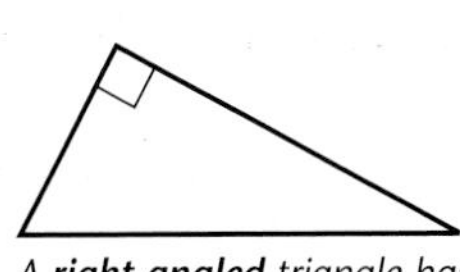

*A **right angled** triangle has one angle of 90°.*

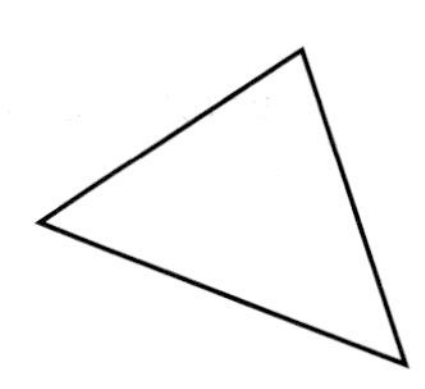

*A **scalene** triangle has no equal sides.*

Triangles are **similar** if

- the corresponding angles are equal, or
- the lengths of corresponding sides are in the same ratio.

*These triangles are **similar**:*

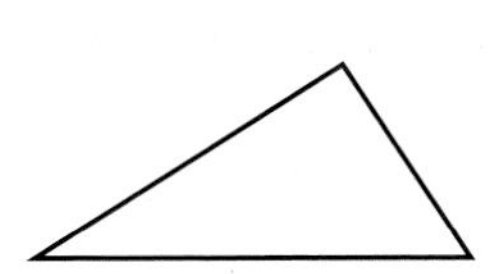

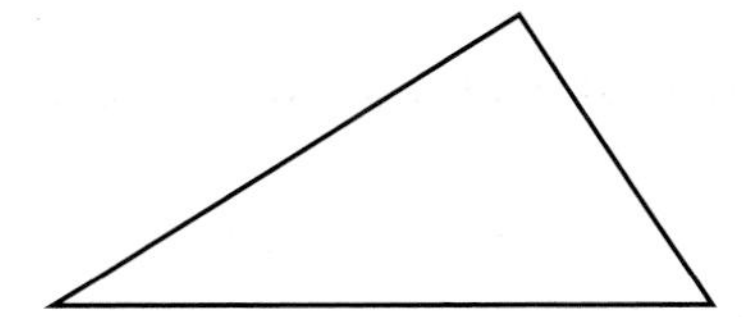

Congruent triangles have the same shape and size, but one may be a reflection or rotation of the other.

*These triangles are **congruent**:*

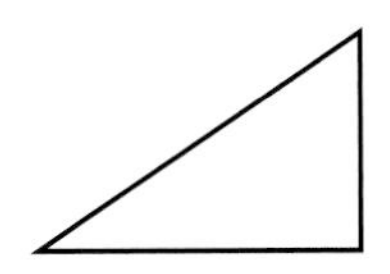

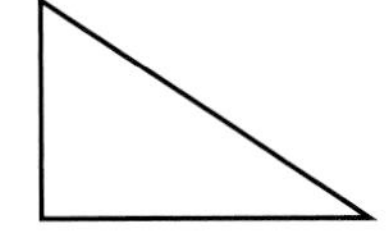

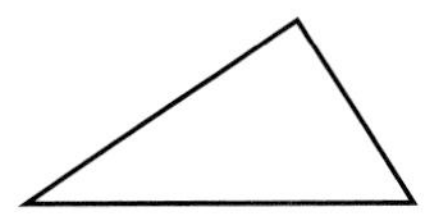

Quadrilaterals

A quadrilateral has 4 sides and angles that add up to 360°

These are special quadrilaterals:

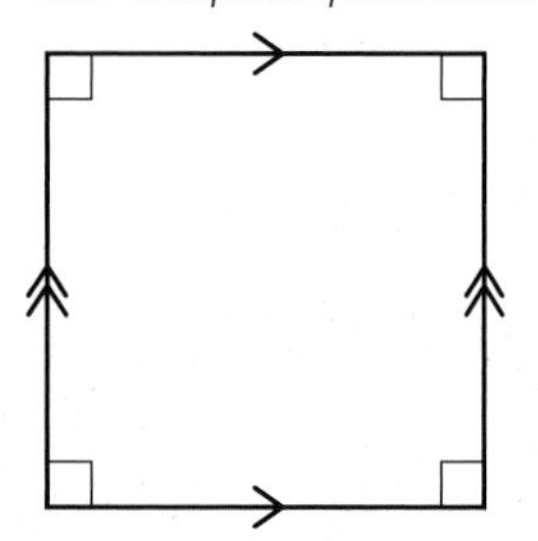

Square *– 4 equal sides, 4 × 90° angles.*

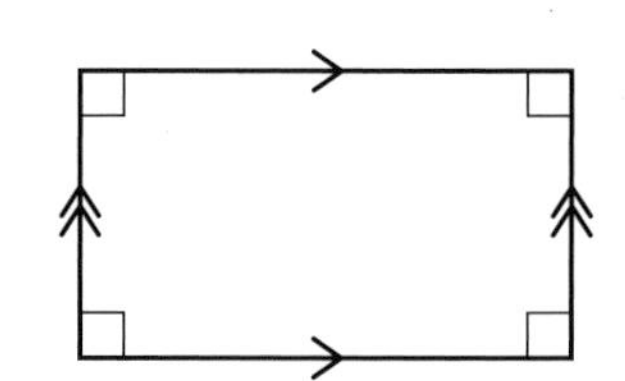

Rectangle *– opposite sides equal, 4 × 90° angles.*

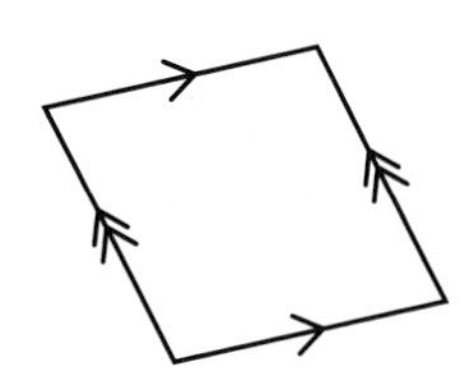

Rhombus *– 4 equal sides, opposite sides parallel, opposite angles equal.*

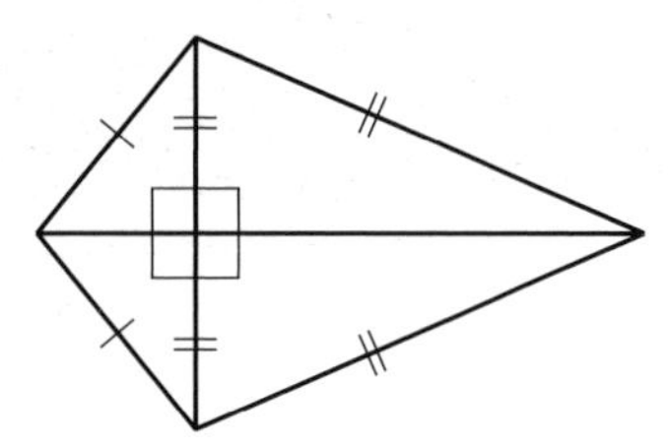

Kite *– 2 pairs of adjacent sides equal.*

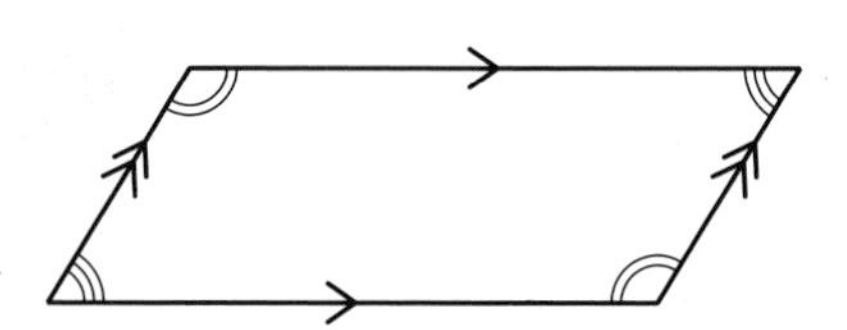

Parallelogram *– both pairs of opposite sides parallel, and equal length. Opposite angles equal.*

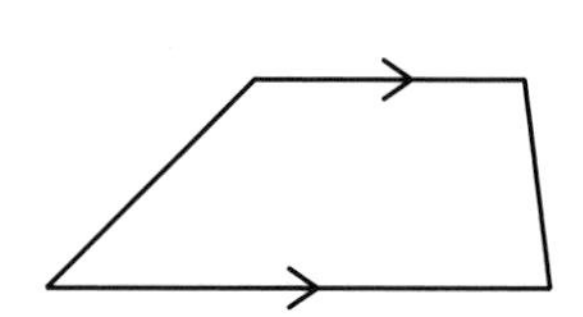

Trapezium *– 1 pair of opposite sides parallel.*

Circles

Circumference – the distance around the edge of a circle.
Chord – a straight line joining two points on the circumference, eg the straight line AB
Diameter – a chord that goes through the centre, eg CD
Radius – a line from the centre to a point on the circumference, eg OC, and OD and DE
Arc – part of a circumference, eg the curve AB.
Tangent – a line touching a circle or curve, eg FG
Segment – an area between a chord and the circumference
Sector – an area between two radii

Hint: SEC TO R
two radii

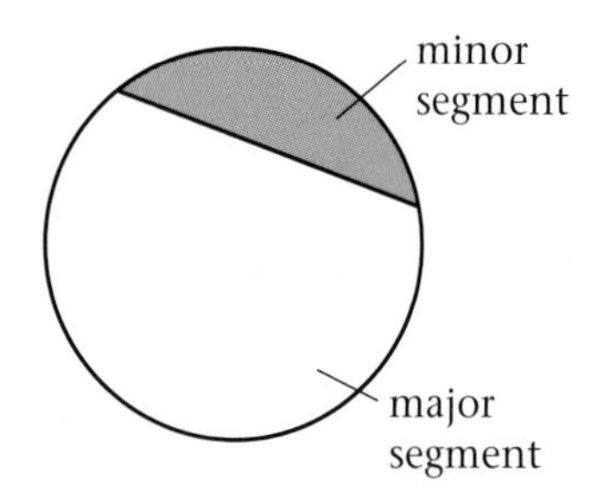

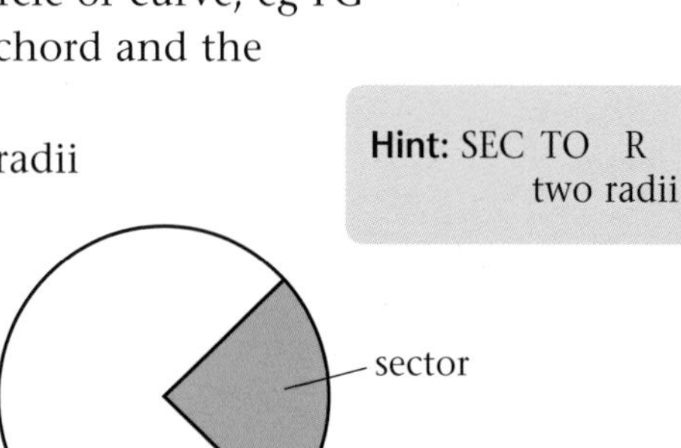

1 Integers

This chapter is about whole numbers (integers), their properties and how to use them when solving problems.

1.1

You need to be able to add and subtract negative numbers. A number line can help you do this.

Example 1

Work out:

a $4 + -3$ **b** $-4 - -2$

c $3 - +2$ **d** $-5 + +3$

e $-3 + -2$

Negative numbers are useful when describing distances above and below sea-level.

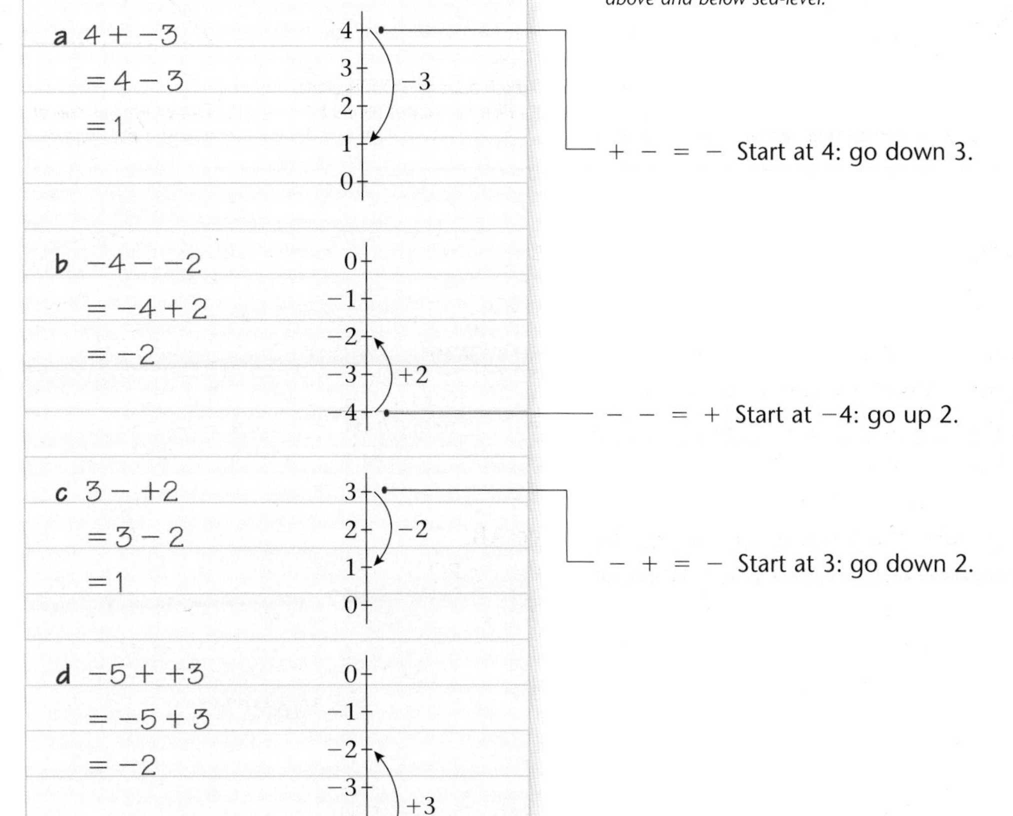

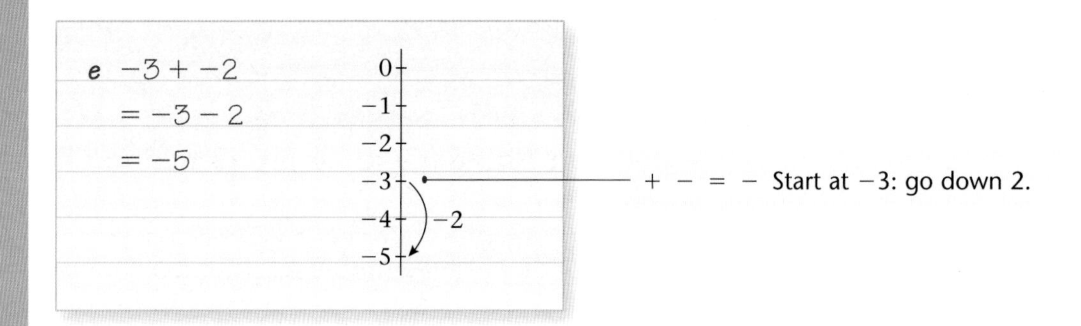

Practice 1A

1 Work out:

a $-6 - -8$ **b** $-10 + 3$ **c** $18 + -2$ **d** $20 - -4$
e $-18 - -4$ **f** $-30 - -4$ **g** $30 + -11$ **h** $50 - -18$
i $-25 + -18$ **j** $-15 - -6$

2 Complete the following tables:

a

2nd number — add on this

start with this — 1st number

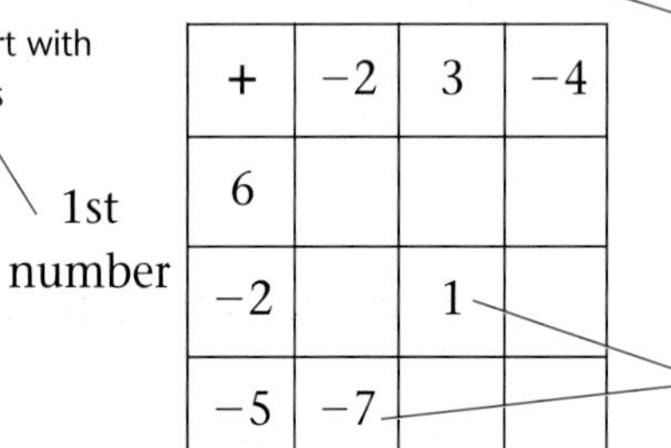

+	−2	3	−4
6			
−2		1	
−5	−7		

these have been done for you

b

2nd number — subtract this

1st number

−	−4	0	5
−3	1		
8			
3			−2

1.2

You need to be able to multiply and divide negative numbers.
This table shows the signs you get:

+	×/÷	+	=	+
+	×/÷	−	=	−
−	×/÷	+	=	−
−	×/÷	−	=	+

Example 2

Work out:

a 6×-4 **b** $-8 \div -2$ **c** $-6 \times +3$ **d** $12 \div -4$

a $6 \times -4 = -24$ — $+ \times - = -$

b $-8 \div -2 = +4$ — $- \div - = +$

c $-6 \times 3 = -18$ — $- \times + = -$

d $12 \div -4 = -3$ — $+ \div - = -$

Practice 1B

1 Work out:

a -3×-4 **b** -4×6 **c** $27 \div -3$ **d** $-21 \div 7$
e $-36 \div -12$ **f** -9×-4 **g** 8×-3 **h** $48 \div -6$
i $-18 \div 3$ **j** -12×4

2 **a**

2nd number; 1st number

×	−3	2	−6
8			
−6			
4			

b

2nd number; 1st number

÷	−2	−3	6
−12			
36			
−18			

1.3 You can work out meaningful approximations of numbers by rounding to a given number of significant figures and estimating.

Example 3

Write the following numbers correct to the number of significant figures (sf) given in the brackets:

a 637 (1 sf) **b** 5878 (3 sf) **c** 6451 (2 sf) **d** 2986 (2 sf) **e** 0.034 (1 sf)

a 637 correct to 1 sf = 600

The first significant figure is the 1st non zero number, counting from the left. Here, the first significant figure is 6.

Use zeros to maintain place value.

b 5878 (3 sf)
5878 correct to 3 sf = 5880

7 is the 3rd sf.
8 is greater than 5 so round up.

c 6451 (2 sf)
6451 = 6500 (2 sf).

4 is the 2nd sf.
This digit is 5 so round up.

d 2986 (2 sf)
2986 = 3000 (2 sf).

9 is the 2nd sf.
8 is greater than 5 so round up.
Think of this as rounding up 29 to give 30.

e 0.034 (1 sf)
0.034 = 0.03 (1 sf)

3 is the 1st sf.

You must keep this zero to maintain the place value of the 3.

Example 4

Estimate the answer to $\dfrac{386 \times 5934}{752}$

$$\frac{386 \times 5934}{752} \approx \frac{400 \times 6000}{800}$$

Rewrite the question, writing all the numbers correct to 1 sf.

$$= \frac{2\,400\,000}{800} = 3000$$

Practice 1C

1 Write the following numbers correct to the number of significant figures given in the brackets:

a 589 (2 sf) **b** 8973 (3 sf)
c 74 531 (2 sf) **d** 6321 (1 sf)
e 237 (2 sf) **f** 63 567 (1 sf)
g 63.25 (3 sf) **h** 6.821 (2 sf)
i 0.037 (1 sf) **j** 0.00281 (2 sf)

Remember:
To find the 3rd (2nd/1st) sf, look at the 4th (3rd/2nd) digit from the left. If it is 5 or more round up, if it is less than 5 round down.

2 Write the following numbers correct to 1 sf:

a 578 **b** 6321 **c** 85
d 5.3 **e** 0.378

3 Write the following numbers correct to 3 sf:

a 78 952 **b** 6342 **c** 78.381
d 6.3452 **e** 0.0 028 465

4 For each of the following questions

i write down a calculation that can be used to give an estimate of the answer
ii work out the value of that estimate.

a 563×29 **b** $8942 \div 283$ **c** $\dfrac{43 \times 319}{58}$

d $\dfrac{7146 \times 392}{37}$ **e** $\dfrac{573 \times 61}{29 \times 49}$

5 There are 1.76 pints in 1 litre. Estimate how many pints there are in 13 litres.

6 There are 2.2 pounds in a kilogram. Estimate how many pounds there are in 5 kilograms.

1.4 You need to know how to carry out calculations involving squares, cubes, square and cube roots.

Example 5

Work out:

a 4^2 **b** 8^3 **c** $\sqrt{25}$ **d** $\sqrt[3]{8}$ **e** $\sqrt[3]{-27}$

a $4^2 = 4 \times 4$
$= 16$

4^2 means 4 squared.
To square a number multiply it by itself.
This can be done on a calculator by pressing [4] [x^2] [=]

b $8^3 = 8 \times 8 \times 8$
$= 512$

8^3 means 8 cubed.
To cube a number multiply it by itself twice. This can be done on a calculator by pressing [8] [x^y] [3] [=]

$\sqrt{25}$ means the square root of 25.
This is the positive square root of 25, because $5 \times 5 = 25$.
This is the negative square root, because $-5 \times -5 = 25$.
Press: [2] [5] [$\sqrt{x}$] [=]

$\sqrt[3]{8}$ means the cube root of 8.
You need to find a number that when multiplied by itself three times = 8.
$2 \times 2 \times 2 = 8$, so 2 is the cube root.
Press: [8] [$\sqrt[3]{}$] [=]

$-3 \times -3 \times -3 = -27$
Press: [2] [7] [+/−] [$\sqrt[3]{}$] [=]

Hint: You need to learn the answers to questions 1–3 for the examination.

Practice 1D

1 Work out:

a 1^2	**b** 3^2	**c** 6 squared	**d** 9 squared	**e** 5^2
f 11^2	**g** 7 squared	**h** 2^2	**i** 10^2	**j** 8 squared
k 4^2	**l** 12^2	**m** 14^2	**n** 13^2	**o** 15 squared.

2 Write down the value of:

a 3^3 **b** 5^3 **c** 4 cubed **d** 10 cubed **e** 2^6

3 Write down the value of:

a $\sqrt{25}$ **b** square root of 36 **c** $\sqrt{16}$

d $\sqrt{196}$ **e** $\sqrt{100}$ **f** $\sqrt{225}$

g $\sqrt{144}$ **h** square root of 49 **i** $\sqrt{9}$

j square root of 121 **k** $\sqrt{169}$ **l** $\sqrt{4}$

m square root of 81 **n** $\sqrt{1}$ **o** $\sqrt{64}$

4 Use a calculator to write down the value of

a 8.2^2 **b** 3.2^3 **c** $\sqrt{1.21}$

d $\sqrt[3]{9.261}$ **e** the negative square root of 1.44 **f** the cube root of -512

g 4.7 squared **h** 6.3 cubed **i** $(-2.3)^3$

j $(-1.3)^2$ **k** the negative square root of 5.29 **l** $\sqrt[3]{-1.331}$

1.5

BIDMAS is a made up word to help you remember the order of operations.

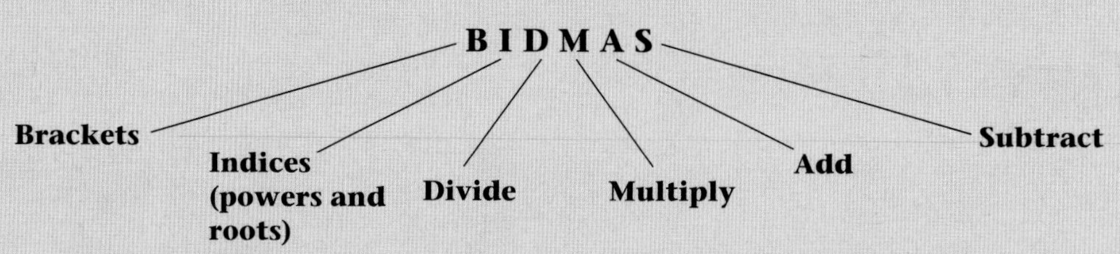

When operations have the same priority, do them in the order that they appear.

Example 6

Work out:

a $\frac{64-16}{2^3}$ **b** $8 \times (3+2) - 5$ **c** $5 + (2^2 + 4^3)$ **d** $\sqrt{\frac{16}{5+11}}$

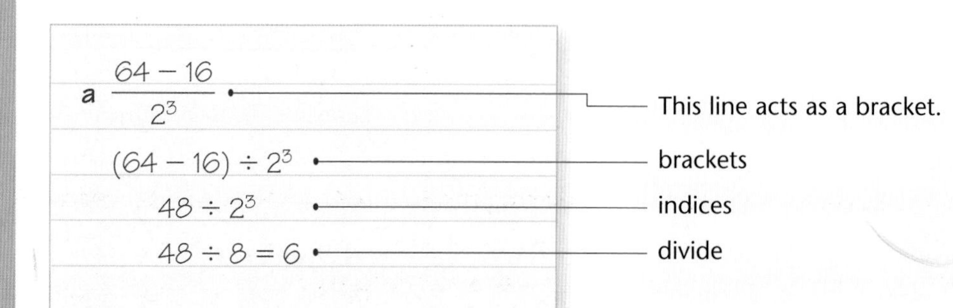

b $8 \times (3 + 2) - 5$ — brackets
$= 8 \times 5 - 5$ — multiply
$= 40 - 5$ — subtract
$= 35$

c $5 + (2^2 + 4^3)$ — brackets; indices
$= 5 + (4 + 64)$ — brackets
$= 5 + 68$ — add
$= 73$

d $16 \div \sqrt{5 + 11}$ — The square root sign acts as a bracket.
$= 16 \div \sqrt{16}$ — indices
$= 16 \div \pm 4$ — divide
$= \pm 4$

Practice 1E

1 Work out:

a $6 + 4 \times 7$

b $9 - 3 \div 2$

c $7 \times (8 - 4)$

d $36 \div (4 \times 3)$

e $36 \div 4 \times 3$

f $(2 + 7)^2$

g $2 + 7^2$

h $\sqrt{9 + 16}$

i $\sqrt{9} + 16$

j $\frac{48 - 12}{3^2}$

k $\frac{\sqrt{57 - 8}}{3^2 - 2}$

l $9 + 3^2 \div (5 - 2)^2 + 1$

2 Make these expressions correct by replacing the * with +, −, ÷, ×. Use brackets if you need to.

a 6 * 8 * 4 = 24

b 3 * 2 * 3 * 2 = 25

c 7 * 7 * 7 * 7 = 49

d 1 * 2 * 3 * 4 = −7

1.6

- The **factors** of a number are whole numbers that divide exactly into the number. The factors always include 1 and the number itself. Eg the factors of 12 are 1, 2, 3, 4, 6, 12.
- **Multiples** of a number are the results of multiplying the number by a positive whole number. Eg the first six multiples of 3 are 3, 6, 9, 12, 15, 18.
- A **prime number** is a whole number greater than 1 which has only two factors: itself and 1. The first eight prime numbers are 2, 3, 5, 7, 11, 13, 17, 19. You will find it useful to learn these.

Example 7

a Find the highest common factor (HCF) of 60 and 18.

b Find the lowest common multiple (LCM) of 3 and 5.

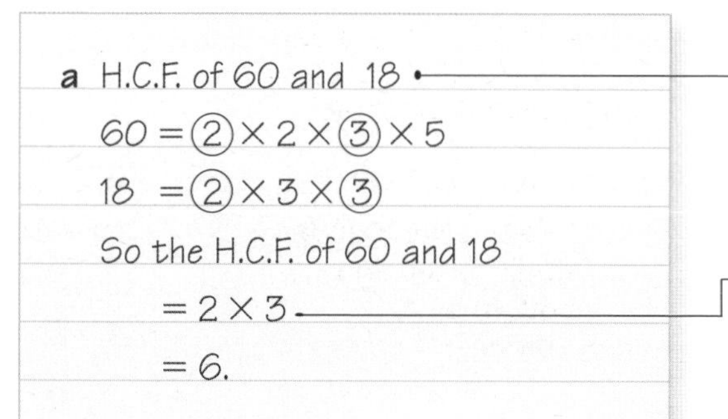

You need to find the highest number that will divide into both 60 and 18 without remainders.

Write each number in prime factor form. both numbers have 2 and 3 as factors.

Multiply the common factors together to give the HCF.

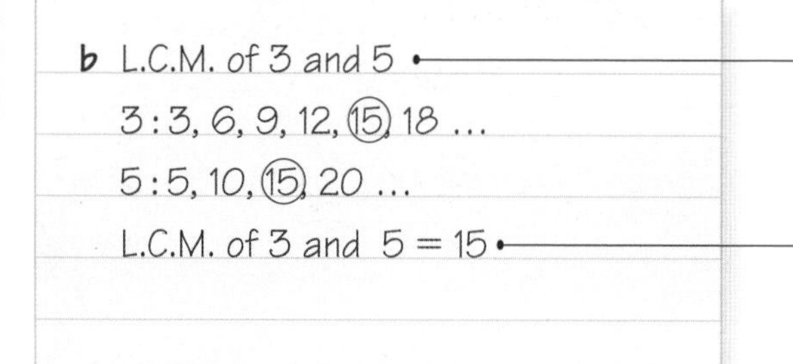

You need to find the lowest number that is a multiple of both 3 and 5.

Write out a list of multiples for each number.

This is the lowest number that appears in both lists.

Practice 1F

1 Write down all the prime numbers between 10 and 30.

2 Find the HCF of:

a 6 and 15 **b** 8 and 30 **c** 16 and 24

d 18 and 48 **e** 15, 25 and 50 **f** 12, 30 and 48

3 Find the LCM of:

a 3 and 7 **b** 4 and 9 **c** 5 and 6

d 2, 3 and 4 **e** 5, 7 and 10 **f** 3, 6 and 15

(S1) (E1)

2 Fractions

This chapter covers the four rules of fractions and how to solve problems involving fractions.

2.1

To write fractions in their simplest form (also known as cancelling) divide top and bottom by their highest common factor (HCF). In any fraction, the number on top is the numerator and the number below the line is the denominator.

$\frac{5}{7}$ — 5 numerator, 7 denominator

Example 1

Write the following fractions in their simplest form: **a** $\frac{15}{20}$ **b** $\frac{12}{20}$ **c** $\frac{32}{48}$

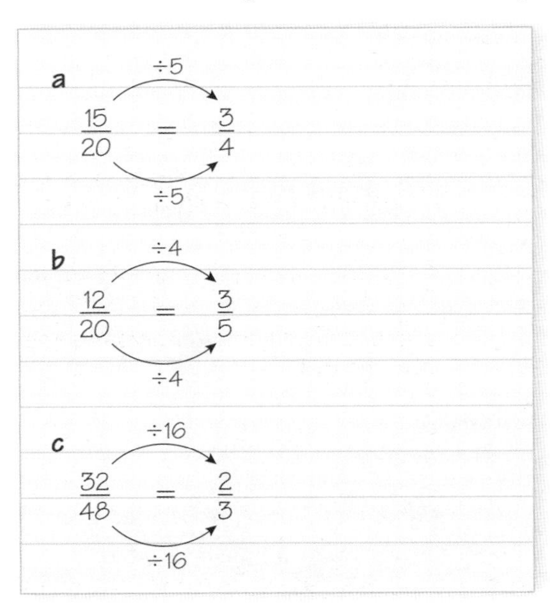

Divide top and bottom by the highest common factor, in this case 5. 5 is the largest number that goes into 15 and 20 exactly.

HCF of 12 and 20 is 4, so divide top and bottom by 4.

HCF of 32 and 48 is 16, so divide top and bottom by 16.

Example 2

Complete this set of equivalent fractions: $\frac{3}{4} = \frac{}{12} = \frac{15}{} = \frac{}{28}$

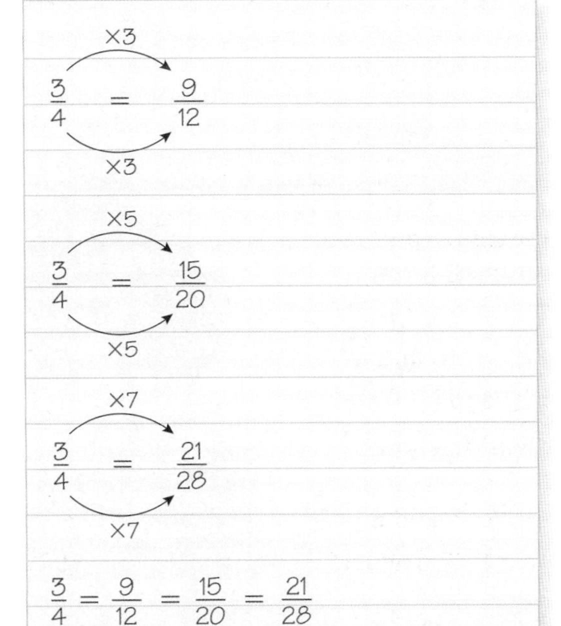

4 has been multiplied by 3 to get 12, so multiply the numerator (3) by 3 to get 9.

3 has been multiplied by 5 to get 15, so multiply the denominator (4) by 5 to get 20.

4 has been multiplied by 7 to get 28, so multiply the numerator by 7 to get 21.

Example 3

Write as improper fractions:

a $2\frac{1}{2}$ **b** $3\frac{2}{5}$ **c** $4\frac{3}{8}$

a $2\frac{1}{2} = 2 + \frac{1}{2}$

$2 = \frac{4}{2}$ — 2 can be written as $2 \times \frac{2}{2} = \frac{4}{2}$

so $2\frac{1}{2} = \frac{4}{2} + \frac{1}{2} = \frac{5}{2}$ — $\frac{5}{2}$ is an **improper fraction**; the numerator is greater than the denominator.

b $3\frac{2}{5} = 3 + \frac{2}{5}$

$3 = \frac{15}{5}$ — 3 can be written as $3 \times \frac{5}{5} = \frac{15}{5}$

so $3\frac{2}{5} = \frac{15}{5} + \frac{2}{5} = \frac{17}{5}$

c $4\frac{3}{8} = 4 + \frac{3}{8}$

$4 = \frac{32}{8}$ — 4 can be written as $4 \times \frac{8}{8} = \frac{32}{8}$

so $4\frac{2}{8} = \frac{32}{8} + \frac{3}{8} = \frac{35}{8}$

Example 4

Change these improper fractions to mixed numbers:

a $\frac{7}{4}$ **b** $\frac{10}{3}$ **c** $\frac{15}{6}$

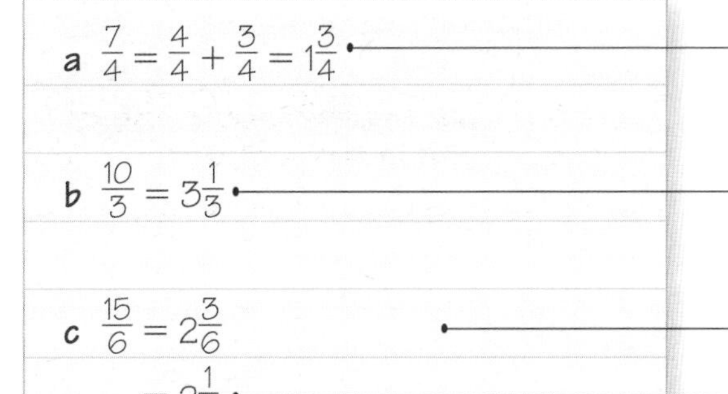

4 goes into 7 once, remainder 3, so $\frac{7}{4} = 1\frac{3}{4}$. This is a **mixed number**.

3 goes into 10 3 times, remainder 1, so $\frac{10}{3} = 3\frac{1}{3}$.

6 goes into 15 twice, remainder 3 so $\frac{15}{6} = 2\frac{3}{6}$.

$\frac{3}{6} = \frac{1}{2}$.

Practice 2A

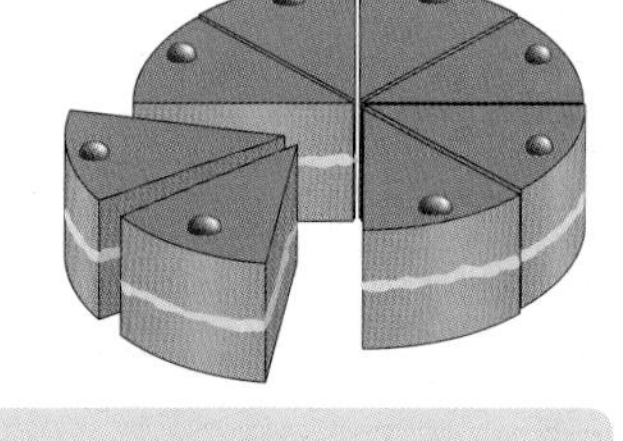

1 Write the following fractions in their simplest form:

a $\frac{6}{8}$ **b** $\frac{6}{9}$ **c** $\frac{18}{24}$ **d** $\frac{20}{24}$

e $\frac{16}{28}$ **f** $\frac{24}{36}$ **g** $\frac{36}{60}$ **h** $\frac{48}{84}$

2 Complete the following pairs of equivalent fractions:

a $\frac{2}{3} = \frac{\ }{6}$ **b** $\frac{4}{5} = \frac{12}{\ }$ **c** $\frac{3}{7} = \frac{\ }{21}$ **d** $\frac{2}{\ } = \frac{12}{18}$

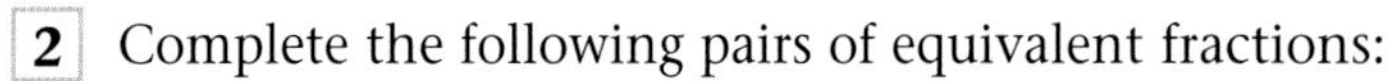

e $\frac{5}{\ } = \frac{20}{28}$ **f** $\frac{3}{4} = \frac{\ }{100}$ **g** $\frac{3}{8} = \frac{15}{\ }$ **h** $\frac{6}{11} = \frac{\ }{33}$

> **Hint:** Equivalent fractions are fractions with the same value, written with different denominators.

3 Write the following mixed numbers as improper fractions:

a $1\frac{1}{5}$ **b** $1\frac{2}{3}$ **c** $2\frac{1}{7}$

d $3\frac{2}{3}$ **e** $4\frac{3}{7}$ **f** $10\frac{2}{3}$

> **Remember:** an improper fraction has a numerator greater than the denominator.

4 Write the following improper fractions as mixed numbers. Simplify your answers if possible.

a $\frac{7}{2}$ **b** $\frac{7}{6}$ **c** $\frac{17}{7}$

d $\frac{10}{4}$ **e** $\frac{16}{6}$ **f** $\frac{20}{8}$

> **Remember:** a mixed number is a mixture of a whole number and a fraction.

2.2 To put fractions in order, you must first rewrite them with the same denominators.

Example 5

Rewrite this list of fractions in ascending order:

$\frac{3}{8}, \frac{7}{24}, \frac{1}{4}, \frac{5}{12}, \frac{5}{6}, \frac{11}{48}$

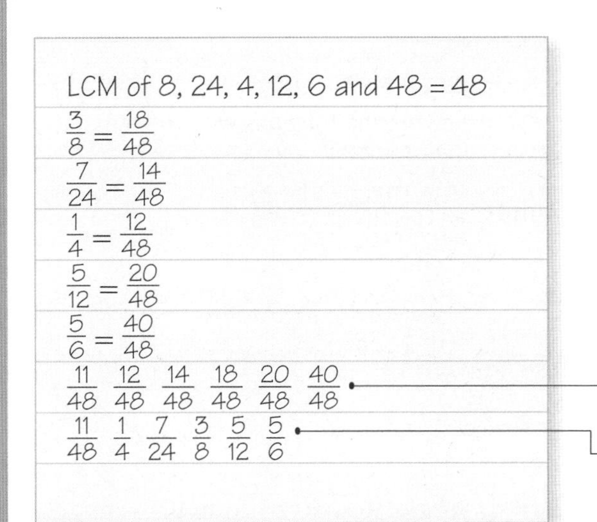

Rewrite the fractions with the same denominator. This will be the LCM of all the denominators.

Put in order of size, smallest first, by comparing the numerators.

Change back to original form.

Practice 2B

1 Rewrite these lists of fractions in ascending order:

a $\frac{11}{14}, \frac{4}{7}, \frac{23}{28}$

b $\frac{1}{2}, \frac{2}{5}, \frac{7}{10}, \frac{1}{4}$

c $\frac{2}{5}, \frac{7}{20}, \frac{4}{15}, \frac{3}{10}$

2 Rewrite these lists of fractions in descending order:

a $\frac{2}{5}, \frac{3}{10}, \frac{7}{20}$

b $\frac{2}{3}, \frac{5}{9}, \frac{13}{18}, \frac{7}{12}$

c $\frac{2}{3}, \frac{3}{4}, \frac{10}{12}, \frac{13}{20}$

(S2)

2.3 To add or subtract fractions, you must first rewrite them with the same denominators.

Example 6

Work out:

a $\frac{2}{3} + \frac{4}{5}$ **b** $2\frac{3}{4} + 1\frac{2}{5}$ **c** $\frac{5}{6} - \frac{1}{4}$ **d** $3\frac{4}{5} - 1\frac{1}{3}$ **e** $4\frac{1}{6} - 1\frac{5}{12}$

a $\frac{2}{3} + \frac{4}{5}$

$\frac{10}{15} + \frac{12}{15}$ — Rewrite the fractions with the same denominator. 15 is the LCM of 3 and 5.

$= \frac{10 + 12}{15}$ — Add the numerators.

$= \frac{22}{15}$ — $\frac{22}{15} = 1$ remainder 7

$= 1\frac{7}{15}$

b $2\frac{3}{4} + 1\frac{2}{5}$

$= 3 + \frac{3}{4} + \frac{2}{5}$ — Add the whole numbers.

$= 3 + \frac{15}{20} + \frac{8}{20}$ — Rewrite the fractions with the same denominator.

$= 3 + \frac{23}{20}$ — $\frac{23}{20} = 1$ remainder 3

$= 3 + 1\frac{3}{20}$

$= 4\frac{3}{20}$

c $\frac{5}{6} - \frac{1}{4}$

$= \frac{10}{12} - \frac{3}{12}$ — Rewrite the fractions with the same denominator. 12 is the LCM of 6 and 4.

$= \frac{10 - 3}{12}$ — Subtract the numerator.

$= \frac{7}{12}$

d $3\frac{4}{5} - 1\frac{1}{3}$

$= 2 + \left(\frac{4}{5} - \frac{1}{3}\right)$ — Subtract the whole numbers $(3 - 1 = 2)$.

$= 2 + \left(\frac{12}{15} - \frac{5}{15}\right)$ — Rewrite the fractions with the same denominator.

$= 2\frac{7}{15}$

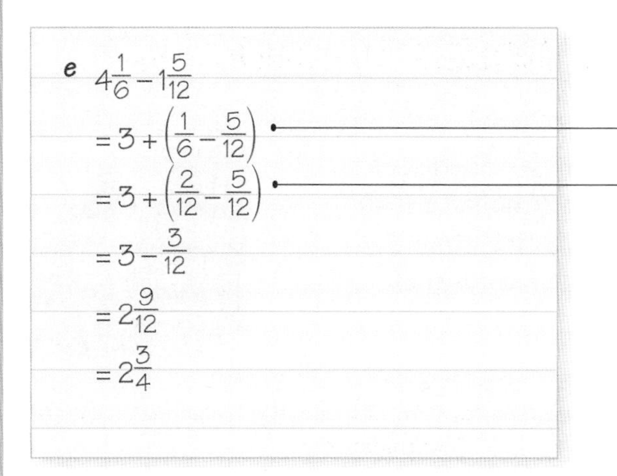

Subtract the whole numbers (4 − 1 = 3).

Rewrite the fractions with the same denominator.

Practice 2C

1 Work out:

a $\frac{1}{6} + \frac{3}{6}$ **b** $\frac{3}{10} + \frac{9}{10}$ **c** $\frac{1}{4} + \frac{1}{8}$ **d** $\frac{2}{5} + \frac{3}{10}$ **e** $\frac{4}{7} + \frac{3}{14}$

f $\frac{2}{5} + \frac{2}{3}$ **g** $\frac{3}{4} + \frac{5}{6}$ **h** $\frac{4}{5} + \frac{3}{8}$ **i** $\frac{5}{9} + \frac{3}{4}$ **j** $\frac{4}{7} + \frac{3}{8}$

2 Work out:

a $\frac{3}{4} - \frac{1}{4}$ **b** $\frac{5}{8} - \frac{3}{8}$ **c** $\frac{2}{5} - \frac{3}{10}$ **d** $\frac{5}{6} - \frac{5}{12}$ **e** $\frac{3}{4} - \frac{5}{8}$

f $\frac{4}{5} - \frac{2}{3}$ **g** $\frac{3}{4} - \frac{1}{6}$ **h** $\frac{7}{8} - \frac{1}{5}$ **i** $\frac{1}{2} - \frac{2}{5}$ **j** $\frac{5}{7} - \frac{1}{3}$

3 Work out:

a $1\frac{1}{2} + \frac{3}{5}$ **b** $2\frac{3}{4} + \frac{5}{8}$ **c** $1\frac{1}{2} + 1\frac{1}{4}$ **d** $2\frac{3}{8} + 3\frac{1}{4}$ **e** $1\frac{1}{2} + 3\frac{2}{3}$

f $2\frac{2}{3} + 4\frac{3}{4}$ **g** $4\frac{1}{2} + 3\frac{5}{6}$ **h** $2\frac{2}{3} + 1\frac{5}{8}$

4 Work out:

a $1\frac{5}{6} - \frac{2}{3}$ **b** $2\frac{1}{2} - 1\frac{1}{4}$ **c** $4\frac{3}{4} - 2\frac{1}{3}$ **d** $3\frac{1}{2} - 1\frac{5}{6}$ **e** $4\frac{1}{4} - 1\frac{3}{5}$

f $5\frac{2}{3} - 2\frac{7}{8}$ **g** $3\frac{1}{8} - 2\frac{2}{3}$ **h** $6\frac{5}{6} - 2\frac{1}{3}$

5 Mrs Bennett buys $1\frac{1}{2}$ kg of potatoes and $\frac{3}{4}$ kg of carrots. Calculate the total weight of the vegetables.

6 A tailor is making a suit. He needs the following length of fabric:

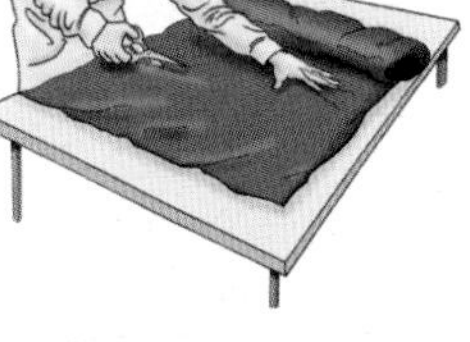

Trousers	$1\frac{1}{2}$ yds
Jacket	$1\frac{3}{4}$ yds
Waistcoat	$\frac{7}{8}$ yds

Calculate the total length of fabric the tailor needs to make the suit.

7 A bottle contains $1\frac{3}{4}$ pts of cooking oil. $\frac{5}{8}$ pt of oil is poured out. How much oil is left in the bottle?

8 The weight of a box including its contents is $5\frac{1}{2}$ kg. The contents weigh $3\frac{5}{8}$ kg. Calculate the weight of the box.

9 Danielle wants to video some programmes. She uses a 5 hour tape. She records a film lasting $2\frac{3}{4}$ hours, a soap opera lasting $\frac{1}{2}$ hr and a documentary lasting $\frac{5}{6}$ hr. How long will be left on the tape if she records these programmes?

2.4 To multiply fractions, multiply the numerators, then multiply the denominators. Then simplify.

Example 7

Work out:

a $\frac{3}{4} \times \frac{2}{5}$ **b** $\frac{3}{5} \times \frac{5}{6}$ **c** $1\frac{1}{2} \times 1\frac{3}{4}$ **d** $\frac{3}{4} \times 5$ **e** $\frac{2}{5}$ of 13

a $\frac{3}{4} \times \frac{2}{5} = \frac{6}{20}$

Multiply the numerators: $3 \times 2 = 6$.
Multiply the denominators: $4 \times 5 = 20$.

$\frac{6}{20} = \frac{3}{10}$

Divide both numerator and denominator by their highest common factor (2), to simplify.

b $\frac{\not{3}^{1}}{\not{5}_{1}} \times \frac{\not{5}^{1}}{\not{6}_{2}}$

$\frac{1}{1} \times \frac{1}{2} = \frac{1}{2}$

$\frac{3}{5} \times \frac{5}{6}$

Cancel the 3s

$\frac{\not{3}^{1}}{5} \times \frac{5}{\not{6}_{2}}$

then cancel the 5s

$\frac{1}{\not{5}_{1}} \times \frac{\not{5}^{1}}{2}$

c $1\frac{1}{2} \times 1\frac{3}{4}$

$\frac{3}{2} \times \frac{7}{4} = \frac{21}{8}$

First write the mixed numbers as improper fractions. Then multiply the numerators and denominators.

$\frac{21}{8} = 2\frac{5}{8}$

Then write as a proper fraction (mixed number).

d $\frac{3}{4} \times 5$

$\frac{3}{4} \times \frac{5}{1} = \frac{15}{4}$ — Write 5 as an improper fraction.

$\frac{15}{4} = 3\frac{3}{4}$ — Write the answer as a proper fraction.

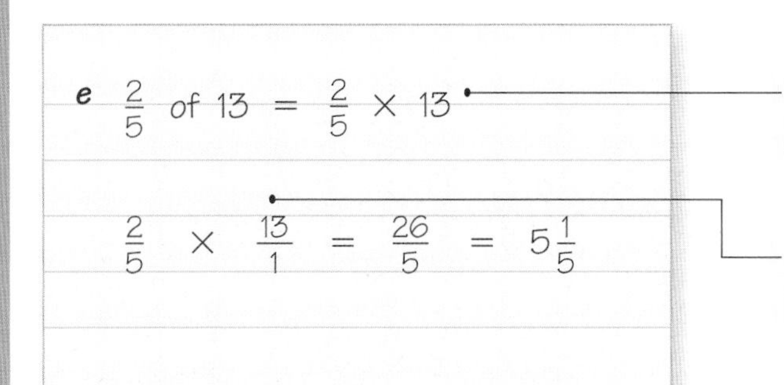

In fraction questions, 'of' means the same as multiply.

Write 13 as an improper fraction.
$2 \times 13 = 26 \quad 5 \times 1 = 5$

Practice 2D

1 Work out:

a $\frac{2}{5} \times \frac{1}{2}$ **b** $\frac{3}{4} \times \frac{5}{6}$ **c** $\frac{3}{8} \times \frac{4}{5}$ **d** $\frac{2}{3} \times \frac{3}{4}$ **e** $\frac{5}{9} \times \frac{6}{15}$

2 Work out:

a $1\frac{1}{2} \times \frac{2}{3}$ **b** $2\frac{3}{4} \times 2\frac{2}{5}$ **c** $1\frac{3}{7} \times 2\frac{4}{5}$ **d** $2\frac{1}{8} \times 3\frac{3}{5}$ **e** $2\frac{1}{4} \times 1\frac{3}{8}$

f $2\frac{1}{4} \times 5$ **g** $6 \times 2\frac{1}{8}$ **h** $3\frac{5}{6} \times 4$

3 Calculate:

a $\frac{1}{4}$ of 8 **b** $\frac{3}{4}$ of 12 **c** $\frac{5}{6}$ of 13 **d** $\frac{4}{5}$ of 21 **e** $\frac{7}{8}$ of 13

4 Suzanna earns £27 working in a supermarket at the weekend.
She saves $\frac{2}{5}$ of her wage. How much does she save?

5 In a maths test, $\frac{3}{8}$ of the 24 marks were on algebra, $\frac{1}{4}$ on statistics and the rest on number. Calculate the number of marks on each topic.

6 A box of chocolates weighs $\frac{3}{4}$ kg. $\frac{2}{5}$ of the chocolates are plain chocolates. How much do the plain chocolates weigh?

2.5

To divide by a fraction, invert it (turn it upside down) and then multiply.

Example 8

Work out:

a $\frac{3}{4} \div \frac{7}{9}$ **b** $\frac{5}{8} \div \frac{3}{4}$ **c** $2\frac{1}{5} \div 1\frac{1}{4}$ **d** $\frac{3}{8} \div 4$

a $\frac{3}{4} \div \frac{7}{9}$

$\frac{3}{4} \times \frac{9}{7} = \frac{27}{28}$ — Change divide to multiply and invert the fraction. Then multiply out.

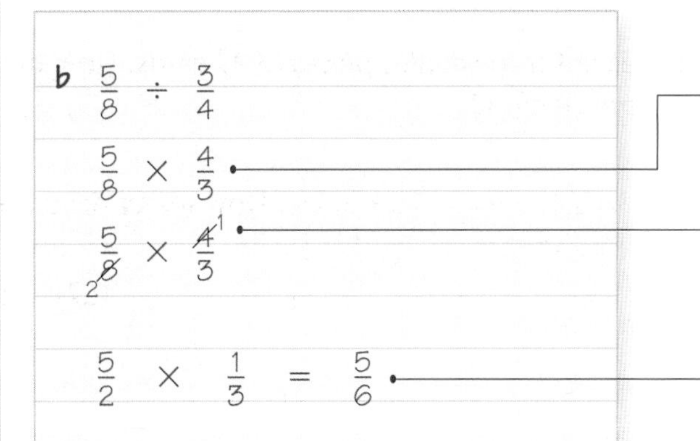

Change divide to multiply and invert the fraction. Then multiply out.

Cancel where possible.

Multiply out: $5 \times 1 = 5$
$2 \times 3 = 6$

c $2\frac{1}{5} \div 1\frac{1}{4}$

$\frac{11}{5} \div \frac{5}{4}$ — Change the mixed numbers into improper fractions.

$\frac{11}{5} \times \frac{4}{5} = \frac{44}{25}$ — Invert $\frac{5}{4}$ and multiply out.

$\frac{44}{25} = 1\frac{19}{25}$ — Write as a proper fraction (mixed number).

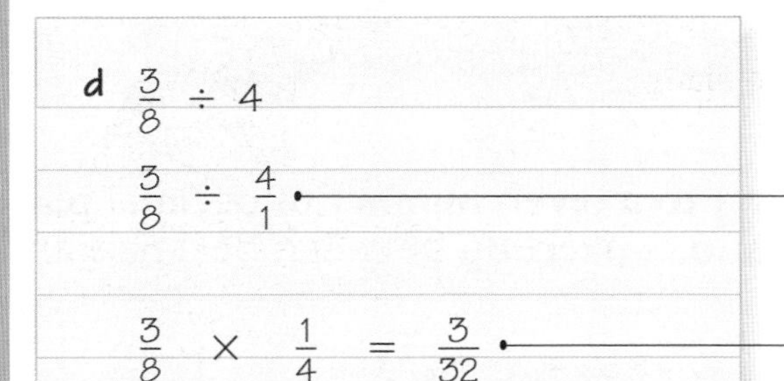

Practice 2E

(Simplify your answers wherever possible.)

1 Work out:

a $\frac{1}{2} \div \frac{1}{5}$ **b** $\frac{2}{3} \div \frac{4}{5}$ **c** $\frac{3}{8} \div \frac{7}{12}$ **d** $\frac{3}{5} \div \frac{7}{10}$ **e** $\frac{5}{8} \div \frac{3}{4}$

2 Work out:

a $1\frac{1}{2} \div 1\frac{1}{4}$ **b** $2\frac{2}{5} \div \frac{7}{10}$ **c** $3\frac{1}{5} \div 2\frac{10}{11}$ **d** $5\frac{1}{3} \div 1\frac{2}{9}$ **e** $1\frac{9}{14} \div 2\frac{1}{7}$

f $\frac{3}{8} \div 6$ **g** $1\frac{5}{6} \div 22$ **h** $5 \div 1\frac{1}{4}$

3 A length of wood is $10\frac{1}{4}$ yards long. It is cut into smaller pieces of $1\frac{5}{8}$ yards. How many pieces of wood can be cut from the larger length?

4 A construction company can lay $2\frac{1}{4}$ miles of railway track a day. How many days will it take the company to lay 30 miles of railway track?

(E2) (E3)

3 Decimals

This chapter tells you about rounding to a given number of decimal places, multiplying and dividing decimals, and converting between decimals and fractions.

(S3)

3.1

To round a number to 1 (2, 3) decimal places (dp), look at the 2nd (3rd, 4th) dp. If it is a 5 or greater, round up. If it is less than 5, round down.

Example 1

Round these numbers to **i** 1 dp **ii** 3 dp

a 3.6825 **b** 4.6398

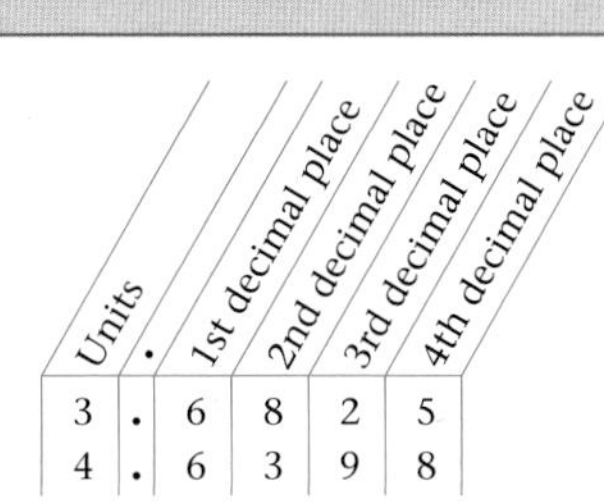

a i 3.6825

3.6825 = 3.7 (1 dp)

1st dp
The second digit is greater than 5, so round up.

ii 3.6825

3.6825 = 3.683 (3 dp)

3rd dp
The fourth digit is 5, so round up.

b i 4.6398

4.6398 = 4.6 (1 dp)

1st dp
The second digit is less than 5, so round down.

ii 4.6398

4.6398 = 4.640 (3 dp)

3rd dp
The fourth digit is greater than 5, so round up. (Think of this as rounding 39 to give 40.)

Keep this zero to show the answer is correct to 3 dp.

Example 2

Three children were told they could share £5 equally between them. One of the children worked out on a calculator that their share would be £1.66666666.

a Explain why this is not a sensible answer. **b** Give a more sensible answer.

a 1.66666666 is the answer a calculator would give but you cannot actually have £1.66666666 because the lowest value of a coin is £0.01.

b £1.67

This answer is more sensible because you can actually have £1.67 in money.

Practice 3A

1 Write the following numbers correct to 1 dp:
a 37.82 **b** 6.843 **c** 132.653 **d** 7.96 **e** 0.38 **f** 0.55

2 Write the following numbers correct to 3 dp:
a 1.4872 **b** 3.9658 **c** 23.2865 **d** 0.0397 **e** 0.0025 **f** 0.1897

3 Write the following numbers correct to
i 2 dp **ii** 1 dp **iii** 3 dp
a 2.3475 **b** 23.8921 **c** 1.3095 **d** 21.0356 **e** 6.2001 **f** 0.9393.

4 Jack earns £20 for 6 hours work. Work out Jack's hourly rate of pay, giving your answer to an appropriate number of dp. Give a reason for your answer.

5 Mr Addison builds a fence using 7 fence panels. The finished fence is 12 m long. Calculate the length of each fence panel. Give your answer to a sensible number of decimal places, explaining the reason for your answer.

6 A forester measures the circumference of a tree in order to find its diameter. He finds the circumference is 93 cm. He works out that the diameter must be 29.60 281 941 cm.

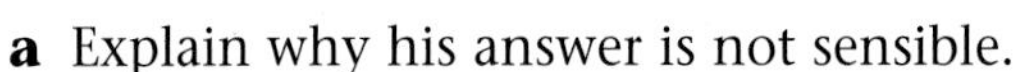

a Explain why his answer is not sensible.
b Give a more sensible answer.

- **When multiplying a decimal by 10, digits move one place to the left of the decimal point.**
- **When dividing a decimal by 10, digits move one place to the right of the decimal point.**
- **When multiplying and dividing by powers of 10 the digits move the same number of places as there are zeros.**

Example 3

Write down the answers to:
a 2.45×10 **b** 3.5321×1000 **c** $132.71 \div 100$ **d** $23.24 \div 10$

a $2.45 \times 10 = 24.5$ — To multiply by 10 move all the digits 1 place to the left.

b $3.5321 \times 1000 = 3532.1$ — To multiply by 1000 move all the digits 3 places to the left.

c $132.71 \div 100 = 1.3271$ — To divide by 100 move all the digits 2 places to the right.

d $23.24 \div 10 = 2.324$ — To divide by 10 move all the digits 1 place to the right.

Example 4

Work out without using a calculator:

a 23.3×1.4 **b** 4.8×0.21

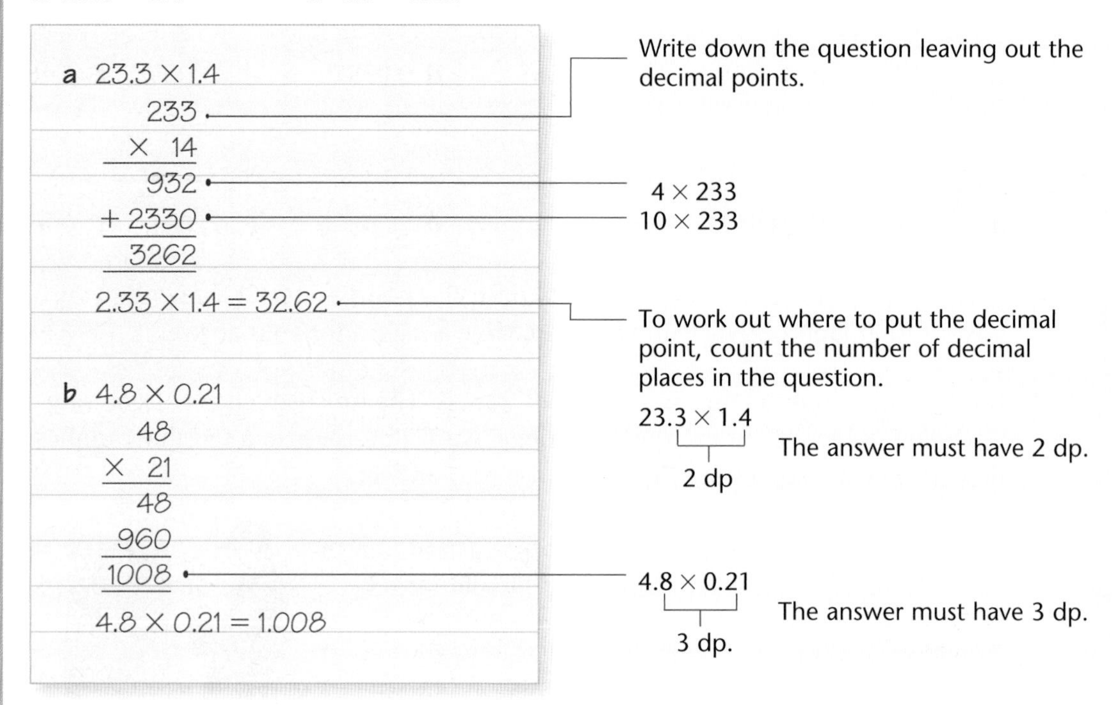

Example 5

Work out without using a calculator:

a $6.35 \div 0.05$ **b** $23.6 \div 0.4$ **c** $3.62 \div 0.2$ **d** $4.06 \div 1.4$

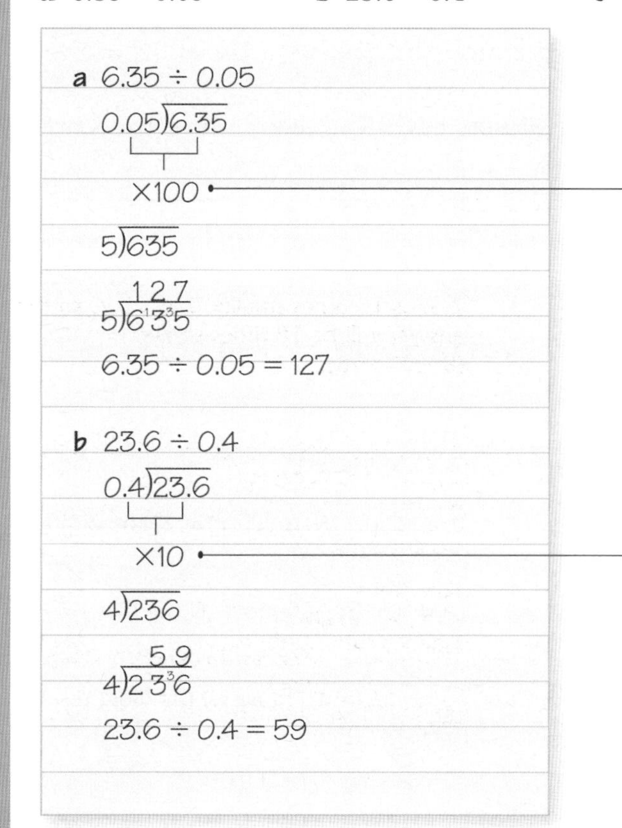

If the number you are dividing by is not a whole number, change it to a whole number.

Multiplying 0.05 by 100 will give a whole number.

You must also multiply 6.35 by 100.

Multiplying 0.4 by 10 will give a whole number. Remember to multiply 23.6 by 10 as well.

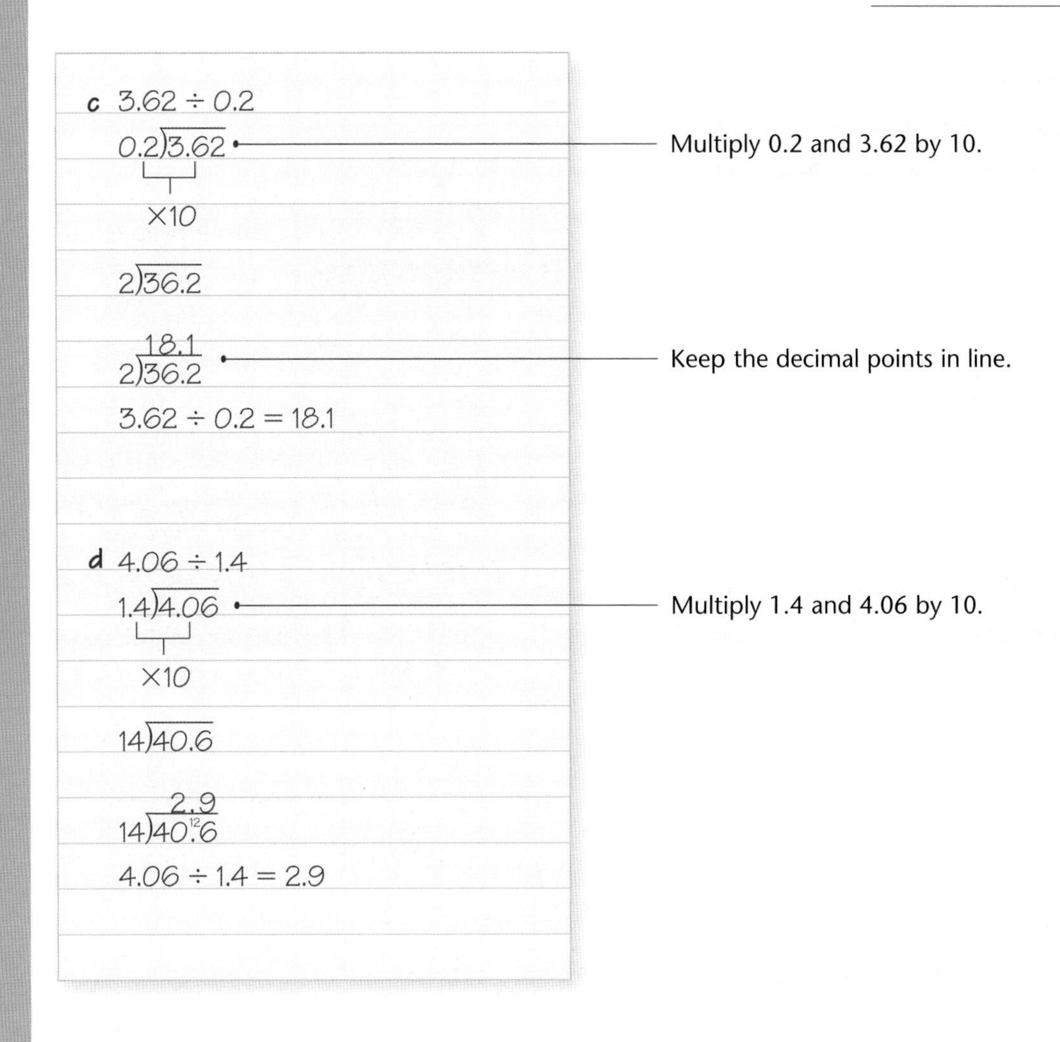

Example 6

$23.7 \times 3.4 = 80.58$

Write down the answers to

a 2.37×3.4 **b** 23.7×340 **c** 0.237×0.34

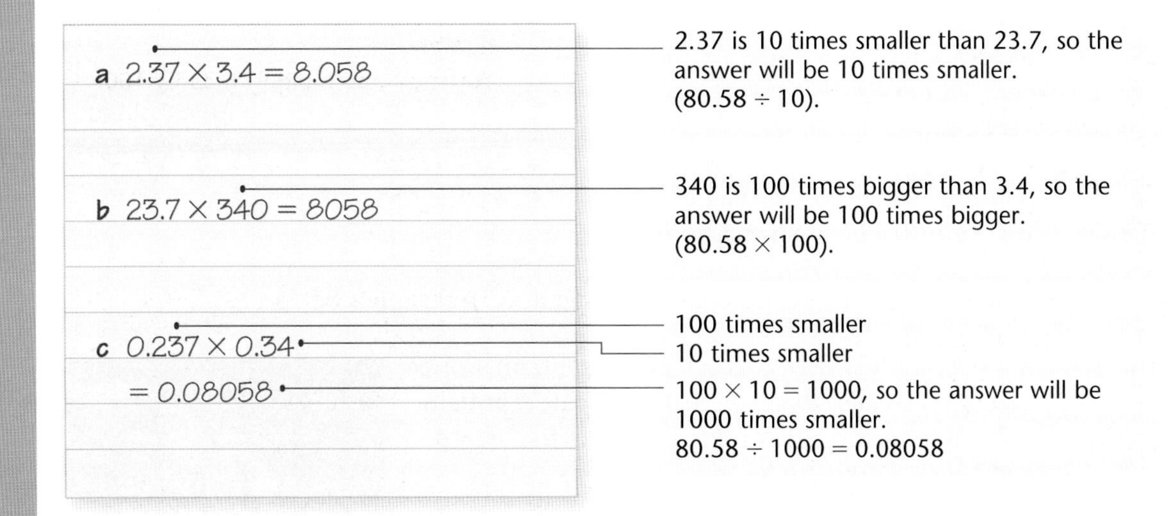

Example 7

$7.285 \div 3.1 = 2.35$

Write down the value of

a $7285 \div 3.1$ **b** $0.7285 \div 0.31$ **c** $728.5 \div 31$

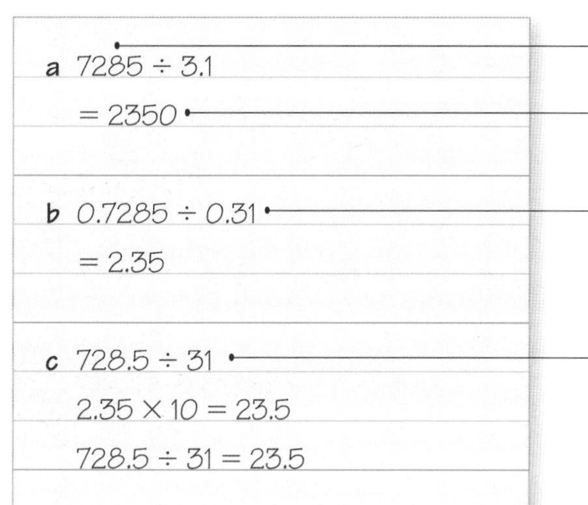

1000 times bigger than 7.285

1000 times bigger than 2.35

As both numbers are 10 times smaller, there will be no difference in the answer.

The first number is 100 times bigger and the second is 10 times bigger. $100 \div 10 = 10$ so the answer will be 10 times bigger.

Practice 3B

Answer the following questions without using a calculator.

1 Multiply each of the following numbers by:
i 100 **ii** 10 **iii** 1000
a 2.34 **b** 3.614 **c** 23.021
d 41.2 **e** 0.0213

Hint: Look at Example 3 page 19

2 Divide each of the following numbers by:
i 10 **ii** 1000 **iii** 100
a 23.4 **b** 37.92 **c** 783.1
d 2.63 **e** 2.01

Hint: Look at Example 4 page 20

3 Work out:
a 3.1×2.6 **b** 3.8×4.1 **c** 97×0.32
d 0.21×0.18 **e** 53.1×2.3 **f** 5.21×1.5
g 0.82×1.5 **h** 6.24×0.37

Hint: Look at Example 5 page 20

4 Work out:
a $42.5 \div 0.5$ **b** $23.8 \div 0.2$ **c** $2.94 \div 0.06$
d $0.236 \div 0.4$ **e** $0.345 \div 0.15$ **f** $67.5 \div 2.5$
g $0.897 \div 0.13$ **h** $8.12 \div 0.014$

5 A book weighs 1.3 kg. Calculate the weight of 100 similar books.

6 A model train is to be made. The model train will be 100 times smaller than the real train. The real train is 7.6 m long and 3.9 m high. Calculate the length and height of the model train.

7 How many stamps costing £0.27 can be bought for £9.18?

8 A packet of sweets weighs 0.113 kg. How much will 9 similar packets weigh?

9 A car will travel 17.3 km on 1 litre of petrol. How far will the car travel on 9.4 litres of petrol?

10 A glass holds 0.25 litres. How many glasses can be filled from a 1.5 litre bottle?

11 $41.2 \times 3.65 = 150.38$

Hint: Look at Example 6 page 21

Write down the value of:

a 4.12×3.65 **b** 412×365 **c** 0.412×3.65 **d** 4.12×0.365

12 $98.42 \div 3.5 = 28.12$

Hint: Look at Example 7 page 22

Write down the value of:

a $9.842 \div 3.5$ **b** $9842 \div 35$ **c** $0.9842 \div 3.5$

3.3 **To change a decimal to a fraction, decide whether the number represents tenths, hundredths, thousandths etc. Then simplify if possible.**

0.256 is 256 thousandths $= \frac{256}{1000} = \frac{32}{125}$

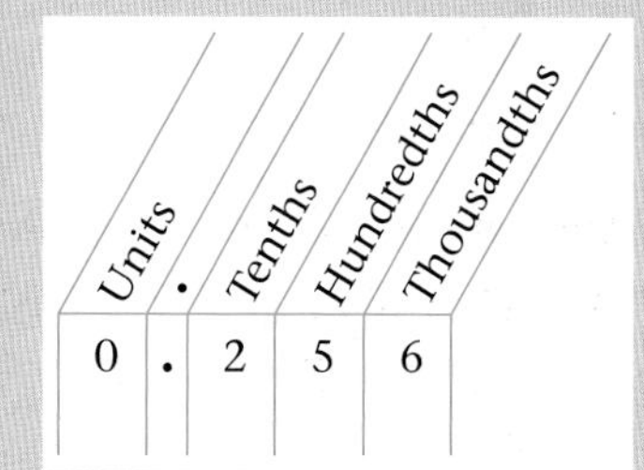

Example 8

Write the following decimals as fractions:

a 0.3 **b** 0.24 **c** 0.385 **d** 0.03

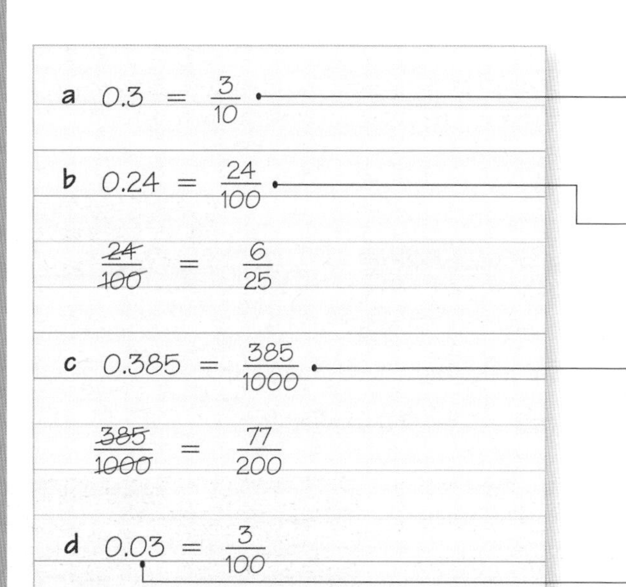

Notice that there are always the same number of zeros in the denominator as there are decimal places in the number.

This can be simplified. Divide top and bottom by 4.

This can be simplified. Divide top and bottom by 5.

Although there is only 1 digit, the number has 2 dp, so the denominator has 2 zeros.

Practice 3C

1 Write the following decimals as fractions:

a 0.5	**b** 0.75	**c** 0.7	**d** 0.9
e 0.85	**f** 0.36	**g** 0.48	**h** 0.125
i 0.625	**j** 0.02	**k** 0.025	**l** 0.008

3.4

To change a fraction to a decimal, divide the numerator by the denominator. This can be done on a calculator if the question allows.

Example 9

Write the following fractions as decimals:

a $\frac{1}{4}$ **b** $\frac{3}{5}$ **c** $\frac{2}{3}$ **d** $\frac{14}{99}$

a $\frac{1}{4} = 1 \div 4 = 0.25$

b $\frac{3}{5} = 3 \div 5 = 0.6$

c $\frac{2}{3} = 0.666666666$ — This is the answer on a calculator.

$= 0.\dot{6}$ — The dot over the 6 shows that the 6 repeats forever. $0.\dot{6}$ is called a recurring decimal.

d $\frac{14}{99} = 0.14141414$

$= 0.\dot{1}\dot{4}$ — A dot is put over both the 1 and the 4 to show that both numbers recur.

Example 10

Rewrite the following list of numbers in descending order:

$\frac{7}{20}$, 0.5, $\frac{2}{5}$, 0.34, 0.3

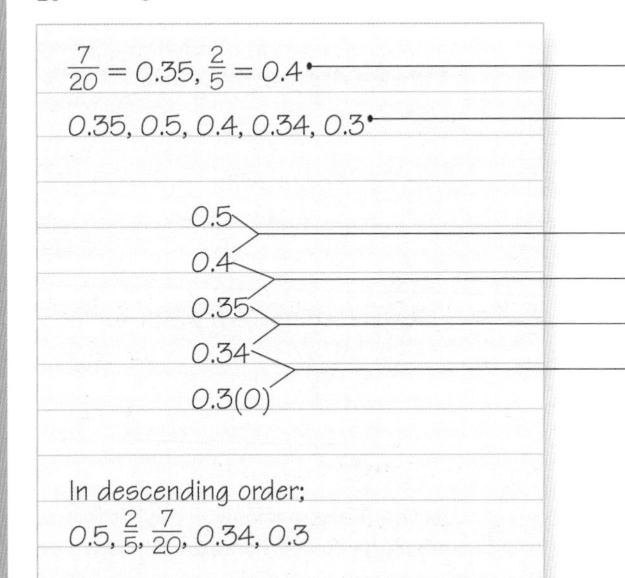

Change the fractions to decimals because decimals are easier to compare.

Rewrite the list.

Compare each number in the list.
1st dp: 5 is bigger than 4.

4 is bigger than 3.

2nd dp: 5 is bigger than 4.

4 is bigger than 0.

Practice 3D

1 Write the following fractions as decimals:

a $\frac{1}{2}$ **b** $\frac{3}{4}$ **c** $\frac{2}{5}$ **d** $\frac{3}{10}$

e $\frac{4}{15}$ **f** $\frac{7}{25}$ **g** $\frac{11}{24}$ **h** $\frac{2}{3}$

i $\frac{4}{11}$ **j** $\frac{41}{100}$ **k** $\frac{17}{125}$ **l** $\frac{5}{6}$

2 Rewrite these lists of numbers in descending order:

a 0.5, 0.45, 0.4, 0.49, 0.04

b $\frac{3}{5}$, 0.64, $\frac{13}{20}$, $\frac{7}{10}$, 0.06

c $\frac{7}{8}$, 0.8, $\frac{15}{16}$, 0.84, $\frac{3}{5}$

d $\frac{2}{5}$, 0.44, $\frac{4}{9}$, $\frac{1}{2}$, 0.404

3.5 **Terminating decimals do not recur. For example, 0.4 and 0.175 are terminating decimals. Only fractions whose denominators have prime factors 2, 5 or 2 and 5 can be written as terminating decimals.**

Example 11

From the list below, state which fractions can be written as terminating decimals and those which are recurring decimals:

$\frac{1}{5}, \frac{5}{12}, \frac{7}{20}, \frac{5}{16}, \frac{2}{14}$.

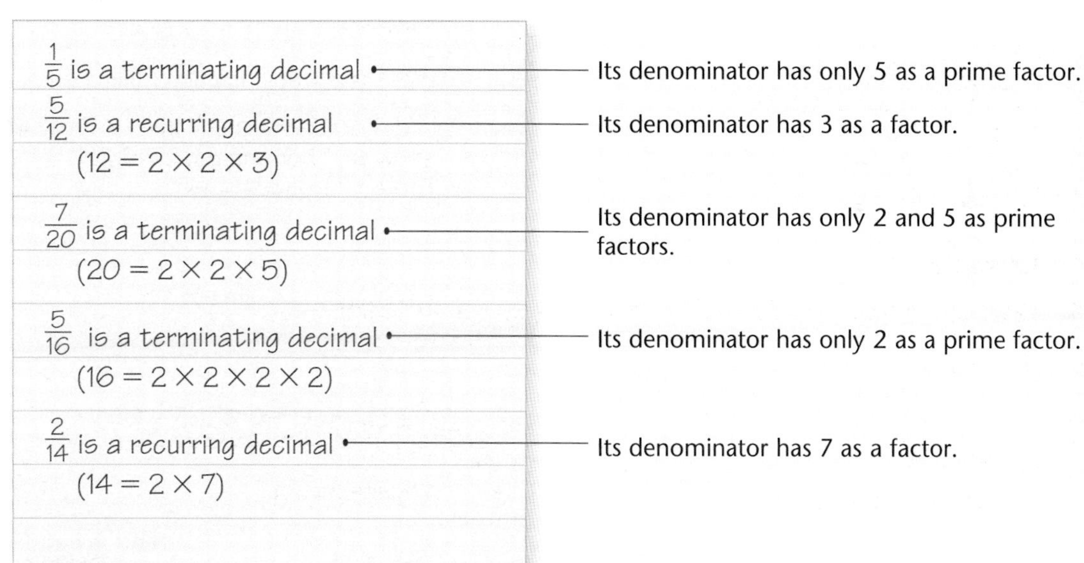

Practice 3E

1 State whether each of the following fractions is a terminating or recurring decimal:

a $\frac{4}{5}$ **b** $\frac{2}{3}$ **c** $\frac{9}{40}$ **d** $\frac{7}{15}$ **e** $\frac{99}{125}$

f $\frac{63}{64}$ **g** $\frac{124}{150}$ **h** $\frac{7}{10}$ **i** $\frac{69}{75}$ **j** $\frac{17}{48}$

In questions 2 and 3, choose different fractions from those in question 1.

2 Write down 3 fractions that can be written as terminating decimals.

3 Write down 3 fractions that can be written as recurring decimals.

3.6 The reciprocal of a number is 1 divided by the number. If the number is a fraction, the reciprocal can be found by inverting.

Example 12

Write down the reciprocals of the following numbers:

a 2 **b** 0.4 **c** $\frac{3}{4}$ **d** $\frac{1}{7}$

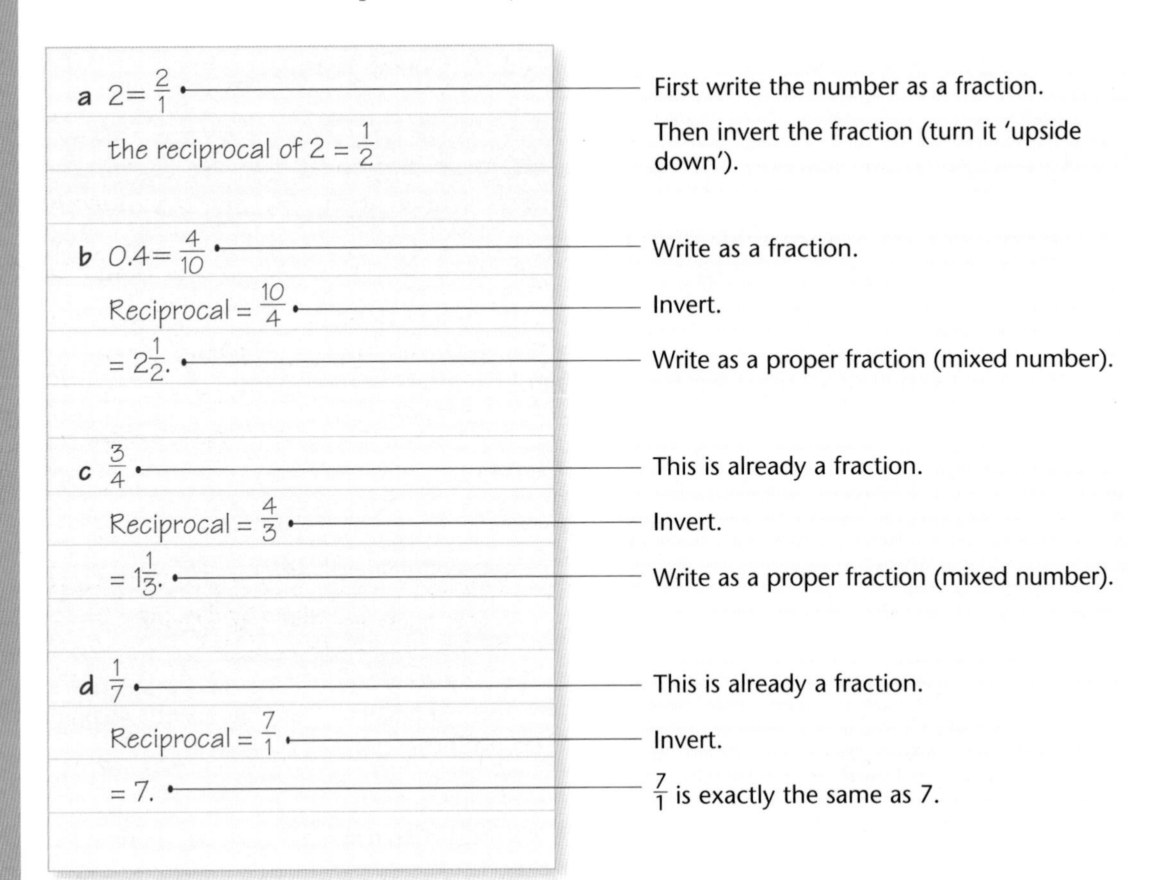

Practice 3F

1 Find the reciprocals of the following numbers:

a 8 **b** 5 **c** 0.3 **d** 0.25

e 0.325 **f** $\frac{1}{9}$ **g** $\frac{1}{20}$ **h** $\frac{7}{8}$

i $\frac{1}{3}$ **j** $0.\dot{6}$

Hint: A number multiplied by its reciprocal equals 1
eg $0.4 \times 2\frac{1}{2} = 1$
$\frac{1}{7} \times 7 = 1$

(E4)

4 Percentages

This chapter covers conversion of percentages to fractions and decimals, and how to work out percentage change.

4.1 **'Percentage' means 'out of a hundred'. Percentages can be written as fractions or decimals.**

Example 1

Write these percentages as **i** fractions **ii** decimals:

a 35% **b** $22\frac{1}{3}\%$

a i $35\% = \frac{35}{100}$ — All percentages can be written as a fraction with a denominator of 100.

$\frac{\not{35}^{7}}{\not{100}_{20}} = \frac{7}{20}$ — Always simplify your answer (see chapter 2).

a ii $35\% = \frac{35}{100}$ — Write the percentage as a fraction. Then convert to a decimal by dividing numerator by denominator.

$\frac{35}{100} = 35 \div 100$

$= 0.35$

b i $22\frac{1}{3}\% = \frac{22\frac{1}{3}}{100}$

$\frac{22\frac{1}{3}}{100} = \frac{67}{300}$ — Multiply top and bottom by 3 to remove the $\frac{1}{3}$.

b ii $22\frac{1}{3}\% = \frac{22\frac{1}{3}}{100}$

$= \frac{67}{300}$ — Multiply top and bottom by 3 to get rid of the fraction.

$= 0.22\dot{3}$ — Check that your answer makes sense: $22\frac{1}{3}\%$ is just less than $\frac{1}{4}$; so is 0.223.

Practice 4A

1 Write the following percentages as fractions:

a 25% **b** 30% **c** 65% **d** 24% **e** 32%
f $17\frac{1}{2}\%$ **g** $42\frac{1}{2}\%$ **h** $33\frac{1}{3}\%$

2 Write the following percentages as decimals:

a 50% **b** 20% **c** 45% **d** 28% **e** 48%
f $37\frac{1}{2}\%$ **g** $2\frac{1}{2}\%$ **h** $66\frac{2}{3}\%$

4.2 Fractions and decimals can be written as percentages.

Example 2

Write the following as percentages:

a 0.32 **b** 0.175 **c** $\frac{11}{20}$ **d** $\frac{23}{40}$

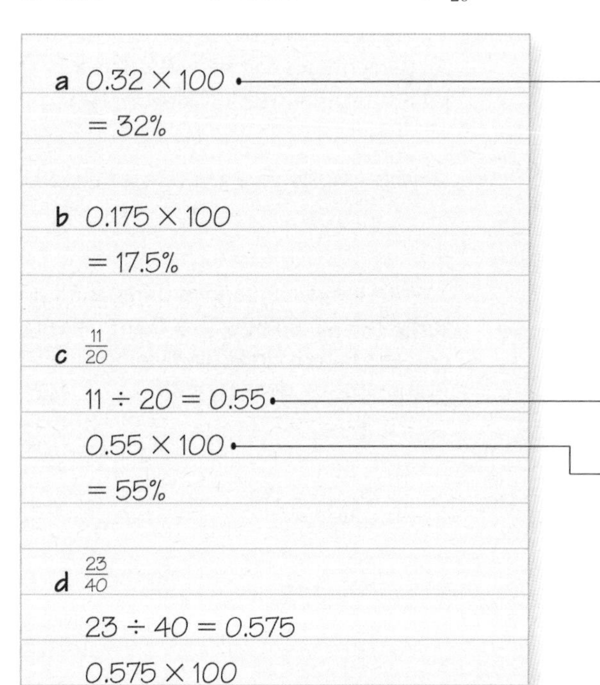

To convert a decimal to a percentage, multiply by 100.

To convert a fraction to a percentage, divide the numerator by the denominator.

Multiply by 100.

Practice 4B

1 Write the following decimals as percentages:

a 0.75 **b** 0.2 **c** 0.85 **d** 0.42 **e** 0.88
f 0.325 **g** 0.825 **h** $0.\dot{3}$

2 Write the following fractions as percentages:

a $\frac{1}{2}$ **b** $\frac{3}{10}$ **c** $\frac{17}{20}$ **d** $\frac{16}{25}$ **e** $\frac{41}{50}$
f $\frac{17}{40}$ **g** $\frac{37}{40}$ **h** $\frac{2}{3}$

3 Suzanna got $\frac{7}{10}$ of the questions in an exercise correct. What percentage of the questions did Suzanna get correct?

4 0.55 of the population of a village are female.
Write down the percentage of the population that are female.

5 Write these numbers in order of size.
Start with the largest number.
0.8 70% $\frac{7}{8}$ $\frac{3}{4}$ **E**

Hint: Write all the numbers as either percentages or decimals first.

6 30%, $\frac{1}{4}$, 0.35, $\frac{1}{3}$, $\frac{2}{5}$, 0.299
Write this list of six numbers in order of size.
Start with the smallest number. **E**

4.3 You can find a percentage of a quantity by changing the percentage to a fraction, then multiplying out.

Example 3

Find **a** 15% of 40 kg **b** $17\frac{1}{2}$% of £250.

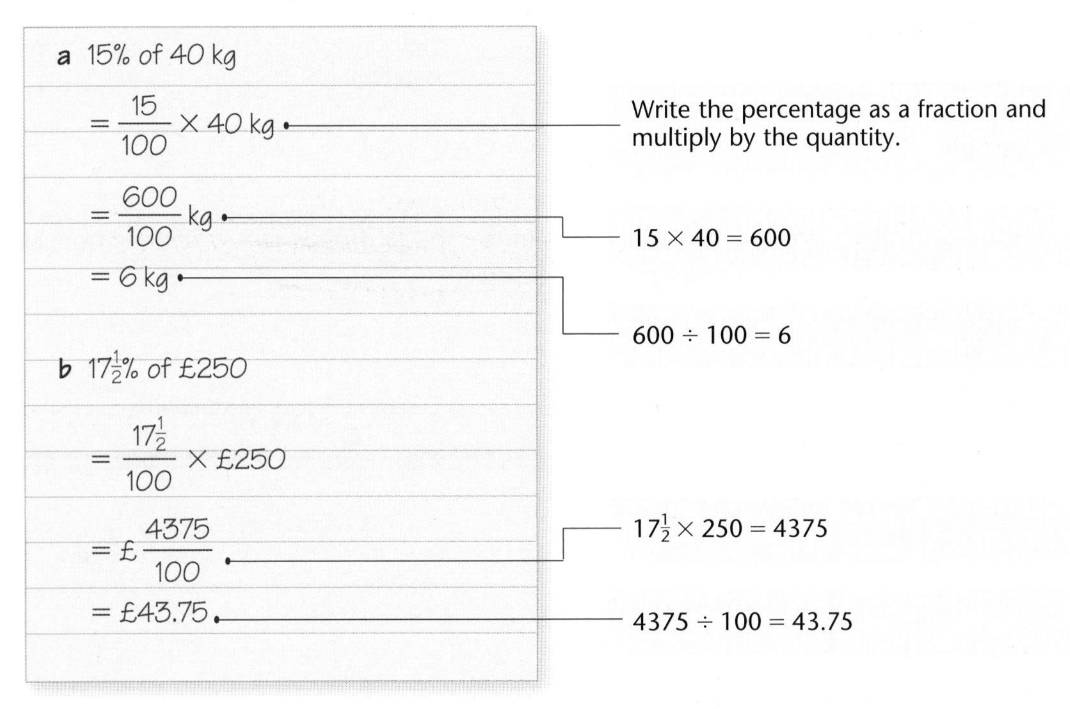

a 15% of 40 kg

$= \frac{15}{100} \times 40$ kg — Write the percentage as a fraction and multiply by the quantity.

$= \frac{600}{100}$ kg — $15 \times 40 = 600$

$= 6$ kg — $600 \div 100 = 6$

b $17\frac{1}{2}$% of £250

$= \frac{17\frac{1}{2}}{100} \times £250$

$= £\frac{4375}{100}$ — $17\frac{1}{2} \times 250 = 4375$

$= £43.75$ — $4375 \div 100 = 43.75$

Practice 4C

1 Calculate:

a 5% of £300 **b** 15% of £7.50 **c** 30% of 60 kg
d 20% of 40 kg **e** 25% of 500 g **f** $22\frac{1}{2}$% of £30
g 24% of 45 m **h** 32% of £6 **i** $33\frac{1}{3}$% of 600 m
j 42% of £7.50

2 Work out 20% of 1800. **E**

3 35% of the employees in a company are under 25.
There are 260 people employed in the company.
How many of the employees are under 25?

4 Work out 45% of £24. **E**

4.4 Quantities can be increased or decreased by a percentage.

Example 4

Peter earns £6.00 an hour working in a garage.
On his eighteenth birthday he receives a 5% increase in his pay.
Work out his new hourly rate.

New rate = 105%
As a decimal 105% = 1.05 — $105\% = \frac{105}{100} = 1.05$
Peter's new hourly rate = £6 × 1.05
= £6.30 — A calculator will show the answer as 6.3. This needs to be written in correct money notation.

Example 5

A sports shop offers 10% discount for all its employees.
The usual selling price of a pair of trainers in the sports shop is £65. Calculate how much an employee would have to pay for these trainers.

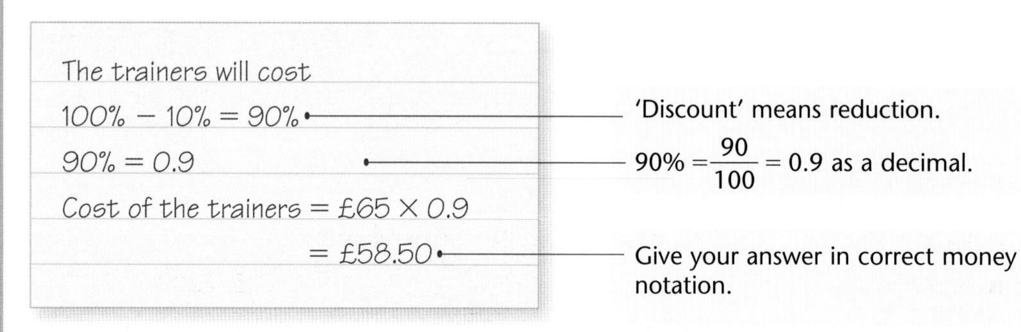
The trainers will cost
100% − 10% = 90% — 'Discount' means reduction.
90% = 0.9 — $90\% = \frac{90}{100} = 0.9$ as a decimal.
Cost of the trainers = £65 × 0.9
= £58.50 — Give your answer in correct money notation.

Practice 4D

1 Mr & Mrs Newman bought their house in 1990 for £50 000. They sold it in 2001 and made a 45% profit.
How much did they sell their house for in 2001?

2 In a sale, all the normal prices are reduced by 15%.
The normal price of a jacket is £42.
Syreeta buys the jacket in the sale. Work out the price of the jacket. **E**

3 Jane is going to buy a computer for £480 + $17\frac{1}{2}$% VAT. Work out the total price, including VAT, that Jane will pay for the computer.

E

4 The selling price of a digital television is the list price plus VAT at $17\frac{1}{2}$%. The list price of a digital television is £786. Work out the selling price.

E

5 The total cost of orange drink for 50 cups is £7.50. Each cup of drink is sold at a 20% profit. Work out the price at which each cup of drink is sold.

E

4.5 You can find one quantity as a percentage of another.

Example 6

14 students from a group of 20 pass an end of unit examination. What percentage of the students passed the examination?

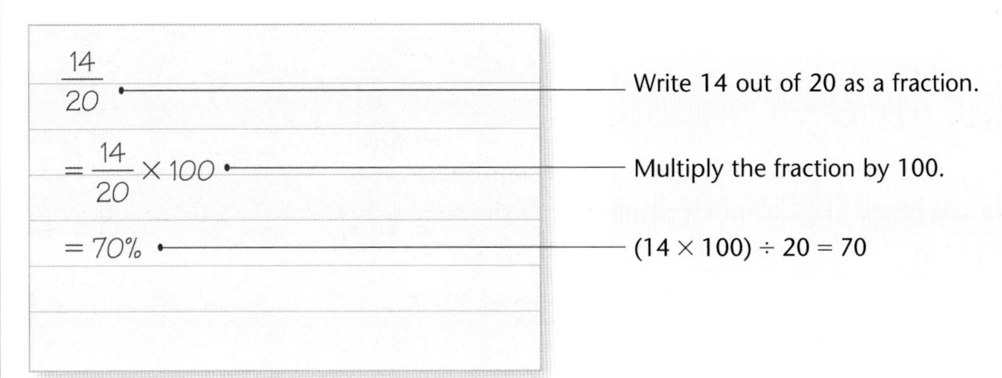

Practice 4E

1 Calculate the first quantity as a percentage of the second:

a 6, 30 **b** £3, £60 **c** £35, £200

d 13 kg, 40 kg **e** 69p, £4.60 **f** 20 cm, 2.5 m

Hint: Express £4.60 in pence.

Hint: Change m to cm first.

2 There are 24 chocolates in a box. 18 of the chocolates in the box are milk chocolates. Work out 18 as a percentage of 24.

3 Jo got 36 out of 80 in an English test. Work out 36 out of 80 as a percentage.

4.6 You can find the percentage change between quantities.

Example 7

In 2002 the cost of a litre of petrol rose from 70p to 77p. Calculate the percentage increase in the price of the petrol.

77p − 70p
= 7p

$\frac{7}{70} \times 100$

= 10% increase

Work out the actual increase.

$\text{Percentage increase} = \frac{\text{increase}}{\text{original cost}} \times 100\%$

Divide the increase by the original cost and multiply by 100, to get the percentage increase.

Example 8

A car was bought for £6500 and sold for £3900. Calculate the percentage loss.

loss = £6500 − £3900
= £2600

$\% \text{ loss} = \frac{2600}{6500} \times 100\%$

= 40%

$\text{Percentage loss} = \frac{\text{loss}}{\text{original cost}} \times 100\%$

Practice 4F

1 The cost of a holiday has risen in price from £750 to £900. Work out the percentage increase.

2 The cost of a CD player has fallen in price from £150 to £105. What is the percentage decrease in price?

3 A greengrocer buys a box of pineapples for £8. He sells the pineapples for £10.80. Work out the greengrocer's percentage profit.

4 Jim buys a car for £8500. He sells the car 3 years later for £5610. Calculate Jim's percentage loss.

5 Sue buys a pack of 12 cans of cola for £4.80. She sells the cans for 50p each. She sells all of the cans. Work out her percentage profit.

E

4.7 You can find the original amount after a percentage change.

Example 9

After a $7\frac{1}{2}$% increase in fares, a train ticket from Bedford to London costs £19.35. Calculate the cost of the ticket before the increase.

£19.35 = $107\frac{1}{2}$% of original price.

Think of original price as 100%
So new price = $(100 + 7\frac{1}{2})$%

Original price = $\frac{£19.35}{1.075}$

= £18

Original price = $\frac{\text{new price}}{\text{percentage increase}}$

1.075 is $107\frac{1}{2}$% as a decimal.

Example 10

During a sale all prices are reduced by 15%.
Graham buys a television in the sale for £212.50.
Calculate how much Graham saved by buying the television in the sale.

£212.50 = 85% of original price

Original price = 100%
Sale price = (100-15)%

Original price = $\frac{£212.50}{0.85}$

0.85 is 85% written as a decimal.

= £250

Saving = £250 − £212.50
= £37.50

Practice 4G

1 A sports centre puts all its membership prices up by 5%. The new membership prices are:

Adult:	£336
Junior:	£178.50
Family:	£672

Calculate the original cost of each type of membership.

2 A shop advertises its prices inclusive of VAT at a rate of 17.5%.
Calculate the cost of these items exclusive of VAT.

Video player = £450
CD player = £120
Games player = £230

Hint: Inclusive means with VAT. Exclusive means without VAT.

3 John sells his car for £4480 making a 20% loss.
Work out how much John lost on the car.

4 In a sale all the normal prices are reduced by 15%.
Winston pays £15.64 for a shirt.
Calculate the normal price of the shirt.

E

5 The selling price of a computer is the list price plus VAT at $17\frac{1}{2}$%.
The selling price of a computer is £1292.50.
Work out the list price of this computer.

E

6 In a sale all the prices are reduced by 30%.
The sale price of a jacket is £28.
Work out the price of the jacket before the sale.

E

4.8 You can work out the effect that combined or repeated percentage changes will have on an amount.

Example 11

The selling price of a DVD player is its list price plus VAT.
The list price of a DVD in 2001 was £780. In 2002 its list price was decreased by 15%.
Calculate the selling price of the DVD in 2002.

New list price of DVD = £780 × 0.85 — New list price is (100 − 15)% of old price. = 85% = 0.85. Keep your answer on your calculator.

New selling price of DVD

= £780 × 0.85 × 1.175 — Cost including VAT is (100 + $17\frac{1}{2}$)%. = $117\frac{1}{2}$% = 1.175

= £779.025 — This is not a sensible answer as you can't actually pay this amount.

= £779 — A shopkeeper would probably give the price to the nearest pound.

Example 12

£750 is invested for 3 years at 8% compound interest which is paid annually.

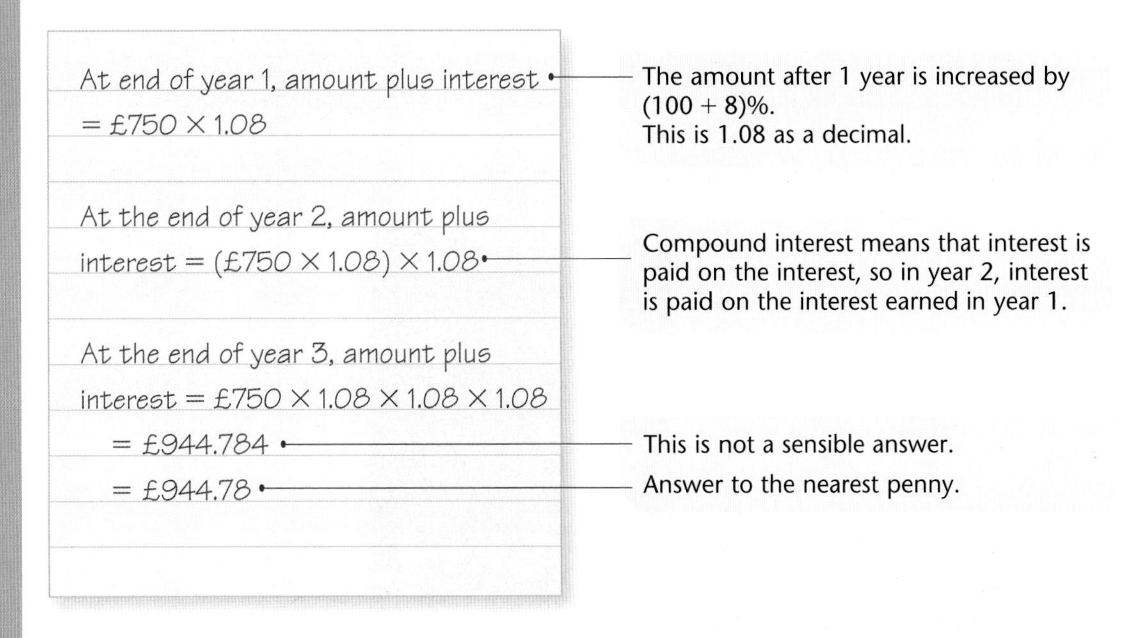

Practice 4H

1 In September the price of airfares rose by 10%. Customers who buy their ticket at least two weeks before departure receive a 20% discount.
In August the cost of the airfare between Birmingham and Belfast was £234.
Calculate the cost of this journey in September, including the discount.

2 Mary earns £130 a week. Mary is told that she is going to have her working time cut by 10% but that she will receive a 5% pay rise. Calculate her new weekly wage.

3 A new lawnmower costs £280. With depreciation, its value is expected to fall each year by 15% of its value at the beginning of the year. What will be the value of the lawnmower in three year's time?

4 £200 is invested for 3 years at 5% per annum compound interest. Work out the total interest earned over the 3 years. *E*

Hint: 'Per annum' means each year.

Hint: It is only the interest you are asked for.

5 £5000 is invested for 3 years at 4% per annum compound interest. Work out the total interest earned over the three years. *E*

6 By the end of each year, the value of a television has fallen by 12% of its value at the start of that year.
The value of a television was £423 at the start of the first year. Work out the value of the television at the end of the third year.

7 A shop buys Indian rugs from a factory.
In July, the cost to the shop of buying a rug was £100.
The shop bought 800 rugs in July.
In August, the cost to the shop of buying a rug increased by 10%.
The number of rugs bought by the shop decreased by 25%.
Find the difference between the total cost to the shop of all the rugs bought in July and the total cost of all the rugs bought by the shop in August. 

(E5)

5 Index notation and standard form

This chapter tells you how to manipulate numbers involving indices and how to express numbers in standard form.

5.1

Index notation is a short way of writing repeated multiplication by the same number.

Example 1

Rewrite these expressions using index notation:

a $4 \times 4 \times 4$

b $3 \times 3 \times 2 \times 2 \times 2 \times 2$

c $7 \times 7 \times 6 \times 6 \times 6 \times 6 \times 6 \times 2$

a $4 \times 4 \times 4 = 4^3$

This is called the **index** or **power**. It tells you to multiply 4 by itself 3 times.

b $3 \times 3 \times 2 \times 2 \times 2 \times 2 = 3^2 \times 2^4$

You can only use index notation for numbers that are the same.

c $7 \times 7 \times 6 \times 6 \times 6 \times 6 \times 6 \times 2$

$= 7^2 \times 6^5 \times 2^1$

$= 7^2 \times 6^5 \times 2$

The 2 has an index of 1 but you don't usually write it down.

Example 2

Evaluate:

a 5^3 **b** $3^3 \times 2^2$ **c** $7^2 \times 2^4 \times 5$

a $5^3 = 5 \times 5 \times 5$

$= 125$

'Evaluate' means 'work out a numerical answer'.

b $3^3 \times 2^2 = 3 \times 3 \times 3 \times 2 \times 2$

$= 108$

If you are not using a calculator, work out 3^3 and 2^2 separately then multiply the two answers together.

c $7^2 \times 2^4 \times 5$

$= 7 \times 7 \times 2 \times 2 \times 2 \times 2 \times 5$

$= 3920$

You could use a calculator for Example 2.

a $5^3 = 125$ — Press [5] [x^y] [3]

The [x^y] button is followed by the index.

b $3^3 \times 2^2 = 108$ — Press [3] [x^y] [3] [×] [2] [x^y] [2]

c $7^2 \times 2^4 \times 5 = 3920$ — Press [7] [x^y] [2] [×] [2] [x^y] [4] [×] [5]

Practice 5A

1 Rewrite these expressions using index notation:

a 4×4 **b** $8 \times 8 \times 8$ **c** $3 \times 3 \times 5 \times 5 \times 5$ **d** $8 \times 8 \times 8 \times 8 \times 2 \times 2 \times 2$

e $9 \times 9 \times 9 \times 7 \times 7 \times 1$ **f** $3 \times 2 \times 2 \times 3 \times 3 \times 3$ **g** $9 \times 9 \times 4 \times 3 \times 4 \times 9$

2 Evaluate, without using a calculator:

a 6^3 **b** 7^2 **c** 2^3

d $2^3 \times 5^2$ **e** $4^2 \times 3^2$ **f** $3^4 \times 10^2$

3 Work out the value of $5^3 \times 3^2$.

4 Use a calculator to evaluate

a $3^4 \times 2^5$ **b** $7^2 \times 9^3$ **c** $2^5 \times 4^3$ **d** $6^3 \times 4^6$

5 Work out the value of

$\sqrt{(4.5^2 - 0.5^3)}$

Write down all the figures on your calculator display.

Hint: To find a square root use the [$\sqrt{\ }$] button.

6 Use your calculator to find the value of

$\sqrt{(47.3^2 - 9.1^2)}$

Write down all the figures on your calculator display.

5.2 **You use the index rules to simplify expressions involving indices.**

$$a^n \times a^m = a^{n+m}$$
$$a^n \div a^m = a^{n-m}$$
$$(a^n)^m = a^{n \times m}$$

Example 3

Simplify

a $2^3 \times 2^4$ **b** $3^2 \times 3^5$ **c** $7^4 \div 7^2$

d $9^5 \div 9$ **e** $\dfrac{3^2 \times 3^4}{3^3}$ **f** $(3^2)^3$

a $2^3 \times 2^4 = 2^{3+4}$
$= 2^7$

To multiply indices of the same number add the indices.

b $3^2 \times 3^5 = 3^{2+5}$
$= 3^7$

c $7^4 \div 7^2 = 7^{4-2}$
$= 7^2$

To divide indices of the same number subtract the indices.

d $9^5 \div 9 = 9^{5-1}$
$= 9^4$

If a number doesn't have an index written down it is 1.

e $\frac{3^2 \times 3^4}{3^3} = \frac{3^{2+4}}{3^3}$
$= 3^6 \div 3^3$
$= 3^{6-3}$
$= 3^3$

f $(3^2)^3 = 3^{2 \times 3}$
$= 3^6$

Practice 5B

1 Simplify

a $3^4 \times 3^3$ **b** $2^4 \times 2^3$ **c** $7^2 \times 7^4$

d $5^6 \times 5^2$ **e** $10^7 \div 10^3$ **f** $\frac{6^4}{6^2}$

g $5^4 \times 5$ **h** $7^6 \div 7$ **i** $\frac{2^5 \times 2^3}{2^3}$

j $\frac{6^4}{6^2 \times 6}$ **k** $\frac{9^4 \times 9^3}{9^2 \times 9}$ **l** $\frac{4^3 \times 4^7}{4^2 \times 4}$

m $(2^4)^2$ **n** $(5^2)^3$

5.3 A number can be written as a product of prime factors.

Example 4

Write 360 as a product of its prime factors.

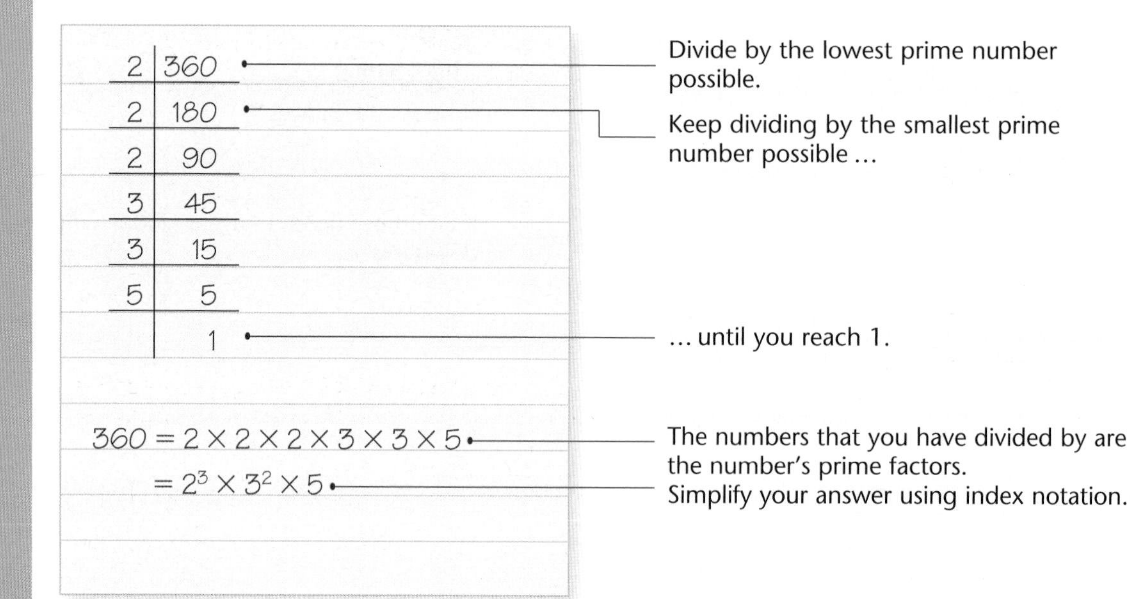

Practice 5C

Hint:
Prime numbers are described in Chapter 1.

1 Write the following as products of their prime factors:

a 24 **b** 48 **c** 96
d 144 **e** 480

2 The number 175 can be written as a product of its prime factors

$175 = 5^2 \times 7$

Write as a product of its prime factors

a 50 **b** 50^2

5.4 Standard form is used as an alternative way of writing very large or very small numbers.

A number is in standard form when it is written like this:

$$a \times 10^n$$

a = a number from 1 up to (but not including) 10

n = an index of 10 where the index is an integer

Example 5

Write the following numbers in standard form.

a 7 800 000 **b** 356 000 **c** 0.003
d 0.000 25 **e** 25×10^4

a 7800000

7800000

$= 7.8 \times 10^6$

An easy way to find the index of 10 is count the number of times the decimal point needs to move.

The index of 10 is positive when the d.p. moves to the left.

b 356000

356000

$= 3.56 \times 10^5$

c 0.003

0.003

$= 3.0 \times 10^{-3}$

The index of 10 is negative when the decimal point moves to the right.

d 0.00025

0.00025

$= 2.5 \times 10^{-4}$

e 25×10^4

This number looks like it is in standard form but 25 is not between 1 and 10.

$25 \times 10^4 = 2.5 \times 10^1 \times 10^4$

$= 2.5 \times 10^{4+1}$

$= 2.5 \times 10^5$

Example 6

Write as ordinary numbers

a 3.2×10^4 **b** 2.6×10^{-5}

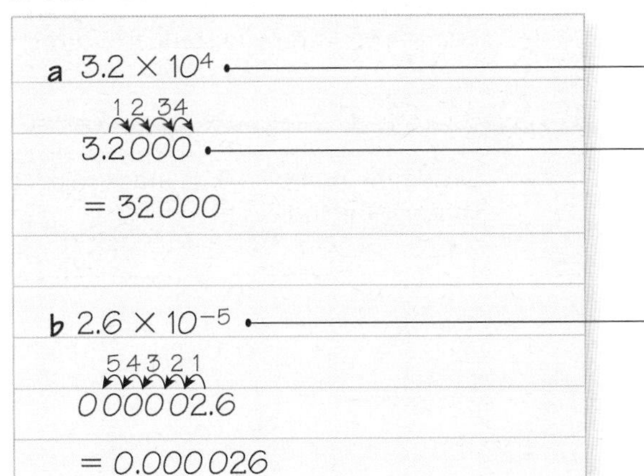

The index tells you how many places to move the decimal point.

For a positive index move the decimal point to the right.

For a negative index move the decimal point to the left.

Practice 5D

1 Write the following numbers in standard form:

a 4800 **b** 370000 **c** 23000 **d** 0.00034
e 0.0025 **f** 0.03 **g** 347.1 **h** 23.78
i 42×10^4 **j** 0.2×10^5 **k** 37×10^{-2} **l** 0.5×10^{-3}

2 Write the following as ordinary numbers:

a 4×10^3 **b** 2.6×10^4 **c** 3.4×10^5
d 2×10^{-3} **e** 5.2×10^{-5} **f** 6.2×10^{-1}

3 **a** Write the number 5.01×10^4 as an ordinary number.
b Write the number 0.0009 in standard form.

5.5 Numbers in standard form can be used in calculations.

Example 7

Work out the following, giving your answer in standard form:

a $(2 \times 10^2) \times (4 \times 10^7)$
b $(4 \times 10^4) \times (3 \times 10^5)$
c $(4 \times 10^6) \div (8 \times 10^3)$

a $(2 \times 10^2) \times (4 \times 10^7)$
$= 2 \times 4 \times 10^2 \times 10^7$ — Rearrange so that the indices of 10 are together.
$= 8 \times 10^{2+7}$ — Multiply these numbers. Use index rules to add the indices of 10.
$= 8 \times 10^9$

b $(4 \times 10^4) \times (3 \times 10^5)$
$= 4 \times 3 \times 10^4 \times 10^5$
$= 12 \times 10^{4+5}$
$= 12 \times 10^9$ — This looks like standard form, but 12 is not between 1 and 10.
$1.2 \times 10^1 \times 10^9 = 1.2 \times 10^{10}$ — The answer is now in standard form.

c $(4 \times 10^6) \div (8 \times 10^3)$
$= (4 \div 8) \times 10^{6-3}$ — Divide the numbers. Use index rules to subtract the indices of 10.
$= 0.5 \times 10^{6-3}$
$= 0.5 \times 10^3$
$= 5 \times 10^2$

Example 8

Use a calculator to calculate $(2.4 \times 10^9) \div (8 \times 10^{-5})$

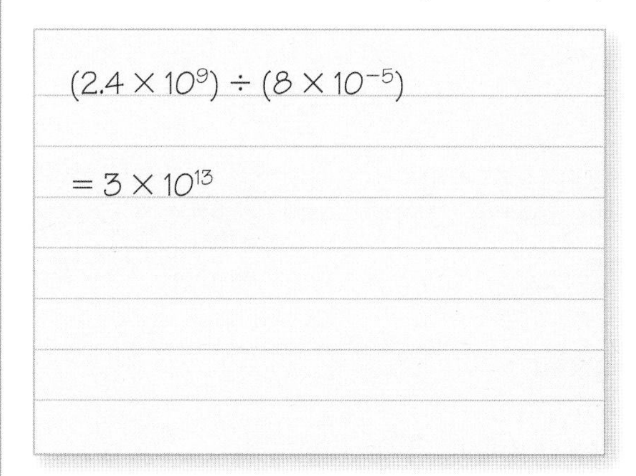

[2] [.] [4] [EXP] [9] [÷] [8] [EXP] [5] [+/−] [=]

Enter the number part of the standard form then press [EXP] (this may be [EE] or [10^x] on your calculator) then the index of 10.

= [3 13]

Do not write down the number as displayed – this is not standard form.

Practice 5E

Do not use a calculator for questions 1–4.

1 Work out the following, giving your answers in standard form:

a $(2 \times 10^4) \times (3 \times 10^5)$
b $(1.7 \times 10^4) \times (3 \times 10^6)$
c $(2 \times 10^4) \times (3 \times 10^{-3})$
d $(8 \times 10^3) \div (4 \times 10^2)$
e $(9 \times 10^{-2}) \div (3 \times 10^5)$
f $(7 \times 10^4) \div (2 \times 10^5)$
g $(7 \times 10^4) \times (8 \times 10^5)$
h $(2.5 \times 10^6) \times (5 \times 10^{-3})$
i $(2.4 \times 10^3) \div (3 \times 10^5)$
j $(3 \times 10^{-4}) \div (4 \times 10^3)$

2 Multiply 4×10^3 by 6×10^5.
Give your answer in standard form. *E*

3 $p = 8 \times 10^3$ $\quad q = 2 \times 10^4$

a Find the value of $p \times q$.
Give your answer in standard form.

b Find the value of $p + q$.
Give your answer as an ordinary number. *E*

4 Work out the value of

$$\frac{3 \times 10^{-6}}{4 \times 10^{-4}}$$

Give your answer in standard form. *E*

5 Calculate, giving your answer in standard form

a $(6.2 \times 10^3) \div (3.2 \times 10^5)$
b $(7.8 \times 10^5) \times (2.3 \times 10^4)$
c $(5.2 \times 10^{-3}) \div (2.1 \times 10^{-4})$
d $(4.9 \times 10^4) \times (3.6 \times 10^{-5})$

6 **a** Write 84 000 000 in standard form.

b Work out

$$\frac{84\,000\,000}{4 \times 10^{12}}$$

Give your answer in standard form. *E*

7 Calculate the value of

$$\frac{(5.98 \times 10^8) + (4.32 \times 10^9)}{6.14 \times 10^{-2}}$$

Give your answer in standard form correct to 3 significant figures.

8 Light travels at a speed of 3×10^8 metres per second.
How long will it take light to travel a distance of 9×10^{23} metres? Give your answer in standard form.

Hint: Time $= \frac{\text{distance}}{\text{speed}}$

9 The mass of a neutron is 5.84×10^{-24} grams.
Calculate the total mass of 4.3×10^6 neutrons.
Give your answer in standard form correct to 3 significant figures.

10 The distance from Earth to the star Proxima Centauri is 4.22 light years.
1 light year $= 9.461 \times 10^{12}$ km.
Work out the distance in kilometers from Earth to the star Proxima Centauri.
Give your answer in standard form correct to 3 significant figures. *E*

(E6) (E7)

6 Ratio and proportion

This chapter covers calculations involving ratio and direct proportion.

6.1 **Two or more quantities can be compared using a ratio. Ratios are always written in the form a : b.**

Example 1

Write these ratios in their simplest form:

a 8 : 4 **b** 100 m : 50 m **c** £1 : 25p

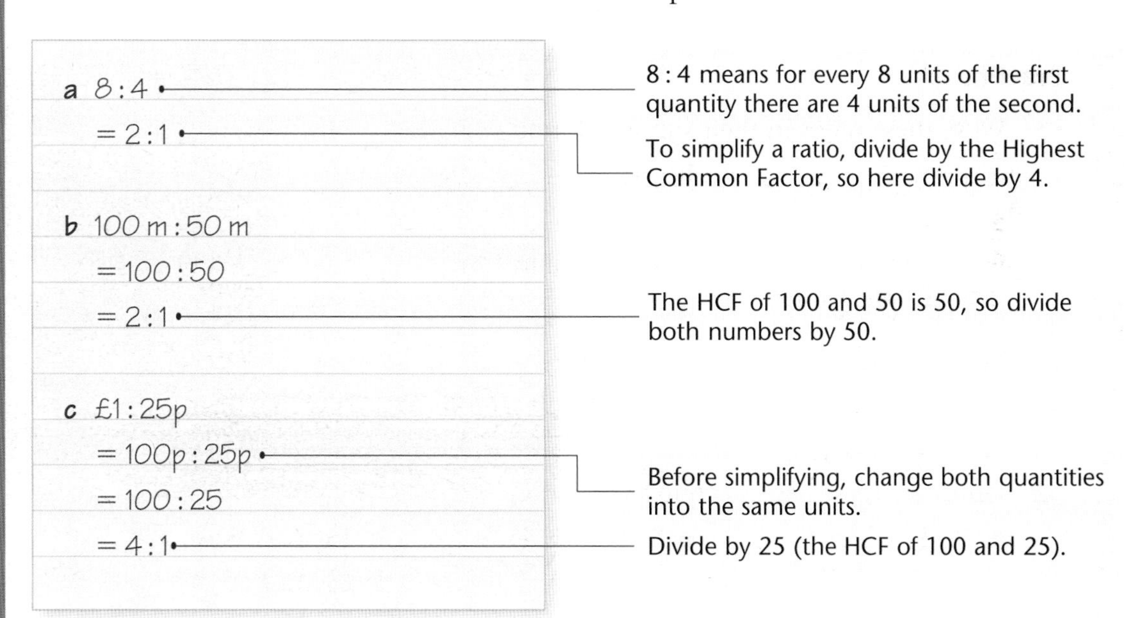

Example 2

a The ratios 10 : 4 and 15 : x are equivalent. Find x.

b The ratio of boys to girls in a college is 2 : 3.
There are 560 boys in the college. Work out the number of girls in the college.

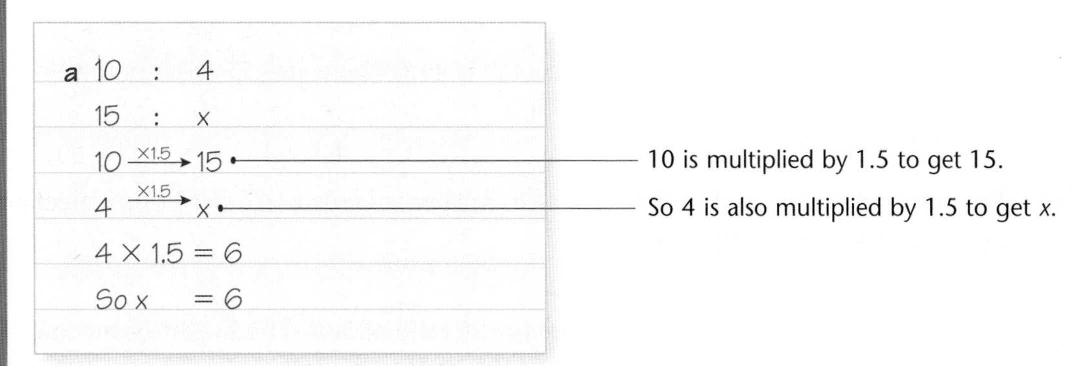

b The ratio of boys to girls is 2 : 3.
It is also 560 : x — Use x for the number of girls.
560 = 2 parts
280 = 1 part
Number of girls
= 3 × 280
= 840

Practice 6A

1 Write these ratios in their simplest form:

a 18 : 9 **b** 36 : 24
c 18 : 9 : 3 **d** 24 : 18 : 6
e £30 : £5 **f** 20 km : 4 km
g 1 km : 250 km **h** £2.50 : 50p
i 3 mins : 30 sec **j** 2 kg : 500 g
k 2 hr : 1 hr : 15 min **l** 2 m : 50 cm : 15 cm

Hint: Remember to change quantities so that the units are the same.

2 Find x for these pairs of equivalent ratios.

a 10 : 5 2 : x **b** 8 : 4 x : 1 **c** 15 : 3 5 : x
d 20 : x 4 : 1 **e** x : 18 2 : 3 **f** 25 : 15 x : 3

3 The ratio of blue paint to yellow paint used to produce a particular shade of green paint is 2 : 3. If 6 litres of blue paint is used to make a quantity of green paint, how much yellow paint should be used?

4 The ratio of sides of a rectangle is 5 : 3. The longer side of the rectangle is 30 cm. Calculate the length of the shorter side.

5 The ratio of men to women in a sports club is 4 : 3. There are 175 women in the club. How many men belong to the sports club?

6 Kim and Donna are paid in the ratio of 6 : 5. Donna earns £4.75 per hour. Calculate how much Kim earns per hour.

7 The heights of two similar triangles are in the ratio of 3 : 2. The height of the larger triangle is 21 cm. Calculate the height of the smaller triangle.

6.2 You can share quantities in a given ratio.

Example 3

Divide £20 in the ratio 3 : 2.

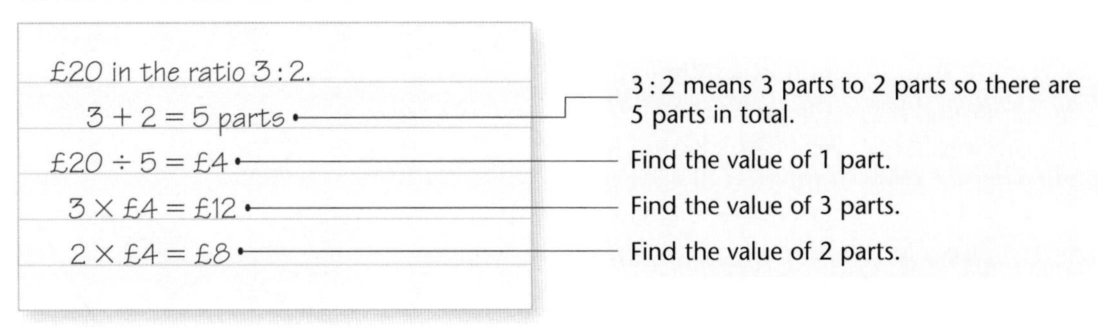

Example 4

Rachael, Davina & Helen share £200 in the ratio 5 : 3 : 2.
How much does Davina receive?

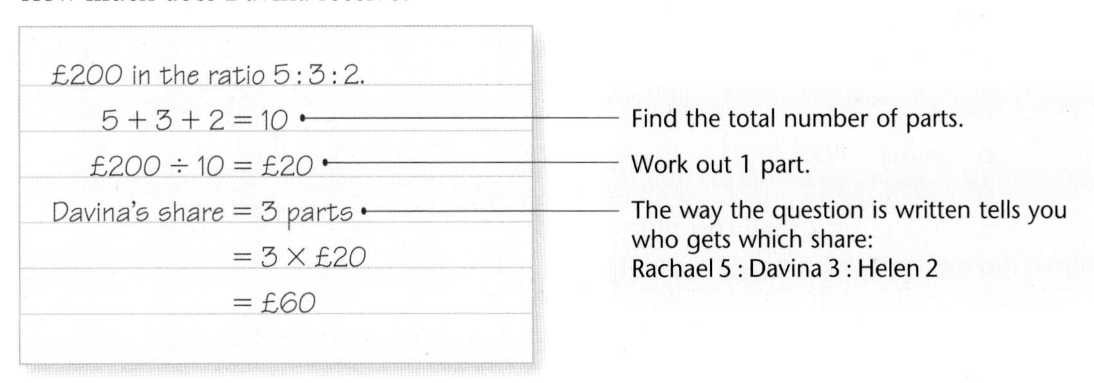

Example 5

$\frac{2}{5}$ of a box of chocolates are plain, the rest are milk chocolates.
What is the ratio of plain chocolates to milk chocolates?

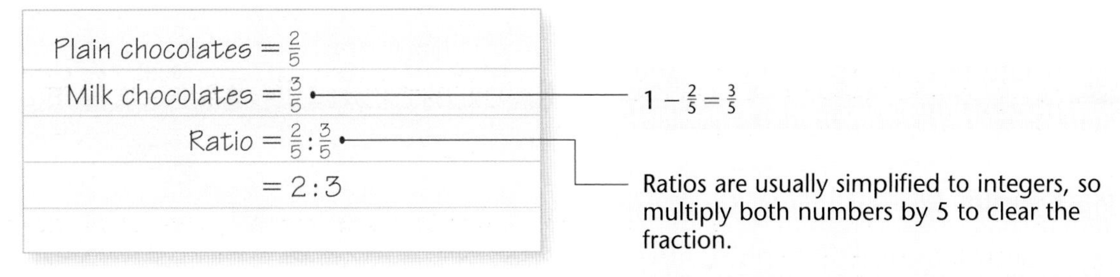

Practice 6B

1 Divide the quantities in the ratios given.

a 45 in the ratio 4 : 5

b 60 in the ratio 5 : 7

c 20 in the ratio 5 : 4 : 1

d £30 in the ratio 3 : 2

e £120 in the ratio 7 : 3 : 2

f 90 cm in the ratio 7 : 5 : 3

2 Tracy and Wayne share £7200 in the ratio 5 : 4.
Work out how much each of them receives.

3 A 500 g loaf of bread is made using plain and wholewheat flour in the ratio 7 : 3.
Calculate how much of each type of flour is used to make the loaf.

4 Rashid has 35 sweets.
He shares them in the ratio 4 : 3 with his sister.
Rashid keeps the larger share.
How many sweets does Rashid keep?

5 A 1 metre length of wood is sawn into three pieces in the ratio 5 : 3 : 2. Calculate the length of each piece of wood.

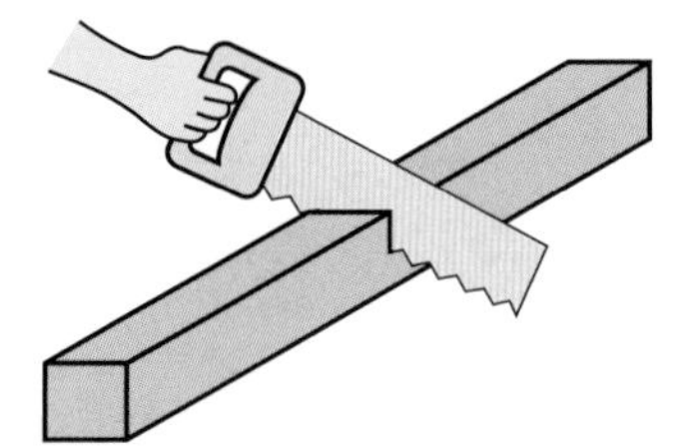

6 Bill gave his three daughters a total of £32.40.
The money was shared in the ratio 4 : 3 : 2.
Jane had the largest share.
Work out how much money Bill gave to Jane.

7 Anna, Beth and Cheryl share the total cost of a holiday in the ratio 6 : 5 : 4.
Anna pays £294.

a Work out the total cost of the holiday.

b Work out how much Cheryl pays.

E

8 $\frac{1}{6}$ of the pages of a book are printed in colour.
The rest of the pages are printed in black and white.
Write down the ratio of colour pages to black and white pages.

9 $\frac{3}{7}$ of the books in a library are fiction. The rest are non-fiction.
Write down the ratio of fiction to non-fiction books in the library.

6.3 Quantities whose ratio stays the same as they increase or decrease are in direct proportion.

Example 6

A car will travel 420 km on 70 litres of petrol.
How far will the car travel on 25 litres of petrol?

$420 \div 70 = 6$ km/ℓ — To find the distance the car will travel on 1 litre of petrol, divide distance by litres.

$6 \times 25 = 150$ km — Multiply by 25 to find the distance travelled on 25 litres.

Example 7

Shortcake biscuit
(makes 18)
8 oz flour
4 oz butter
2 oz sugar

Calculate the amount of each ingredient needed to make 27 biscuits.

Ratio = 18 : 27 or 2 : 3.

The recipe is for 18 biscuits, but you need 27.
Simplifying the ratio will make it easier to work with.

Ratio of flour = 8 : x

$$\frac{2}{3} = \frac{8}{x}$$

x = 12

12 oz flour

As the quantities are in direct proportion to the number of biscuits, you can use equivalent ratios to solve the problem.
8 is 2×4, so x is $3 \times 4 = 12$

Ratio of butter = 4 : x

$$\frac{2}{3} = \frac{4}{x}$$

x = 6

6 oz butter

4 is 2×2, so x is $3 \times 2 = 6$

Ratio of sugar = 2 : x

$$\frac{2}{3} = \frac{2}{x}$$

x = 3

3 oz sugar

Practice 6C

1 16 books cost £24. Work out the cost of
a 1 book **b** 13 books.

2 Dress fabric costs £14.70 for 5 metres. Work out the cost of
a 1 metre of fabric **b** 13 metres of fabric.

3 The cost of 5 metres of wire is £3.
What is the cost of 8 metres of the same wire? *E*

4 It takes 20 litres of orange drink to fill 50 cups.
Work out how many litres of orange drink are needed to fill 60 cups. *E*

5 The cost of 15 pens is £2.45.
Work out the cost of 9 pens.

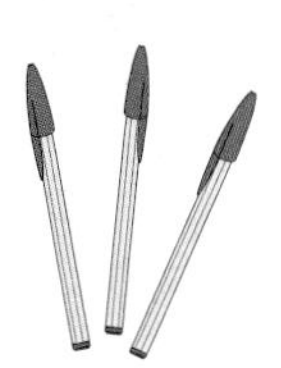

E

6 £15 can be exchanged for €25.
Eric changes £320. How many euros should he receive?

7 Robert used these ingredients to make 24 buns.

100 g of butter
80 g of sugar
2 eggs
90 g of flour
30 ml of milk

Robert wants to make 36 similar buns.
Write down how much of each ingredient he needs for 36 buns.

E

8 This is a recipe for making a Tuna Bake for 4 people.

TUNA BAKE
Ingredients for 4 people
400 g tuna
400 g mushroom soup
100 g grated cheddar cheese
4 spring onions
250 g breadcrumbs

Work out the amounts needed to make a Tuna Bake for 10 people.

E

7 Algebra

Algebra is used to describe relationships between quantities. In algebra you use letters to represent numbers. **(S4)**

7.1 You can simplify expressions by collecting like terms.

Example 1

Simplify these expressions

a $3a - 2b + 4a + 5b$ **b** $3a^2b + 5ab^3 - a^2b + 2ab^3 - ab^3$

a $3a - 2b + 4a + 5b$

$= 3a + 4a - 2b + 5b$ — Rewrite the expression with the like terms next to each other.

$= 7a + 3b$ — $-2b + 5b = 3b$

b $3a^2b + 5ab^3 - a^2b + 2ab^3 - ab^3$

$= 3a^2b - a^2b + 5ab^3 + 2ab^3 - ab^3$ — a^2b and ab^3 are not like terms: $a^2b = a \times a \times b \quad ab^3 = a \times b \times b \times b$

$= 2a^2b + 6ab^3$ — $5ab^3 + 2ab^3 - ab^3 = 6ab^3$

$3a$ and $4a$ are called **like terms**. Like terms have the same letters.

Practice 7A

1 Simplify these expressions:

a $2x + 3y - x + 2y$ **b** $4p - 3q + 2p + 5q$

c $3c + 5d - c + 6d + 4c$ **d** $5a - 7b - 4a - 2b + 12b$

e $5ef + 7gh - 9ef + 2gh$ **f** $5r - 3s - 7s + 2r - s$

g $7t^2 - 9t^2 + 10t^2 - t^2$ **h** $3x^2y + 4x^2y^2 - 3x^2y + x^2y^2$

i $-2cd^3 + 2c^2d - 3cd^3 + 2c^2d + 6cd^3$ **j** $3ab^3 - 2a^2b + 5ab^3 + 6a^2b - 3a^2b$

Hint: x^2y and x^2y^2 are not like terms.

7.2 Removing brackets from an expression is called multiplying out the brackets.

Example 2

Multiply out the brackets in these expressions:

a $2(a + 3)$ **b** $9(3b + 4c - 2d)$

a $2(a + 3)$
$= 2 \times a + 2 \times 3$
$= 2a + 6$

Multiply the term outside the bracket by both terms inside the bracket:
$2 \times a = 2a$
$2 \times 3 = 6$
So $2(a + 3) = 2a + 6$

b $9(3b + 4c - 2d)$
$= 27b + 36c - 18d$

You can sometimes write down the answer straight away.

Practice 7B

1 Multiply out the brackets:

a $2(a + 3)$
b $3(2p + 3)$
c $-4(x - 3)$
d $-4(2x + 3)$
e $7(4 + 5x)$
f $-(3 + x)$
g $3(4x + 5y)$
h $6(5x - 7)$
i $-2(5 + 2y - 3w)$
j $9(3b + 4c - 2d)$

2 Simplify:

a $5x + 2(3x + 1)$
b $3a + 4(a - 2)$
c $4(3m + 2) + 5$
d $2(3x - 5) + 12$
e $2(2x - 1) + 3(x + 4)$
f $3x + 7 - 2(2x + 5)$
g $-2(2x + 1) + 7(x + 3)$
h $3(2a + b) + 4(5a + 2b)$
i $4(p + 3q) - 2(3p + 2q)$
j $4(4a + 5) - 2(3a - 2)$

Hint: multiply both terms inside the bracket by -2.

7.3

When exploring sequences you can find:
- **a rule to find the next term.**
- **a rule to find the *n*th term.**

Example 3

For the sequence 3, 8, 13, 18, 23, …

a Work out the next two terms.
b Write down a rule to find the next term.
c Write down an expression to find the *n*th term.

a The next two terms are 28 and 33

To go from one term to the next you add 5.

b

Term no.	1	2	3	4	5
Term	3	8	13	18	23

The rule to find the next term is 'add 5'.

Write out the term numbers and the values of the terms in the sequence first. Look for the rule that takes you from one term to the next.

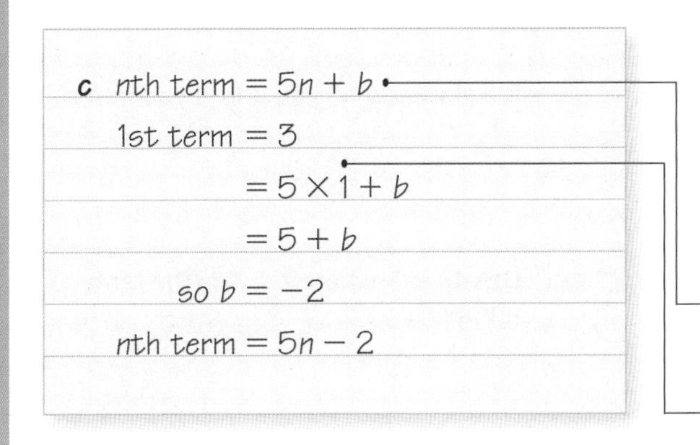

c nth term $= 5n + b$

1st term $= 3$

$= 5 \times 1 + b$

$= 5 + b$

so $b = -2$

nth term $= 5n - 2$

The general rule for the sequence will look like:

$$n\text{th term} = an + b$$

where a and b are numbers. The value of a is the difference between the terms.

Substitute a value to find b.

For the 1st term, $n = 1$.

Practice 7C

1 For each sequence

i work out the next two terms

ii work out an expression to find the nth term.

a 4, 7, 10, 13, …　**b** 2, 3, 4, 5, …　**c** 4, 8, 12, 16,…

d 3, 7, 11, 15, 19, …　**e** 7, 13, 19, 25, 31, …　**f** 1, 4, 7, 10, 13, …

g 21, 15, 9, 3, −3, …　**h** 4.5, 2.0, −0.5, −3.0, …　**(E8)**

7.4

When exploring a pattern of shapes you can find:

- **a rule for the next term.**
- **a rule for the *n*th term.**

Example 4

Here is a pattern of shapes made from matchsticks:

  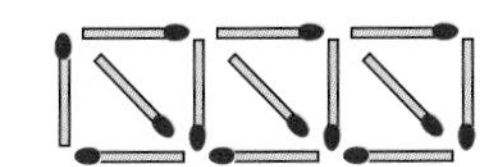

a Draw the next pattern in the sequence.

b Copy and complete this table of results:

Pattern (term no)	1	2	3	4	5
Number of matchsticks (term)	5	9	13		

c Find an expression for the number of matchsticks in the nth term.

d How many matchsticks would you need to make the 46th term?

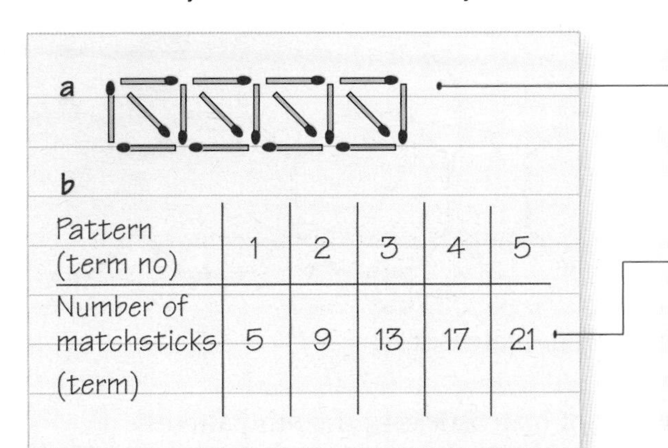

a

b

Pattern (term no)	1	2	3	4	5
Number of matchsticks (term)	5	9	13	17	21

Copy the pattern for the 3rd term and put the extra matchsticks in to continue the pattern to make the 4th term. The number of matchsticks is 17.

You add 4 to the first term to get 2nd term, then 4 to the 2nd term to get the 3rd and so on

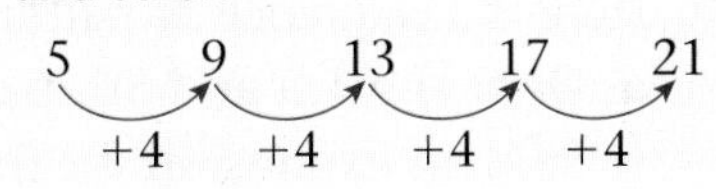

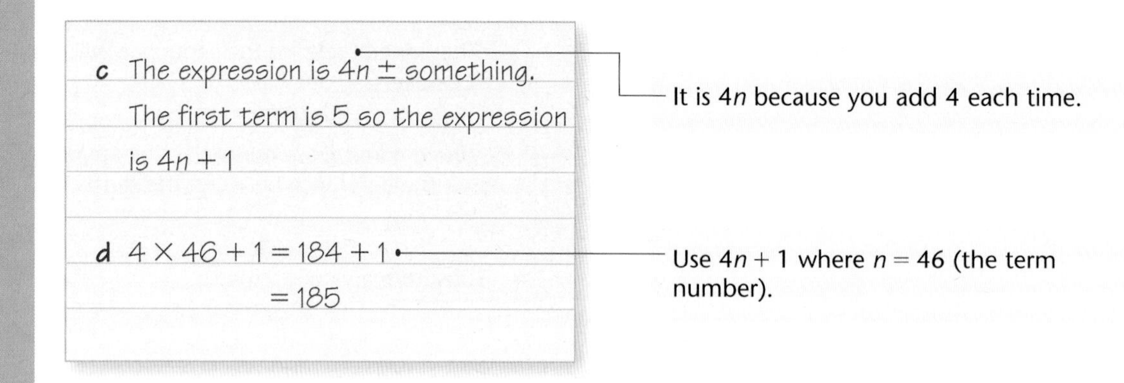

c The expression is $4n \pm$ something.
The first term is 5 so the expression
is $4n + 1$

It is $4n$ because you add 4 each time.

d $4 \times 46 + 1 = 184 + 1$
$= 185$

Use $4n + 1$ where $n = 46$ (the term number).

Practice 7D

1

Here is a sequence of patterns made from matchsticks.

a Draw the next pattern in the sequence.
b Find an expression for the number of matchsticks in the *n*th pattern.
c Write down the number of matchsticks in the 38th term.
d Work out the number of matchsticks in the 26th pattern.

2

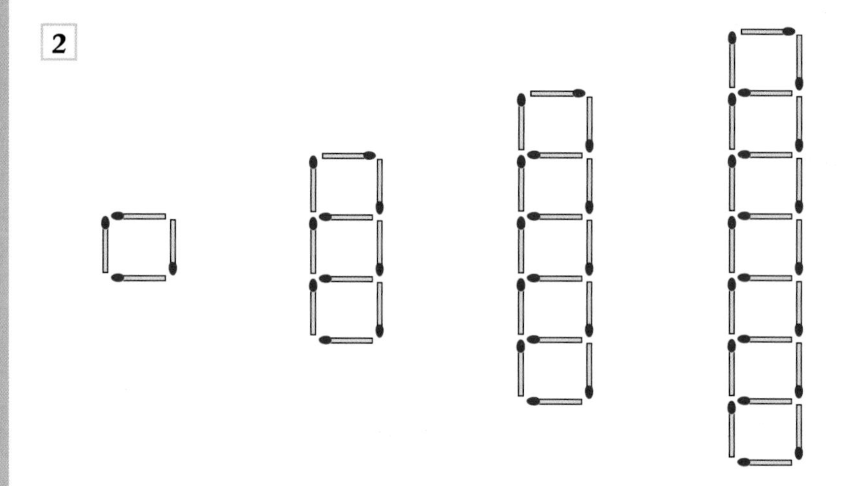

Here is a pattern of squares.

a Write down the first 5 terms in the sequence of squares.
b Work out the expression for the *n*th term of the number of squares in the *n*th pattern.

3

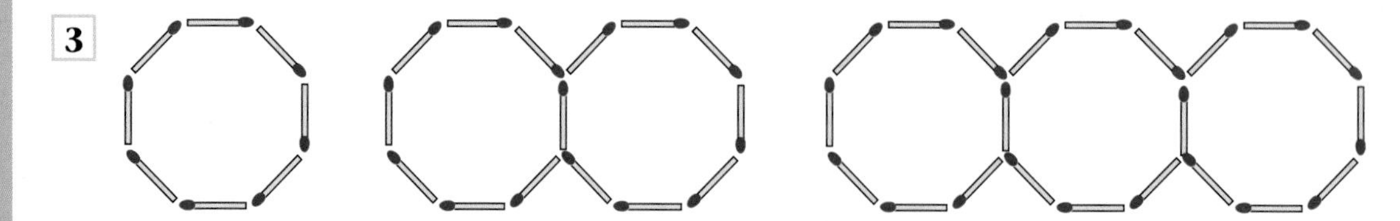

Here is a sequence of patterns made from matchsticks.

a Draw the 4th and 5th pattern.
b Find an expression for the number of matchsticks in the *n*th pattern.
c How many matchsticks would you need to make the 50th pattern?

4 Here are some patterns made of crosses.

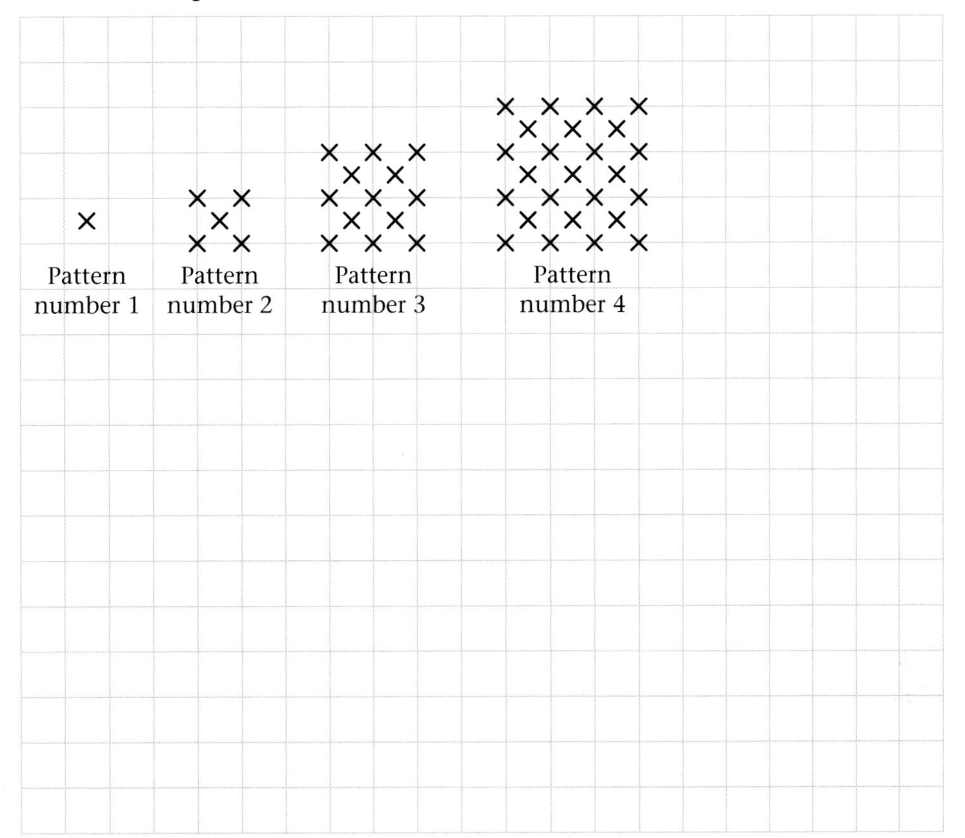

a Draw Pattern number 5.

b Complete the table.

Pattern number	1	2	3	4	5
Number of crosses	1	5	13		

c Work out the number of crosses in Pattern number 8.

E

7.5 You can simplify expressions by using rules of indices (or powers):

$$a^n \times a^m = a^{n+m}$$
$$a^n \div a^m = a^{n-m}$$
$$(a^n)^m = a^{n \times m}$$

Example 5

Simplify these expressions:

a $x^4 \times x^3$ **b** $3y^2 \times 2y$ **c** $p^5 \div p^3$

d $2x^{-4} \div 4x^{-5}$ **e** $(x^2)^3 \times 3x^2$ **f** $(5d)^2 \div d^{-3}$

a $x^4 \times x^3$

$= x^{4+3}$ — Use the rule $a^n \times a^m = a^{n+m}$ to simplify the power.

$= x^7$

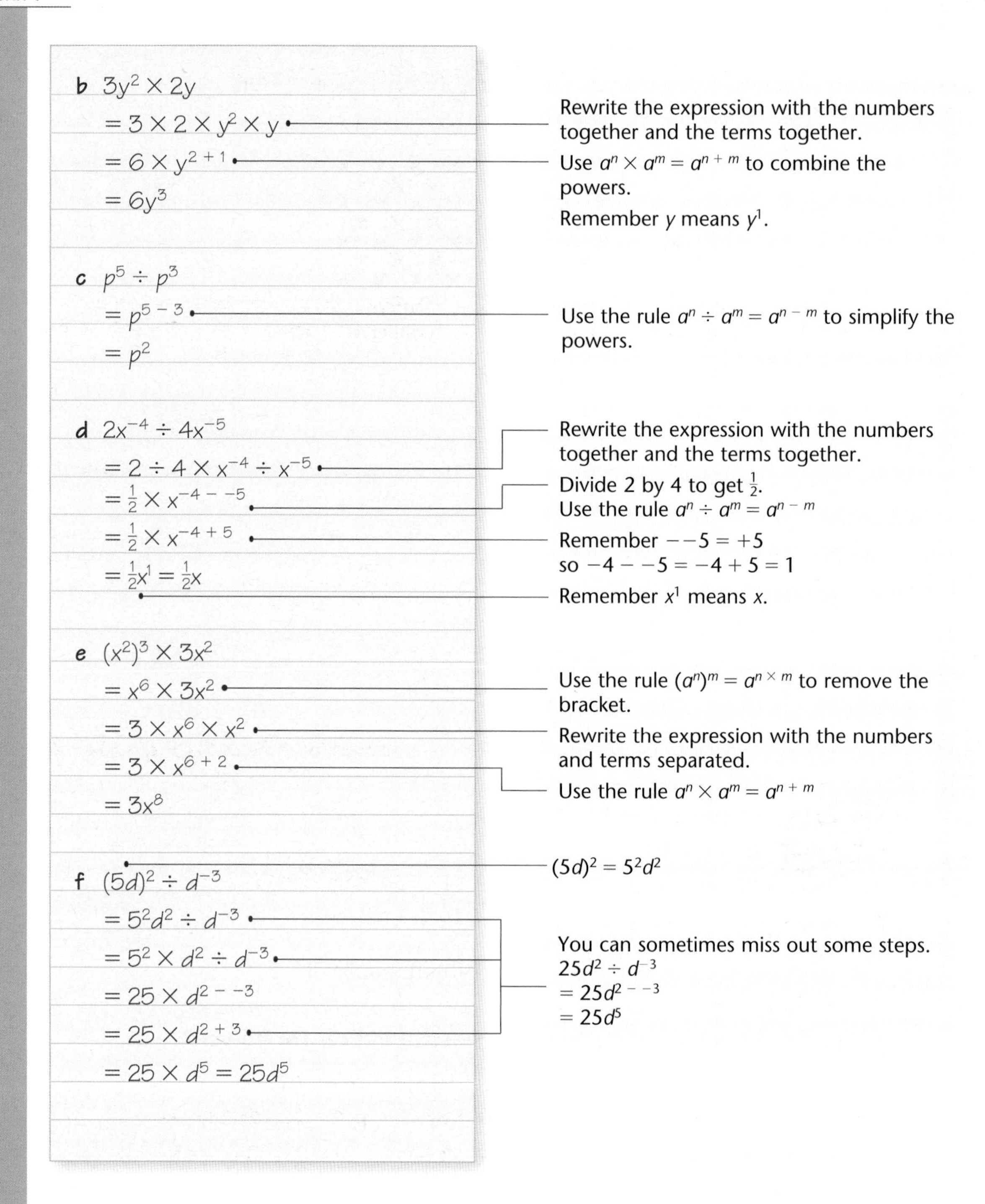

Practice 7E

1 Simplify the expressions:

a $b \times b$ **b** $d^2 \times d^5$ **c** $e^7 \div e^2$ **d** $(y^2)^3$

e $x^4 \div x^7$ **f** $6c^2 \div 2c$ **g** $3p^3 \times 2p^2$ **h** $(3x^2)^2$

i $x^{-7} \times x^2$ **j** $3b^5 \div b^2$ **k** $3d^{-2} \div 6d^{-3}$ **l** $(3y)^2 \div 2y^3$

m $(4x)^2 \div x^{-5}$ **n** $(a^{-2})^{-3}$ **o** $10e^2 \times 5e^3$ **p** $(4x^2y)^3$

7.6 Multiplying out brackets is used to expand and simplify expressions.

Example 6

Expand and simplify:

a $(y + 2)(y + 3)$ **b** $(2b - 2)(3b + 5)$ **c** $(3x + 2)^2$

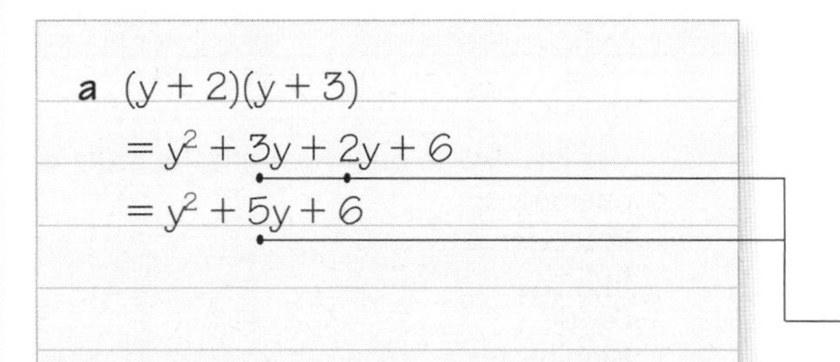

You must multiply each term in each bracket by each term in the outer bracket. You can think of this as

$(y + 2)\ (y + 3)$

$3y$ and $2y$ are like terms so combine them.

b $(2b - 2)(3b + 5)$

$= 6b^2 + 10b - 6b - 10$

$= 6b^2 + 4b - 10$

You use the rule $a^m \times a^n = a^{m+n}$ to combine the indices
$b \times b = b^{1+1} = b^2$

c $(3x + 2)^2$

$= (3x + 2)(3x + 2)$

$= 9x^2 + 6x + 6x + 4$

$= 9x^2 + 12x + 4$

Rewrite $(3x + 2)^2$ *as* $(3x + 2) \times (3x + 2)$

Combine the like terms.

Practice 7F

1 Expand and simplify:

a $(a + 1)(a + 2)$ **b** $(b - 2)(b + 3)$ **c** $(x + 3)(x - 2)$

d $(2x + 1)(x + 3)$ **e** $(3x + 2)(x - 3)$ **f** $(3d - 2)(2d - 1)$

g $(3t + 2)(5t - 2)$ **h** $(x + 2)(x - 2)$ **i** $(2a - 3)(a + 5)$

j $(3e - 5)(2e - 1)$ **k** $(2p + 1)(3p - 8)$ **l** $(3 - x)(2 - 4x)$

m $(d - 5)(3d - 4)$ **n** $(3t - 2)(t + 4)$ **o** $(3p - 1)(3p + 1)$

p $(4 + x)^2$ **q** $(2a + 5)^2$ **r** $(3y - 1)^2$

s $(5p + 2)^2$ **t** $(3x - 4)^2$

8 Formulae and expressions

This chapter covers how to work out values using formulae and how to simplify expressions.

8.1 You can work out the values of expressions and formulae by substituting values.

Example 1

Work out the value of these algebraic expressions using the values given.

a $\dfrac{4a - b}{2c}$ when $a = 5$, $b = 6$, $c = 7$

b $\dfrac{3p^2 - 2q}{q}$ when $p = 5$, $q = 15$

Use BIDMAS to help you with the order of operations:
Brackets
Indices
Divide
Multiply
Add
Subtract

a $\dfrac{4a - b}{2c}$ — The line acts as a bracket, so work out the top first. Apply BIDMAS to the top and bottom separately.

$4a = 4 \times 5 = 20$ — Multiply.

$4a - b = 20 - 6 = 14$ — Subtract.

$2c = 2 \times 7 = 14$ — Work out the bottom.

$\dfrac{4a - b}{2c} = \dfrac{14}{14} = 1$ — Divide.

b $\dfrac{3p^2 - 2q}{q}$ — Bracket, the dividing line acts as a bracket so work out the top first, using BIDMAS.

$p^2 = 5 \times 5 = 25$ — Indices.

$3p^2 = 3 \times 25 = 75$ — Multiply.

$2q = 2 \times 15 = 30$ — Multiply.

$3p^2 - 2q = 75 - 30 = 45$ — Subtract.

$\dfrac{3p^2 - 2q}{q} = \dfrac{45}{15} = 3$ — Divide; now divide the top by the bottom.

(S5)

Practice 8A

Work out the values of these algebraic expressions using the values given:

1 $2a + 3b$ $(a = 3, b = 5)$

2 $4c - 2d$ $(c = 5, d = 3)$

3 $3x - 2y$ $(x = 4, y = -1)$

4 $3d^2 - 2e$ $(d = 5, e = 9)$

5 $2p^2 + 7q$ $(p = 3, q = -4)$

6 $m^2 - 2n^2$ $(m = 7, n = 3)$

7 $3k - 2l + m^2$ $(k = 1, l = 8, m = 2)$

8 $2(3a + 2b)$ $(a = 2, b = 5)$

9 $b^2 + 4d$ $(b = -2, d = 3)$

10 $(2x - 3y)^2$ $(x = 2, y = 3)$

11 $\frac{p^2 - 3q}{4}$ $(p = 10, q = 8)$

12 $\frac{3a^2 - 2b}{c}$ $(a = 4, b = 6, c = 3)$

13 $3x + 2y$, $x = -4, y = 5$ **E**

14 $\pi(R^2 - r^2)h$, $\pi = \frac{22}{7}, R = 4, r = 0.5, h = 2\frac{3}{4}$ 

For questions 15 and 16, give your answer correct to 3 sf.

15 $2\pi\sqrt{\frac{l}{g}}$, $\pi = 3.14, l = 1.4, g = 9.8$ **E**

16 $\frac{a - 3c}{a - c^2}$, $a = 19.9, c = 4.05$ **E**

Hint: The square root acts as a bracket so work out the square root first.

8.2 A formula is a way of describing a rule or a relationship using algebraic expressions. You can find values by substituting into a formula.

Example 2

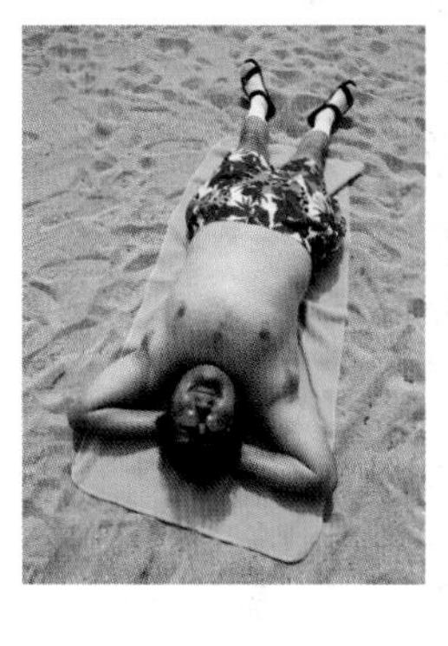

The formula to change degrees Fahrenheit (°F) to degrees Celsius (°C) is

$$C = \frac{5}{9}(F - 32)$$

Work out the temperature in Celsius when $F = 86°$

$C = \frac{5}{9}(F - 32)$

$F - 32 = 86 - 32 = 54$ — Brackets.

$\frac{5}{9}(54) = \frac{5}{9} \times 54 = 30$ — Multiply (the bracket means multiply).

So $C = 30$

The temperature is 30 °C

Use BIDMAS to help with the order of operations.

Practice 8B

1 $v = u + at$

Calculate the value of v when

a $u = 0, a = 5, t = 3$

b $u = 10, a = -10, t = 5$

c $u = 12.5, a = 5.2, t = 7$

Give your answer correct to 2 dp.

2 The area of a trapezium is given by the formula

$$A = \frac{1}{2}(a + b)h$$

Find the value of A when

a $a = 2, b = 8, h = 3$ **b** $a = 3.2, b = 6, h = 4.5$ **c** $a = 8.3, b = 9.4, h = 1.9$

3 The formula to change Celsius (°C) to Fahrenheit (°F) is

$$F = \frac{9C}{5} + 32$$

Work out the temperature in °F when C is:

a 18°C **b** 100°C **c** −5°C

4 The cost, in pounds (£), of hiring a car is given by the formula

$$C = 0.12m + 15$$

C is the cost of hiring the car and m is the number of miles travelled.
A customer hires a car and travels 315 miles. Use the formula to work out the total cost of hire.

5 A formula used in physics is

$$v^2 = u^2 + 2as$$

Calculate the value of v when $u = -15$, $a = 2.8$ and $s = 86$.

Give your answer correct to 2 dp.

Hint: Work out $u^2 + 2as$ first then square root to find v.

6 Kinetic energy is calculated using the formula

$$K = \frac{1}{2}mv^2$$

Calculate the value of K when $m = 30$ and $v = 9.16$.

Give your answer correct to 2 dp. **(E9)**

8.3 A formula is a mathematical way of expressing a rule.

Example 3

Mary sells p pork pies and q coffees every day for d days.
Write down a formula T for the total number of pork pies and coffees she sells in d days.

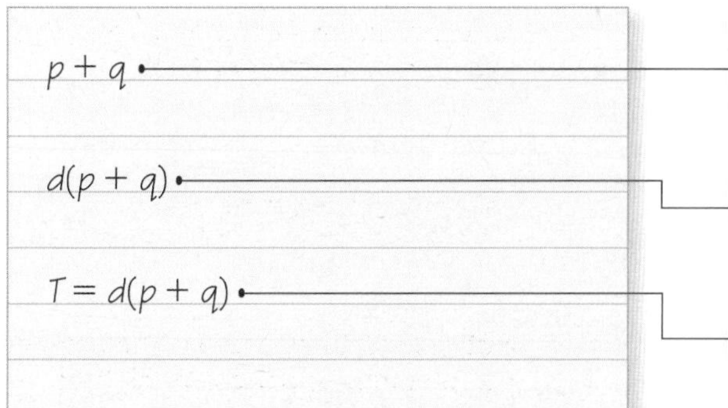

The number of pork pies and coffees sold every day.

Multiply the number each day by d.

This is the rule (formula) for d days.

Practice 8C

1 Write down a formula for the area of a triangle.
Use b for the length of the base
h for the perpendicular height
a for the area.

2 Peter earns p pounds per hour.
Write down a formula for T the total earned in n hours.

3 Choc bars cost 27 pence each.
Write down a formula for the cost, C pence, of n choc bars.

4 Write down a formula for the total cost, C pounds, of x chairs at £12 each and y tables at £45 each.

5 Write in symbols the rule
'To find S, double v and add u squared'.

6 Hanging baskets cost £15 each and tubs cost £8 each.
Write down a formula for the cost, C pounds, of b hanging baskets and t tubs.

7 Kate's age and Andrew's age add up to 59.
Andrew is 3 years older than Kate.
Write down a formula for Andrew's age A in terms of Kate's age K.

8.4 Factorising expressions helps to make expressions simpler.

Example 4

Factorise each of these expressions:

a $3a + 9$ **b** $3x^2 + 9x$ **c** $6a^2b + 9ab^2$

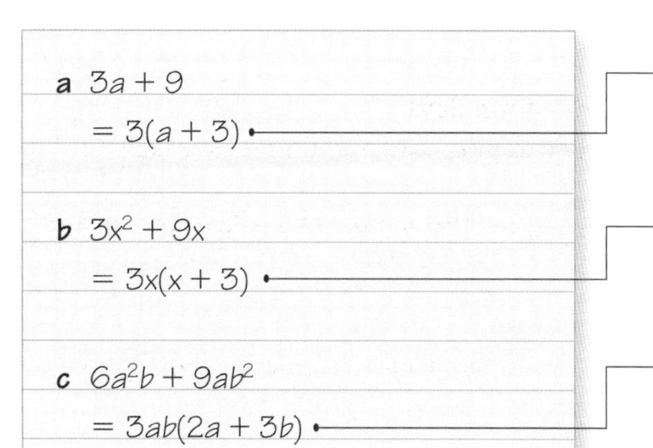

3 is a common factor of $3a$ and 9 so take it outside the bracket.

$3x$ is a common factor of $3x^2$ and $9x$ so take $3x$ outside the bracket.

Take $3ab$ outside the bracket.

Notice in all three answers there are now no common factors inside the bracket.

Practice 8D

Factorise these expressions:

1 $2a + 8$

2 $7c - 28$

3 $12x + 15$

4 $8y - 12$

5 $3a - 9b$

6 $7s + 21t$

7 $3a^2 + 6a$

8 $4x^2 - 10x$

9 $12m^2 + 6m$

10 $14x^2 - 21x$

11 $3x^2 - xy$

12 $4ab - 10a$

13 $r^2s - 2rs$

14 $9d - 36e$

15 $15x^2 - 21x$

16 $21a - 35a^2$

17 $8g - 24g^2$

18 $8x^3 - 2x^2$

19 $11a^2 + 33a$

20 $12ab^2 - 4b^2$

21 $6at^2 - 15t$

22 $c^2d - cd$

23 $2ab - 6a^2$

24 $8x^2y - 20xy^2$

9 Equations

This chapter shows you how to solve equations to find an unknown value, and how to solve simultaneous equations.

9.1

You can solve linear equations by using a balancing method. You need to get the unknown on one side on its own.

Example 1

Solve these equations:

a $x + 4 = 9$ **b** $5 - y = 8$

c $6x - 3 = 27$ **d** $2 - \frac{3}{4}x = -4$

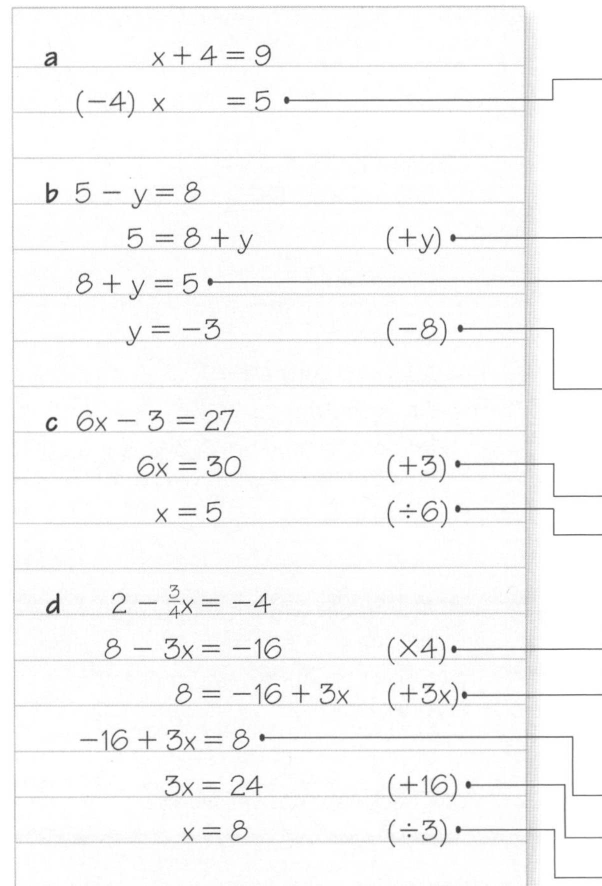

To keep the equation balanced, you must do the same to both sides. The note (-4) tells the examiner that you are subtracting 4, and helps you do the calculation.

Add y to both sides to make y positive.

Write the equation the other way round so that the unknown is on the left. It makes it easier to solve.

Take 8 from both sides to get the unknown y on its own.

Add 3 to each side.

Divide both sides by 6.

Multiply each term by 4 to clear the fraction.

Add $3x$ to both sides to make $3x$ positive.

Write the equation the other way round.

Add 16 to both sides.

Divide both sides by 3.

Practice 9A

Solve these equations:

1 $x + 3 = 7$ **2** $x - 5 = 9$ **3** $a + 3 = 2$

4 $p + 4 = 8$ **5** $5 + t = 7$ **6** $w + \frac{2}{3} = 3$

7 $7 - y = 4$ **8** $3 - x = 5$ **9** $6 - b = 2$

10 $6 - d = 2\frac{1}{2}$ **11** $12 - e = 21$ **12** $9 - m = 5$

13 $4x + 3 = 11$ **14** $6x - 5 = 13$ **15** $8x + 5 = 21$

16 $\frac{1}{2}x + 3 = 5$ **17** $3x - 2 = 16$ **18** $4x + 1 = 15$

19 $5 - 2x = 9$ **20** $3 - 4x = 11$ **21** $9 - 2y = 14$

22 $\frac{1}{2} - 3x = 5$ **23** $7 - 3a = -20$ **24** $11 - 2p = 15$

(S6)

9.2 To solve equations with brackets, first remove the brackets by multiplying out.

Example 2

Solve these equations:

a $2(x - 3) = 4$ **b** $3(2x + 1) - 2(x + 4) = 7$

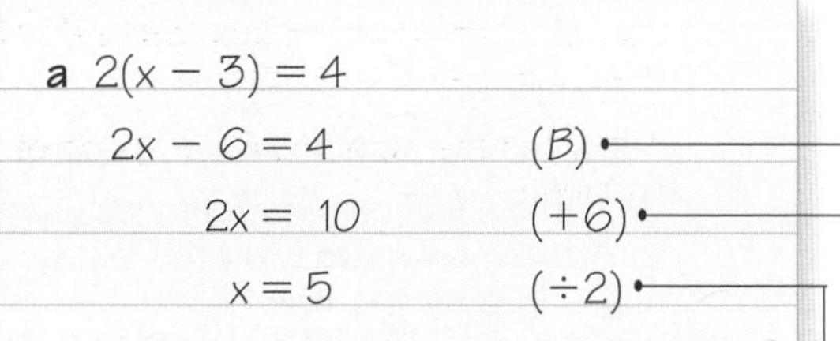

b $3(2x + 1) - 2(x + 4) = 7$

$6x + 3 - 2x - 8 = 7$ (B)

$4x - 5 = 7$

$4x = 12$ (+5)

$x = 3$ (÷4)

Check $3(6 + 1) - 2(3 + 4)$

$= 3 \times 7 - 2 \times 7$

$= 7 =$ *RHS* ✓

Multiply out the bracket. It is normally best to do this first.

Now use the balancing method as in Example 1.

Add 6 to both sides.

Divide both sides by 2.

In more difficult equations, it is a good idea to check your answer works.

Remember when multiplying out the bracket $- \times + = -$

Collect like terms (you met this in Chapter 8).

Check your answer works.

Practice 9B

Solve these equations:

1 $2(x + 3) = 10$ **2** $3(x - 1) = 12$

3 $5(2 - x) = 5$ **4** $4(2x + 1) = 28$

5 $5(3a - 2) = 20$ **6** $3(5 - 2y) = 48$

7 $4(5a + 3) = 32$

8 $4(x - 3) + 2(x + 5) = 10$

9 $2(3x + 1) + 3(2x - 3) = 31$

10 $2(3x - 5) = 2(x + 2)$

11 $7(9 - a) = 5(2a - 1)$

12 $5(2y - 1) - 3(3y + 4) = 3$

9.3

You need to be able to solve equations with *x* on both sides, and equations with fractions.

Example 3

Solve these equations:

a $5x + 2 = 3x - 4$

b $3(2b - 1) = 4b + 5$

c $\frac{2}{3}(x + 3) = 3$

a $5x + 2 = 3x - 4$

$2x + 2 = -4$ $(-3x)$ — Collect all the terms in x on one side of the equation.

$2x = -6$ (-2) — Collect all the numbers on one side of the equation.

$x = -3$ $(\div 2)$

	LHS	RHS
Check:	$5 \times -3 + 2$	$3 \times -3 - 4$
	$-15 + 2$	$-9 - 4$
	-13	-13

✓

Substitute $x = -3$ into each side to check.

The check works.

b $3(2b - 1) = 4b + 5$

$6b - 3 = 4b + 5$ (B) — Expand brackets.

$2b - 3 = 5$ $(-4b)$ — Subtract $4b$ from both sides.

$2b = 8$ $(+3)$ — Add 3 to both sides, so that only terms in b are on the left.

$b = 4$ $(\div 2)$ — Now divide, to get b.

	LHS	RHS
Check:	$3(8 - 1)$	$4 \times 4 + 5$
	3×7	$16 + 5$
	21	21

✓

The check works.

c $\frac{2}{3}(x + 3) = 3$

$\frac{2}{3}x + 2 = 3$ — $\frac{2}{3} \times 3 = 2$

$2x + 6 = 9$ $(\times 3)$ — Multiply by 3 to clear the fractions.

$2x = 3$ (-6)

$x = 1\frac{1}{2}$ $(\div 2)$

Check: $\frac{2}{3}(1\frac{1}{2} + 3)$

$\frac{2}{3}(4\frac{1}{2})$

3 ✓

You may wish to use a calculator for this check.

Practice 9C

Solve these equations:

1 $3x + 5 = 2x - 3$

2 $5x - 2 = 3x + 5$

3 $3 - 2y = 2y - 9$

4 $3x - 1 = 5x - 9$

5 $2(3a - 2) = 4a + 5$

6 $2(3p + 5) = 9p - 2$

7 $2m - 3 = 3(m - 6)$

8 $4y - 7 = 7y - 4$

9 $\frac{3}{4}(x + 4) = 1$

10 $\frac{1}{2}(4 - 8y) = 5 - 3y$

9.4

You can solve simultaneous equations using graphs or by the method of elimination. Simultaneous equations are equations of the type $x + 2y = 3$, $2x + 3y = 8$

You need to find a common solution for *x* and *y*.

Example 4

Solve the simultaneous equations

$2x + y = 5$ and $x - y = 1$

a by drawing graphs **b** algebraically

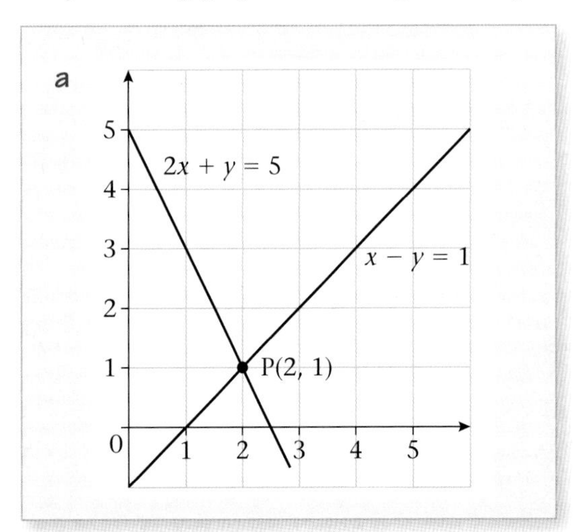

The point P(2, 1) where the lines cross, is the solution of the pair of equations.

This is the solution of the simultaneous equations $2x + y = 5$, $x - y = 1$

There is more about how to draw straight line graphs in Chapter 10.

b
$2x + y = 5$ ①
$x - y = 1$ ②
$3x = 6$ ① + ②
$x = 2$ ÷3
Put $x = 2$ into ①
$4 + y = 5$
$y = 1$

Check $x = 2$, $y = 1$
in ① $4 + 1 = 5$ ✓
in ② $2 - 1 = 1$ ✓
So $x = 2$, $y = 1$

First label each equation

Notice adding ① and ② eliminates y. So work out ① + ②

Substitute $x = 2$ into either ① or ② to find the value of y.

$x = 2$ and $y = 1$ works in both equations so $x = 2$, $y = 1$ is the correct solution.

Example 5

Solve the simultaneous equations

$3x - 2y = 6$
$4x + y = 19$

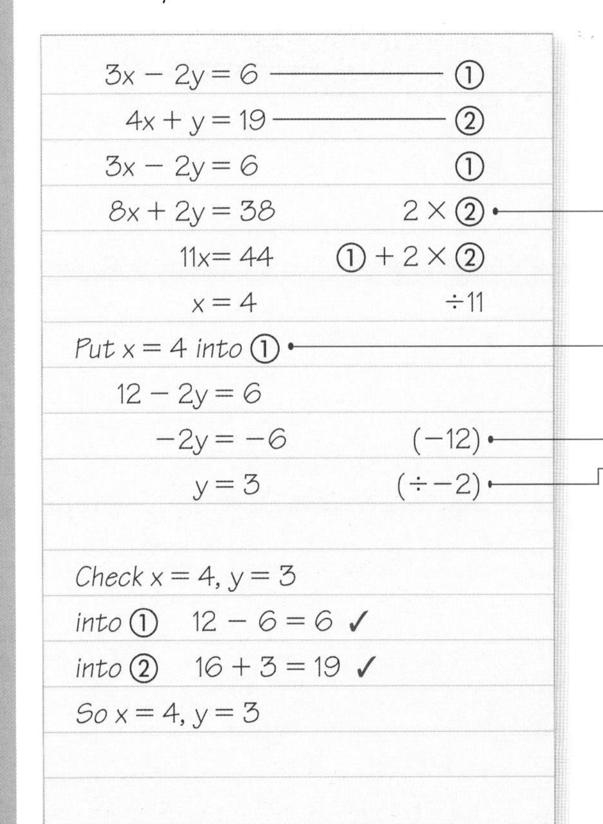

Look for a way to eliminate x or y.

Notice multiplying equation ② by 2 gives $2y$. So adding equation ① to 2 times equation ② eliminates y.

This will find the value of y.

Subtract 12 from both sides.

Remember − ÷ − = +

Always check your answer.

Practice 9D

1 Solve the following simultaneous equations
i graphically **ii** algebraically

a $2x + y = 9$
$3x - y = 6$

b $5x - 2y = 8$
$x + 3y = 5$

c $4x + y = 11$
$2x + y = 7$

2 Solve the simultaneous equations:
$3a + 4b = 2$
$5a + b = 9$

Hint: When a specific method is not required, use an algebraic method.

3 Solve the simultaneous equations:

a $2x + 5y = -1$
$6x - y = 5$ *E*

b $4x + y = 4$
$2x + 3y = -3$ *E*

c $2x + 6y = 17$
$3x - 2y = 20$ *E*

(E10)

10 Linear and quadratic graphs

This chapter shows you how to describe positions using coordinates, how to draw linear and quadratic graphs, and how these graphs may be used in real-life situations.

10.1 You can describe positions on a grid using coordinates.

Example 1

a Plot the following points on a coordinate grid: A(2, 5), B(−1, 4), C(5, −2), D(−3, −5).
Use a grid for x and y values from −6 to +6.

b Find the mid-point of the line segments (i) AB (ii) BD.

The mid-point of the line segment means half-way between the points.

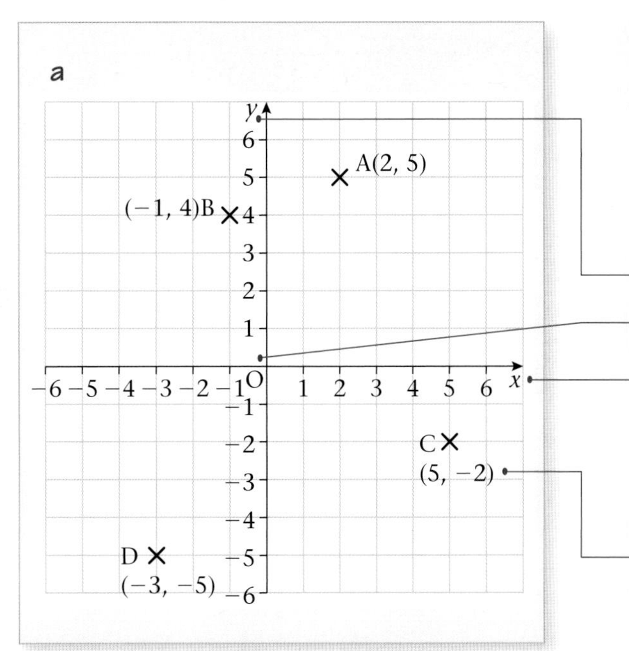

Label this axis y.

O is the origin.

Label this axis x.

The first number is the x value, the second number is the y value.

b i $\frac{x_1 + x_2}{2} = \frac{2 + -1}{2} = \frac{1}{2}$

$\frac{y_1 + y_2}{2} = \frac{5 + 4}{2} = 4\frac{1}{2}$

So mid-point of AB is $(\frac{1}{2}, 4\frac{1}{2})$

ii $\frac{x_1 + x_2}{2} = \frac{-1 + -3}{2} = \frac{-4}{2} = -2$

$\frac{y_1 + y_2}{2} = \frac{4 - 5}{2} = \frac{-1}{2} = -\frac{1}{2}$

So mid-point of BD $= (-2, -\frac{1}{2})$

To find the mid-point use the rule $\left(\frac{x_1 + x_2}{2}, \frac{y_1 + y_2}{2}\right)$ for the pair of coordinates (x_1, y_1) and (x_2, y_2).

$-1 + -3 = -1 - 3$
$= -4$

Practice 10A

Draw a coordinate grid for values of x and y from 10 to -10.

1 Plot the following coordinates
A(1, 5), B(3, 9), C(−4, 7), D(−1, −9)
E(0, −7), F(3, −2), G(−3, −3), H(7, −2)

2 Find the mid-point of the line segments
a AB **b** BD **c** EH **d** HF **e** CF **f** EC **g** CG **h** AG

10.2

You can use straight line graphs to represent relationships between two variables which lie on a straight line.

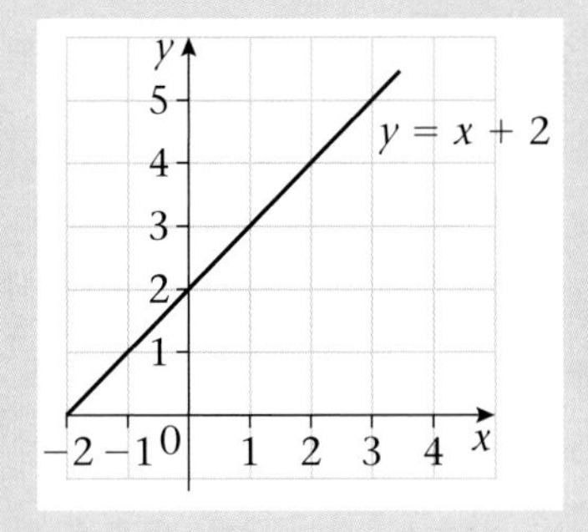

Example 2

a Complete the table of values for $y = 2x + 1$:

x	−3	−2	−1	0	1	2	3
y			−1				

b Draw the graph of $y = 2x + 1$.

c Use your graph to find the values of
i y when $x = 2.5$ **ii** x when $y = -1.5$.

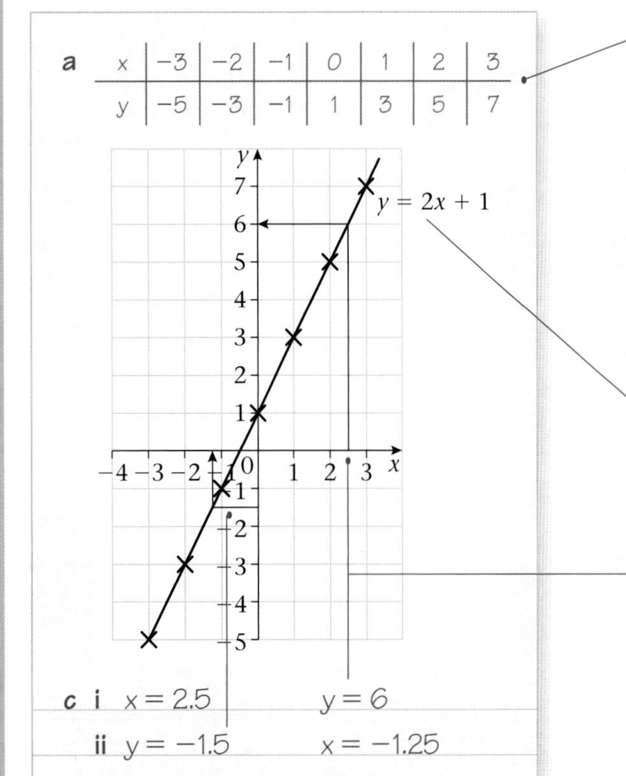

Substitute the values of x into $y = 2x + 1$ to find y.

So when $x = -3$, $y = 2 \times -3 + 1 = -5$
$x = -2$, $y = 2 \times -2 + 1 = -3$
and so on.

You need x values from -3 to $+3$ and y values from 7 to -5.

Plot each point on the grid and join them up with a straight line (use a ruler).

Label the line.

Draw a vertical line from $x = 2.5$ to the line and read off the y value ($y = 6$).

Draw a horizontal line from $y = -1.5$ and read off the x value (-1.25).

Example 3

The equations of five straight lines are:

$y = 3x - 2$, $y = x + 3$, $y = 4x - 1$, $y = 3x + 5$, $y = 2x + 3$

Two of the lines go through the point (0, 3).

a Write down the equation of these lines.
Two of the lines are parallel.

b Write down the equations of these lines.

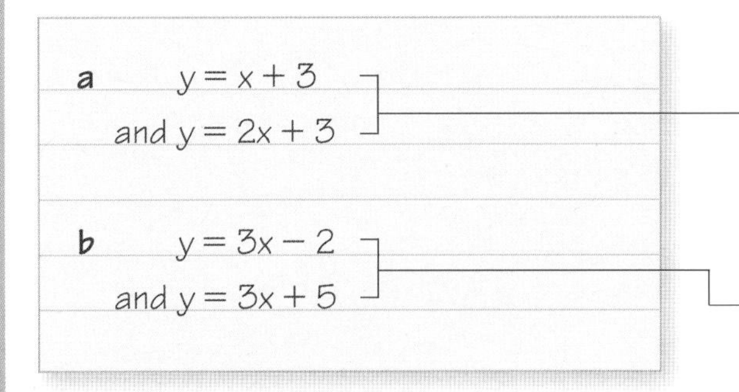

$y = mx + c$ is the general equation of a straight line.

c is the intercept of the y axis ie $(0, c)$. So in these two equations $c = 3$ and they both pass through (0, 3).

In $y = mx + c$, m is the gradient.
In both these equations the gradient is the same so they are parallel.

Example 4

Find the equation of the line shown:

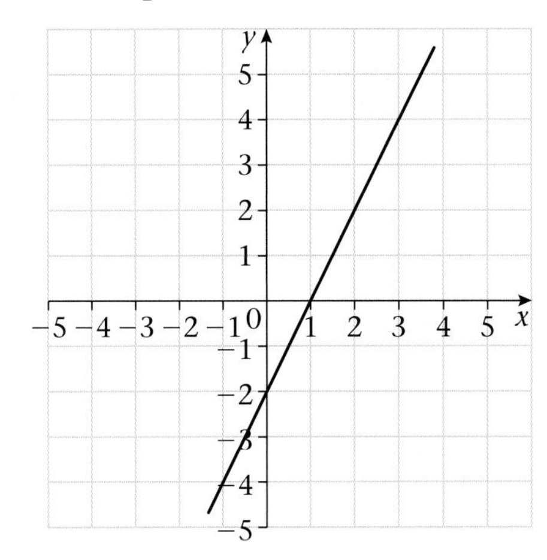

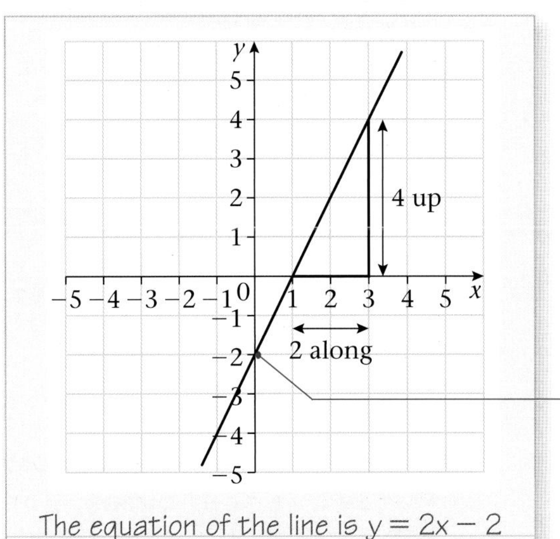

m is the gradient.

So $m = \frac{4 \text{ (up)}}{2 \text{ (along)}}$
$= 2$

Compare to $y = mx + c$.

This is $(0, -2)$ so $c = -2$.

Practice 10B

1 **a** Copy and complete the table of values for $y = 2x + 3$:

x	-3	-2	-1	0	1	2	3
y		-1					

b Draw the graph of $y = 2x + 3$.

c Use your graph to find

i the value of y when $x = 1.5$

ii the value of x when $y = -0.5$.

E

2 **a** Copy and complete the table of values for $y = 3x - 1$:

x	-3	-2	-1	0	1	2	3
$y = 3x - 1$	-10		-4			5	

b Draw the graph of $y = 3x - 1$.

c Use your graph to find the value of x when $y = 6.5$.

E

3 The equations of five straight lines are:

$y = 2x + 3$, $y = \frac{1}{2}x - 3$, $y = 4x + 3$, $y = 2x - 7$, $y = 3x - 4$

a Which lines go through the point (0, 3)?

b Which lines are parallel?

4 Find the equation of the straight line for:

a

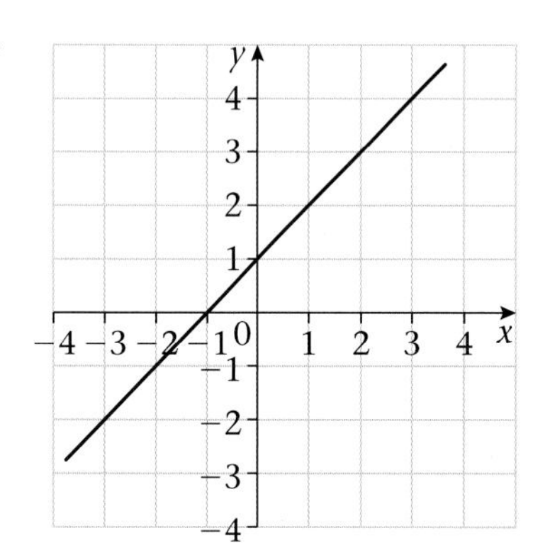

b

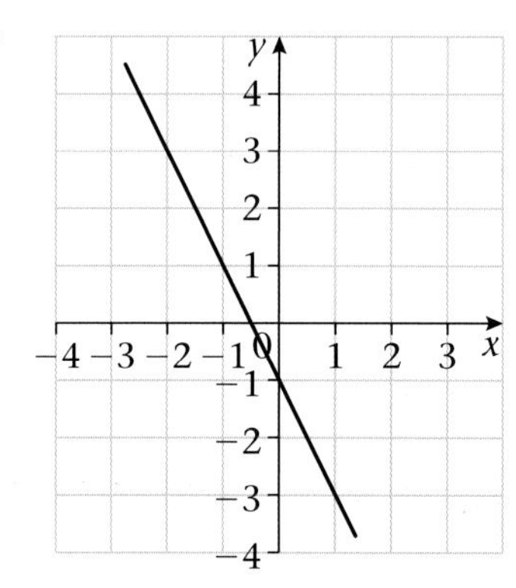

5 Write down the gradient and y-intercept for these equations of a straight line:

a $y = 3x + 2$

b $y = 5x - 3$

c $y = -\frac{1}{2}x + 3$

d $y = -4x - 3$

e $y = 5 - 2x$

f $2y = 6x - 3$

Hint: For **f**, rewrite the equation in the form $y = mx + c$, to find a value for m.

10.3 Straight line graphs describe real-life situations, eg distance–time graphs, conversion graphs.

Example 5

This graph can be used to convert between Celsius and Fahrenheit.

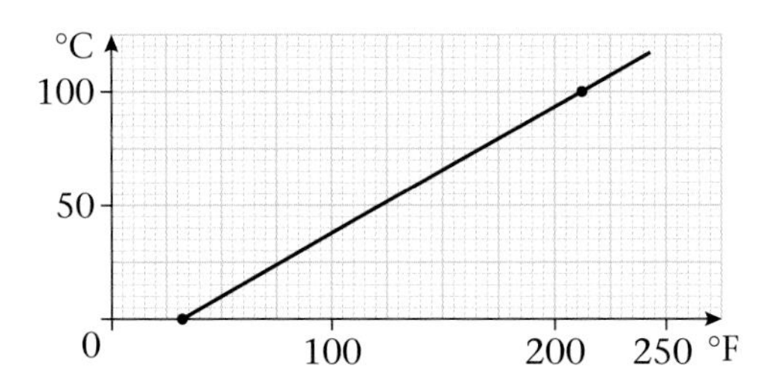

Use the graph to convert

a 40°C to °F **b** 176°F to °C.

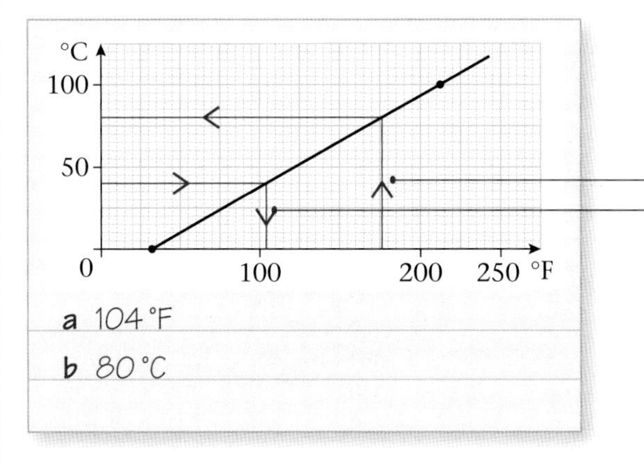

Draw lines on the graph to read off the conversions.

Example 6

Here is the distance–time graph for Susan's car journey.

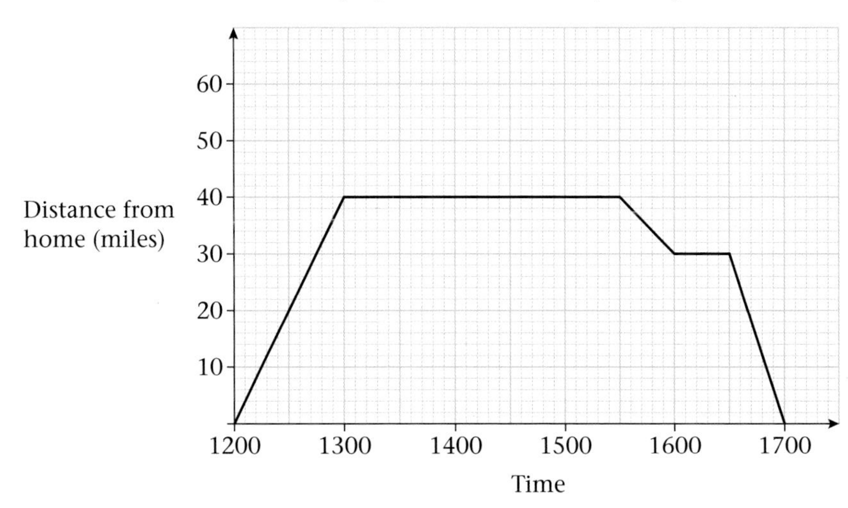

a Work out Susan's distance from home at 1230.

b Describe Susan's journey between 1200 and 1530.

c Work out Susan's average speed between 1630 and 1700.

a 20 miles — Draw lines on the graph to read this value.

b Between 1200 and 1300, she travels 40 miles at a constant speed. — The graph is a straight line and sloping.

Between 1300 and 1530 Susan is not moving (she has arrived at her destination). — This is shown by a straight horizontal line.

c Average speed $= \frac{\text{distance}}{\text{time}}$

$= \frac{30}{0.5}$

$= 60$ m.p.h.

This may help you remember the rule.

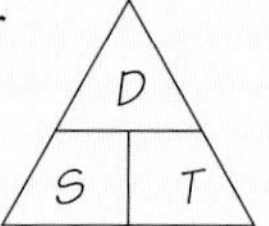

Remember to convert 30 minutes to 0.5 hours.

Practice 10C

1 £1 = 0.6 euros.

a Draw a conversion graph for amounts up to £200.

b Use your conversion graph to convert

i £68 into euros **ii** 100 euros into pounds.

2 A train travelled 430 km from London to Durham.
The graph shows the train's journey from London as far as York.

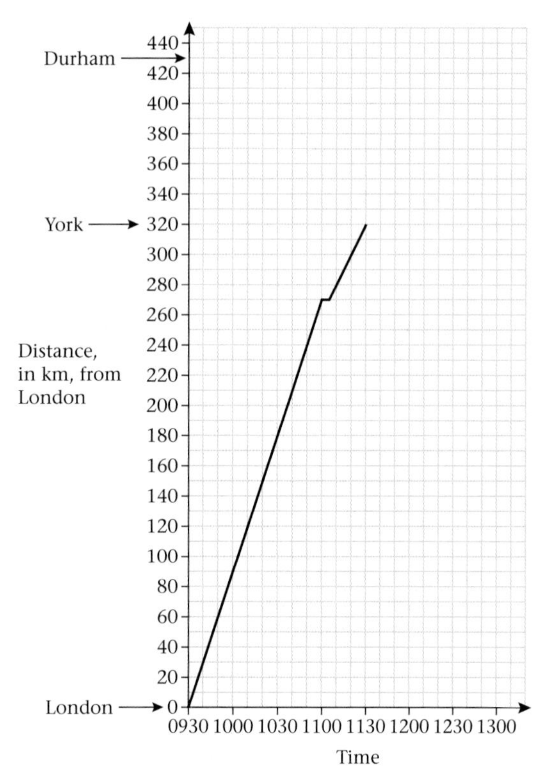

On the way to York, the train stopped at Doncaster.

a Write down the distance of Doncaster from London.

b Work out the average speed at which the train travelled from London to Doncaster.

The train stopped at York for 10 minutes. It then went on to Durham at a steady speed. It did not stop between York and Durham. It reached Durham at 1230.

c Copy and complete the graph of the train's journey to Durham.

3 David went for a ride on his bike. He rode from his home to the lake. The travel graph shows this part of his trip.

a Find David's average speed between 1300 and 1500.

b What happened to David between 1500 and 1600?

David started to travel back to his home at 1600.

He travelled at a speed of 10 miles per hour for half an hour. He remembered he had left his water bottle at the lake. He immediately rode back to the lake at 10 miles per hour. He picked up his water bottle and immediately travelled back home at 20 miles per hour.

c Copy and complete the travel graph.

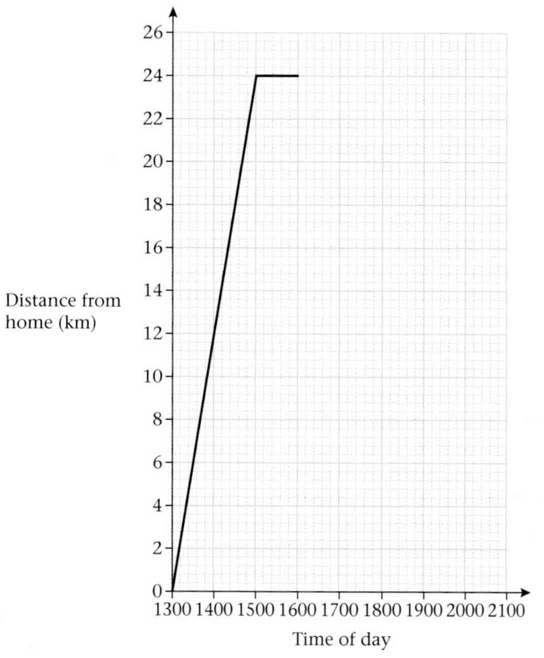

E

4 Ken and Wendy go from home to their caravan site. The caravan site is 50 km from their home. Ken goes on his bike. Wendy drives in her car. The diagram shows information about the journeys they made.

a At what time did Wendy pass Ken?

b Between which two times was Ken cycling at his greatest speed?

c Work out Wendy's average speed for her journey.

d Describe what Ken is doing between 1130 and 1200.

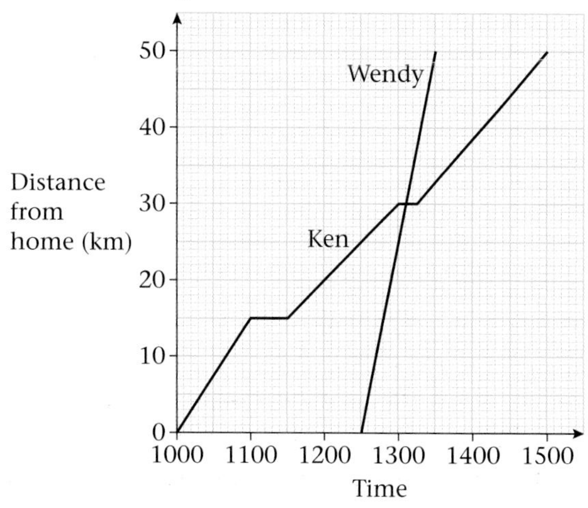

E

5 Elizabeth went for a cycle ride.
The distance–time graph shows her ride.

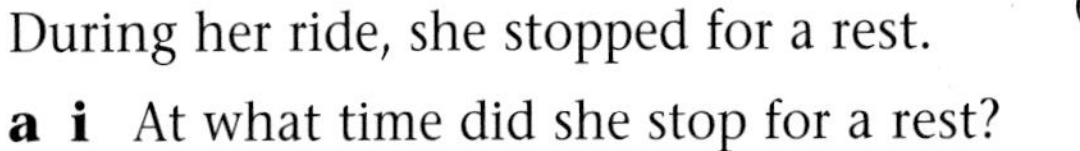

She set off from home at 1200 and had a flat tyre at 1400.
During her ride, she stopped for a rest.

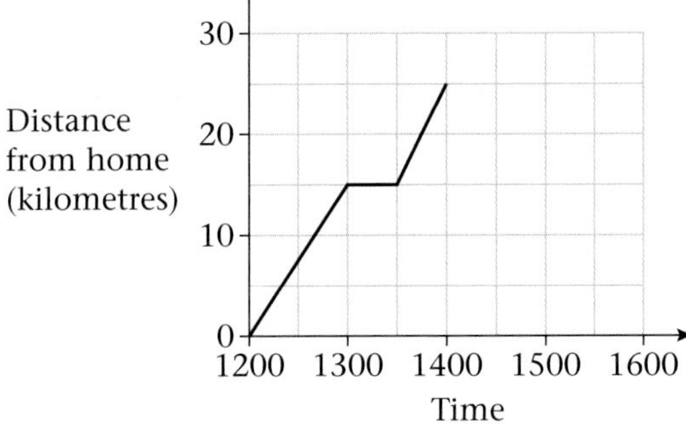

a i At what time did she stop for a rest?

ii At what speed did she travel after her rest?

It took Elizabeth 15 minutes to repair the flat tyre. She then cycled home at 25 kilometres per hour.

b Copy and complete the distance–time graph to show this information.

10.4

You will need to draw graphs of quadratic equations. The general equation of a quadratic is $y = ax^2 + bx + c$. Notice the highest power of x is 2 ie x^2.

Example 7

a Draw the graph of $y = x^2 - 2$ for values of x from -3 to $+3$.

b Write down the minimum value of y and the value of x for this point.

c Label the line of symmetry.

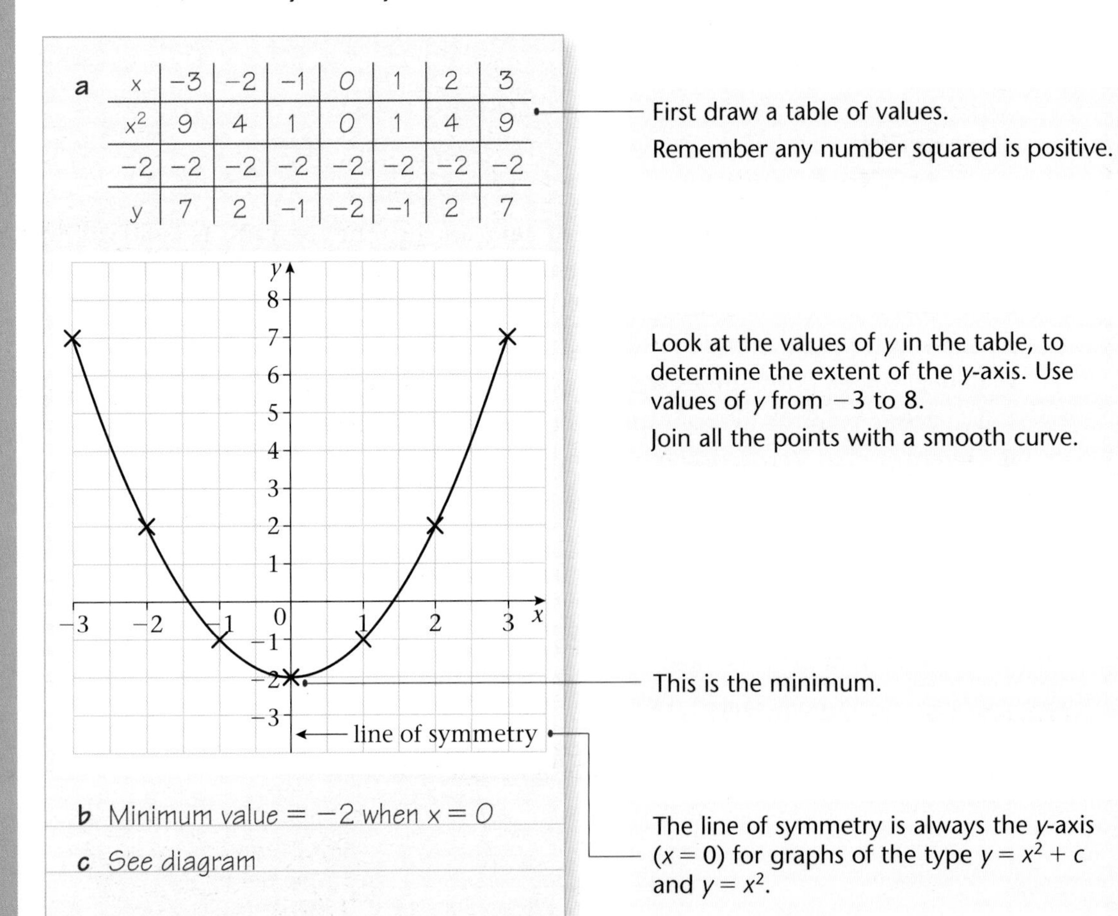

a

x	−3	−2	−1	0	1	2	3
x^2	9	4	1	0	1	4	9
−2	−2	−2	−2	−2	−2	−2	−2
y	7	2	−1	−2	−1	2	7

b Minimum value = −2 when x = 0

c See diagram

First draw a table of values.
Remember any number squared is positive.

Look at the values of y in the table, to determine the extent of the y-axis. Use values of y from −3 to 8.
Join all the points with a smooth curve.

This is the minimum.

The line of symmetry is always the y-axis ($x = 0$) for graphs of the type $y = x^2 + c$ and $y = x^2$.

Example 8

a Draw the graph of $y = x^2 + 3x - 4$ taking values of x from -5 to 2.
b Draw on the graph the line of symmetry and label its equation.
c Write down the co-ordinates of the minimum value.
d Use your graph to solve $x^2 + 3x - 4 = 0$.

a

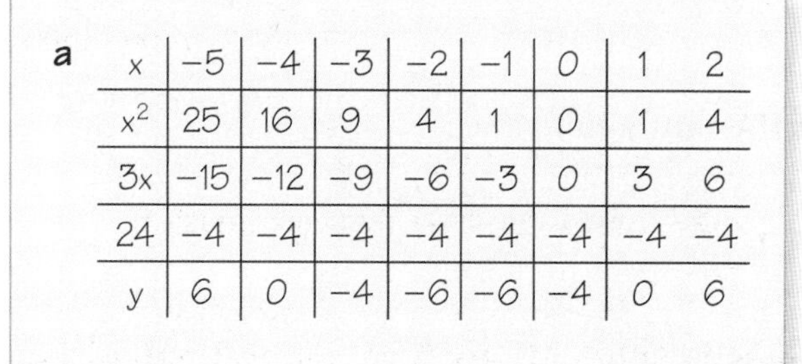

x	−5	−4	−3	−2	−1	0	1	2
x^2	25	16	9	4	1	0	1	4
3x	−15	−12	−9	−6	−3	0	3	6
24	−4	−4	−4	−4	−4	−4	−4	−4
y	6	0	−4	−6	−6	−4	0	6

Choose values of y from 6 to -6 for the graph.

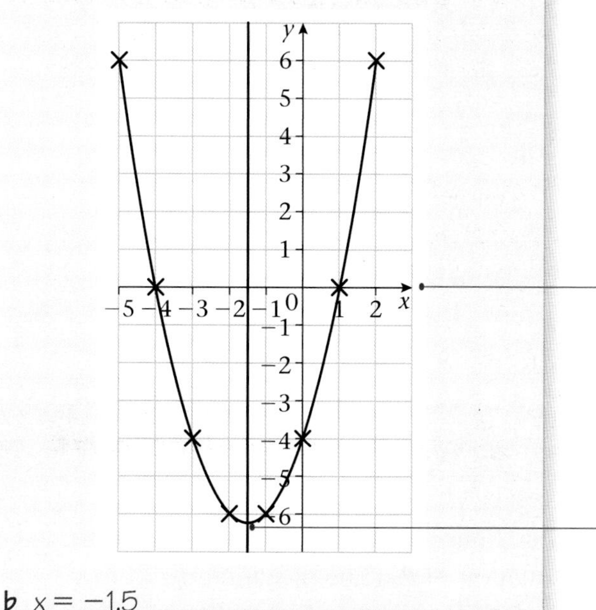

Notice the shape is again a U.

Here $x^2 + 3x - 4 = 0$. Read off the x coordinates for the solutions to the equation.

Draw a smooth curve below -6 to join these two points of symmetry.

b $x = -1.5$

c Minimum value $= (-1.5, -6.3)$

From the graph, the value is about -6.3 (the true value is -6.25 by calculation).

d $x = -4$ or $x = 1$

There are two solutions to $x^2 + 3x - 4 = 0$.

Practice 10D

1 Draw the graph of the following quadratic equations.
Use values of x from -3 to $+3$.

a $y = x^2 + 2$ **b** $y = x^2 - 5$

c $y = x^2 + 3x$ **d** $y = \frac{1}{2}x^2 + 1$

e $y = 3 - x^2$

Hint: For **e** the general shape is an upside down U-shape.

2 **a** Copy and complete the table of values for $y = 2x^2$:

x	−3	−2	−1	0	1	2	3
y	18				2	8	

b Draw the graph of $y = 2x^2$.

c Use your graph to find:

i the value of y when $x = 2.5$

ii the value of x when $y = 12$.

d Write down the coordinates of the minimum value.

e Use your graph to solve
$5 = 2x^2$

Hint: Draw a line at $y = 5$.

3 **a** Copy and complete the table of values for $y = x^2 + 3x + 1$:

x	−4	−3	−2	−1	0	1
y				−1		

b Draw the graph of $y = x^2 + 3x + 1$.

c Use your graph to solve $0 = x^2 + 3x + 1$.

4 Draw the graph for each of the following equations.

a $y = x^2 - 4x + 1$ for values of x from −1 to 5

b $y = 3x^2 + 4x - 3$ for values of x from −4 to +2

c $y = 4 - x^2$ for values of x from −4 to +4

d $y = 5 + 3x - 2x^2$ for values of x from −2 to 4.

Hint: A maximum value occurs when the equation contains $-x^2$.

For each part also:

i write down the minimum or maximum value

ii draw in and write down the equation of the line of symmetry

iii solve each equation for values of $y = 0$ and $y = -2$.

(E11)

11 Quadratics and subject of the formula

This chapter covers factorising and solving quadratic equations, and shows you how to change the subject of a formula.

11.1

You can factorise quadratic expressions of the type $x^2 + bx + c$ where b and c are numbers. Factorising is the reverse process of removing the brackets.

Example 1

Factorise

a $3x^2 - 12x$ **b** $x^2 + 3x + 2$ **c** $x^2 - 5x - 6$ **d** $x^2 - 16$

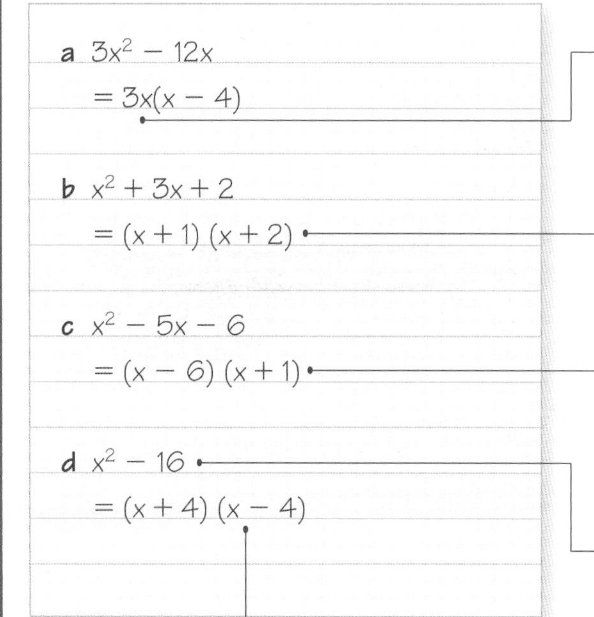

Take $3x$ outside the bracket as it is the highest common factor (HCF).

Find two numbers that:
- multiply to give 2 and
- add to give 3.

They are 1 and 2.

Find two numbers that:
- multiply to give −6
- add to give −5

They are −6 and 1.

Find two numbers that:
- multiply to give −16
- add to give 0

They are +4 and −4.

This is known as the difference between two squares.
In general $x^2 - a^2 = (x + a)(x - a)$

Practice 11A

Factorise

1 **a** $x^2 + 3x$ **b** $2x^2 + 10x$ **c** $5x^2 + x$
d $6x^2 - 2x$ **e** $6x^2 + 15x$ **f** $4x - 6x^2$

2 **a** $x^2 + 5x + 4$ **b** $x^2 + 6x + 9$ **c** $x^2 + x - 12$
d $x^2 - 2x - 15$ **e** $x^2 + 5x - 6$ **f** $x^2 - 3x - 18$

3 **a** $x^2 - 4x + 3$ **b** $x^2 - 5x + 4$
c $x^2 - 7x + 12$ **d** $x^2 - 3x + 2$

4 **a** $x^2 - 9$ **b** $x^2 - 25$
c $x^2 - 4$ **d** $x^2 - 36$

5 **a** $3x^2 - 27$ **b** $2x^2 - 8$
c $5x^2 - 20$ **d** $3x^2 - 75$

Hint: In question 5 look for a common factor first.

6 **a** $x^2 + 6x - 7$ **b** $x^2 - 6x + 8$ E **(E12)**

11.2 You can solve quadratic equations by factorisation.

Example 2

a Solve $x^2 + 5x + 4 = 0$

b Solve $x^2 - 2x - 8 = 0$

a $x^2 + 5x + 4 = 0$

$(x + 1)(x + 4) = 0$ — First factorise the quadratic into two brackets by finding two numbers whose sum is 5 and product is 4.

then $(x + 1) = 0$ or $(x + 4) = 0$ — Either bracket equals zero.

$x = -1$ or $x = -4$ — Now solve the linear equations.

Check $x = -1$ — You can check your answer by substituting $x = -1$ or $x = -4$ into the equation.

$(-1)^2 + 5 \times (-1) + 4$

$= +1 - 5 + 4$

$= 0$ ✓

$x = -4$

$(-4)^2 + 5 \times (-4) + 4$

$= +16 - 20 + 4$

$= 0$ ✓

b $x^2 - 2x - 8 = 0$

$(x - 4)(x + 2) = 0$ — Factorise ($-4 \times +2 = -8$ and $-4 + 2 = -2$)

then $x - 4 = 0$ or $x + 2 = 0$ — Set both brackets equal to zero.

$x = 4$ or $x = -2$ — Solve the linear equations.

Practice 11B

Solve the following quadratic equations:

> **Hint:** The brackets may be the same.

1 $x^2 + 5x - 6 = 0$

2 $x^2 + 2x + 1 = 0$

3 $x^2 - x - 20 = 0$

4 $x^2 + 7x + 10 = 0$

5 $x^2 - 2x - 15 = 0$

6 $x^2 - x - 12 = 0$

7 $x^2 + 8x + 15 = 0$

8 $x^2 + 2x - 15 = 0$

9 Solve the equations:
a $x^2 - 6x + 8 = 0$
b $x^2 + 6x - 7 = 0$

E

11.3

You need to be able to rearrange formulae such as: $A = \frac{1}{2}(a + b)h$. A is called the 'subject of the formula'. It appears on its own on one side of the formula.

Example 3

Rearrange each formula to make a the subject of the formula.

a $W = a + b$

b $A = \frac{1}{2}(a + b)h$

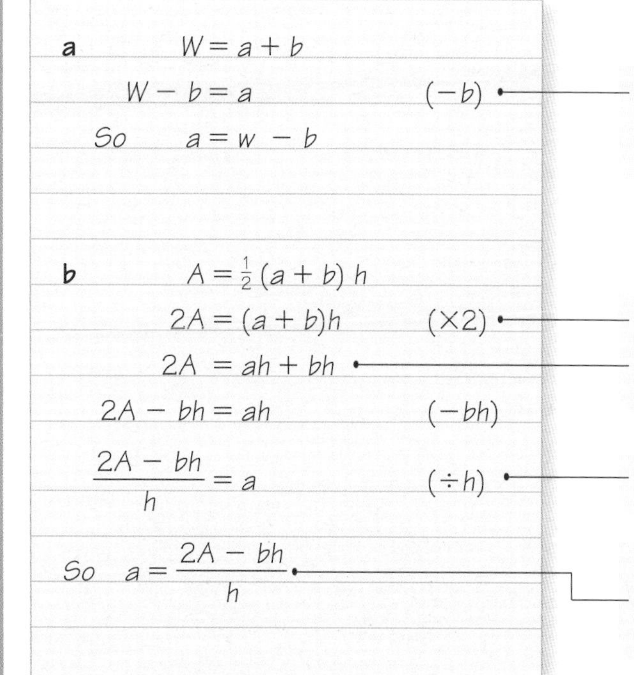

Subtract b.
The subject of the formula is normally written on the left hand side.

Multiply by 2 to clear the fraction.

Clear the brackets by multiplying them out.

So a is on its own on one side.

There are other ways to rearrange the formula but you will always get the same answer.

Practice 11C

In questions 1–10 make the letter in the brackets the subject of the formula.

1 $P = 4s$, (s)

2 $A = lw$, (w)

3 $P = 2a + b$, (a)

4 $v = u - at$, (t)

5 $V = \frac{D}{T}$, (T)

6 $V = \frac{1}{3}\pi r^2 h$, (h)

7 $m = \frac{a + b + 2c}{2}$, (b)

8 $y = \frac{2}{5}x - 7$, (x)

9 $y = 5(x - 2)$, (x)

10 $A = \frac{1}{2}(a + b)h$, (b)

11 Make t the subject of the formula

$$v = u + 10t$$

12 Make x the subject of the formula

$$y = \frac{x^2 + 4}{5}$$

Hint: Remember to square root x^2 to get x on its own.

13 Make u the subject of the formula

$$v^2 = u^2 + 2as$$

12 Inequalities

This chapter shows you how to solve inequalities, and how to represent inequalities using graphs. **(S7)**

12.1

An inequality is an expression in which the left hand side is not necessarily equal to the right hand side, eg $x + 2 < 3x - 1$.
You need to remember these symbols:

- **$>$ means 'greater than'**
- **$<$ means 'less than'**
- **$\geqslant$ means 'greater than or equal to'**
- **$\leqslant$ means 'less than or equal to'**

Example 1

a Show each of these inequalities on a number line:

i $x > 4$ **ii** $-2 < x \leqslant 3$

b Write down the inequalities shown on these number lines:

i

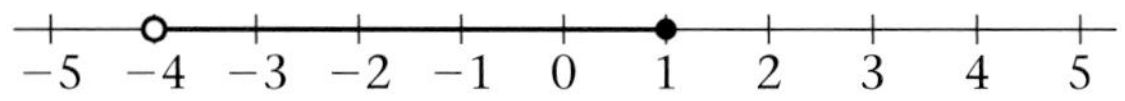

ii

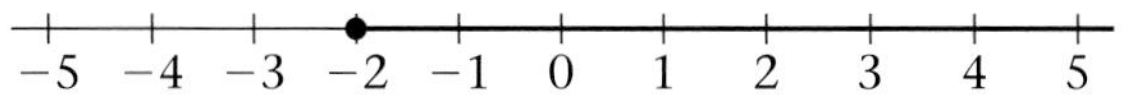

a i

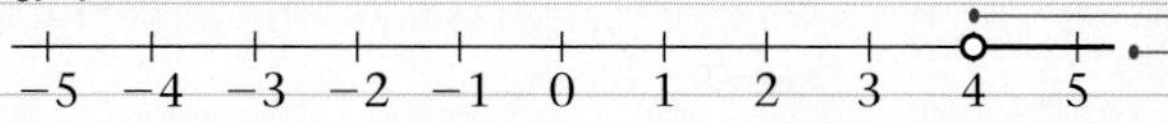

The open circle shows that x cannot equal 4.

The bold line shows that x extends from 4 and beyond.

ii

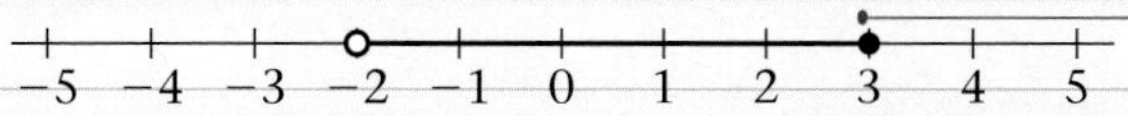

The solid circle shows that x can equal 3.

b i $-4 < x \leqslant 1$

You could also write this as $x > -4$ and $x \leqslant 1$.

ii $x \geqslant -2$

Practice 12A

1 Show each of these inequalities on a number line

a $x > -2$ **b** $x \leqslant 4$

c $x \geqslant 0$ **d** $x > 1$ and $x < 3$

e $o < x \leqslant 5$ **f** $5 \geqslant x \geqslant -2$

g $-4 \leqslant x < 1$ **h** $0 \leqslant x < +4$

2 Write down the inequalities shown on these number lines.

a 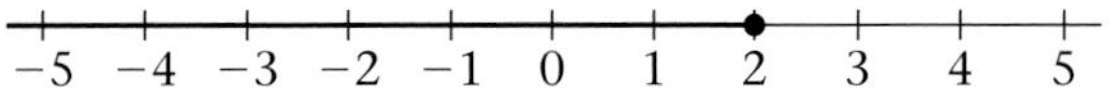

b 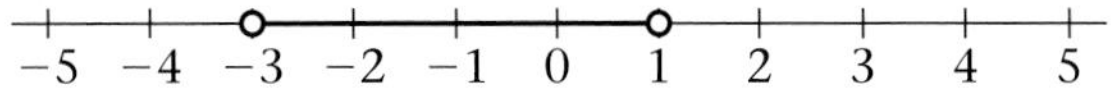

c 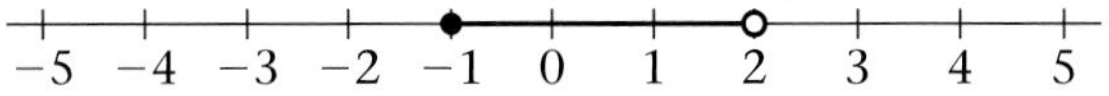

d 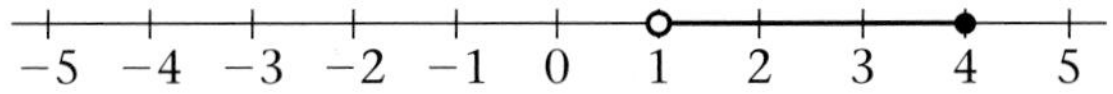

12.2 **You can solve inequalities in a similar way to linear equations, *except* if you multiply or divide by a negative number you must change the direction of the inequality sign.**

Example 2

Solve the inequalities:

a $5x - 3 \geqslant 2x + 3$ **b** $2x + 1 \leqslant 5x - 2$

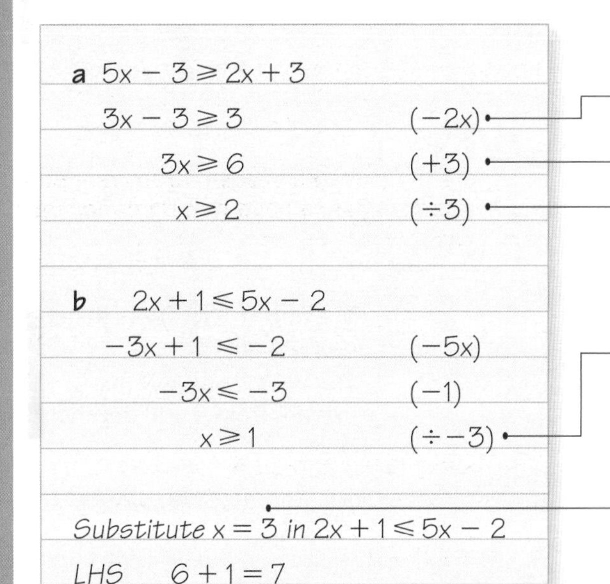

a $5x - 3 \geqslant 2x + 3$

$3x - 3 \geqslant 3$ $(-2x)$ — Take $2x$ from both sides to get x's on one side.

$3x \geqslant 6$ $(+3)$ — Add 3 to both sides.

$x \geqslant 2$ $(\div 3)$ — Divide both sides by 3 to get single x.

b $2x + 1 \leqslant 5x - 2$

$-3x + 1 \leqslant -2$ $(-5x)$

$-3x \leqslant -3$ (-1)

$x \geqslant 1$ $(\div -3)$ — Remember to change the direction of the inequality sign.

Substitute $x = 3$ *in* $2x + 1 \leqslant 5x - 2$ — Choose a value for $x \geqslant 1$ (say 3) to check your answer.

LHS $6 + 1 = 7$

RHS $15 - 2 = 13$

So $7 \leqslant 13$ ✓ — The inequality is correct so $x \geqslant 1$ is likely to be correct.

Practice 12B

1 Solve these inequalities:

a $3x \geqslant 8$ **b** $2x < -4$ **c** $8x < 2$

d $2x + 1 > 10$ **e** $3x - 2 \leqslant 13$ **f** $x + 3 > 2x - 5$

g $5x - 2 \geqslant 4x + 7$ **h** $3x - 7 < 15 + 7x$ **i** $4x + 3 \geqslant 7x - 6$

j $5x + 1 \leqslant 2x + 7$ **k** $2x + 3 \leqslant 8$ **E** **l** $7y > 2y - 3$ **E**

12.3 You can show inequalities on a coordinate grid.

Example 3

Shade a region which satisfies the inequalities
$x \geqslant 2$ and $4y + 2x < 8$.

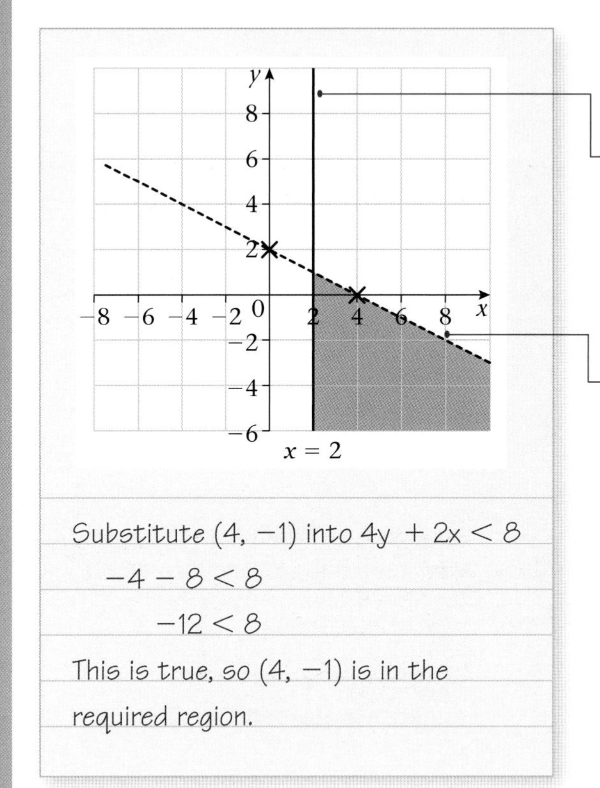

Use a coordinate grid.

Draw the line $x = 2$.
The line is solid as 2 is included in the region.
(If you need to, refer back to graphs in Section 10.2.)

Draw the line $4y + 2x < 8$.
The line is broken to show the line is not part of the region.

To find out which region to shade, test a point: choose a point that is not on either line, such as $(4, -1)$.

Practice 12C

1 Shade a region to represent each inequality. Use a coordinate grid.

a $y + x > 1$
b $6x - 5y \geqslant 15$
c $2x + 3y \leqslant 6$
d $y < 2y - 1, y < 2$
e $6x - 3y \geqslant 9, x < 4$
f $x > 0, 2x + 4y > 10$
g $y > 0, 3x - 2y < 12$
h $4 - x \leqslant 2y, x > 0$
i $x > 2y + 1, x < y + 4$
j $3x + y > 12, x > 8 - 2y$

13 Angles

Angles are to do with change of direction and are measured in degrees or turns. This chapter covers angle facts and bearings. **(S8)**

13.1 A straight line is a half turn or 180° and a full turn is 360°.

Example 1

Work out the named angle in the diagrams:

a

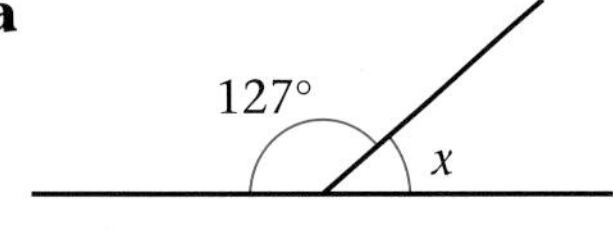

b

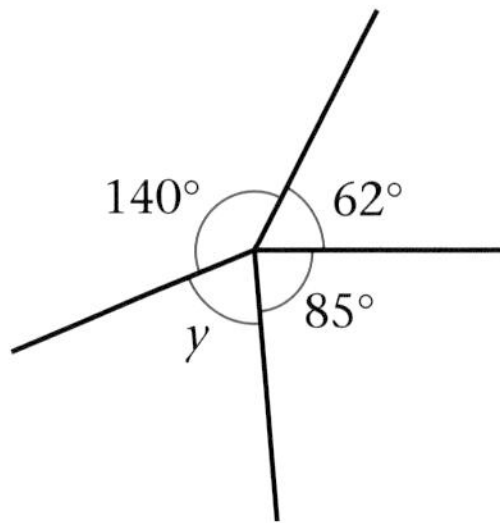

c

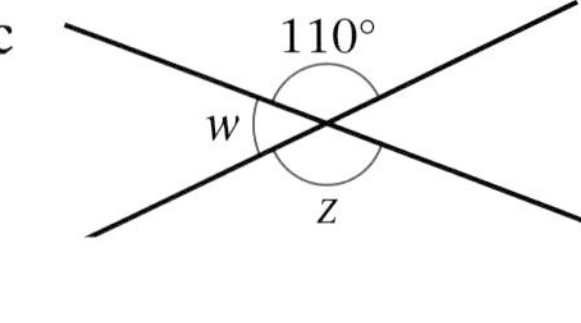

a $127 + x = 180$ — Angles on a straight line add up to 180°. Find x using the balancing method (Section 9.1).

$x = 180 - 127$

$x = 53°$

b $140 + 62 + 85 + y = 360$ — Angles at a point add up to 360°.

$287 + y = 360$

$y = 360 - 287$

$y = 73°$

c $z = 110°$ — Opposite angles are equal.

$w = 180 - 110$ — Angles on a straight line add up to 180°.

$w = 70°$

Practice 13A

For each question, work out the value of the marked angle.
Give reasons for your answers.

1

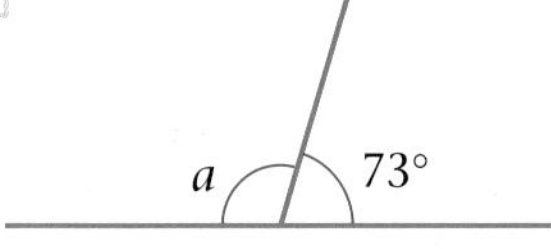

2

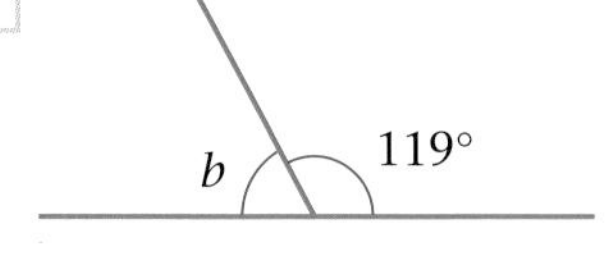

3

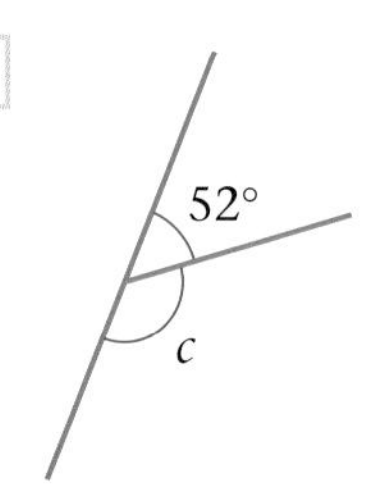

4

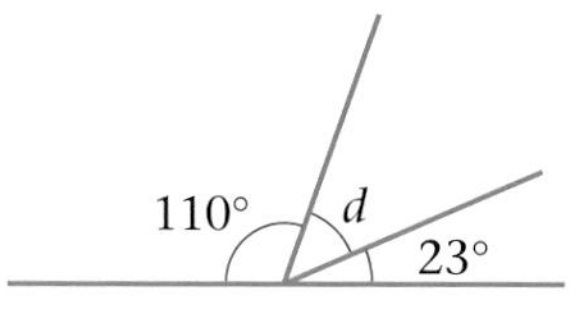

5

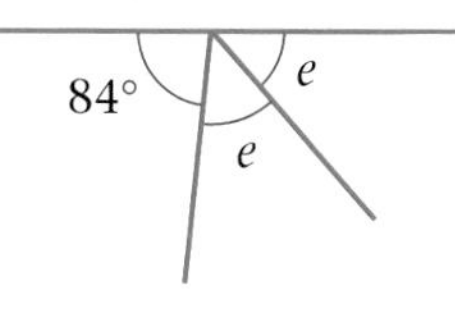

6

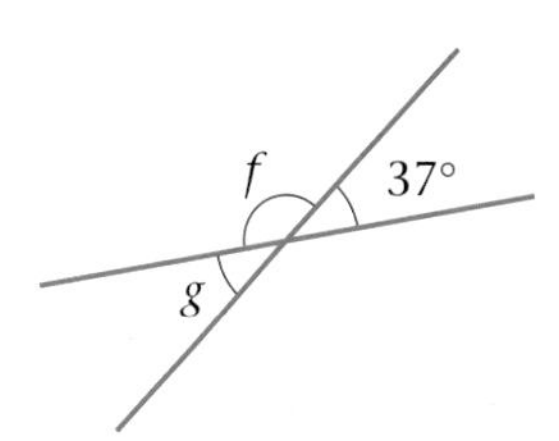

7

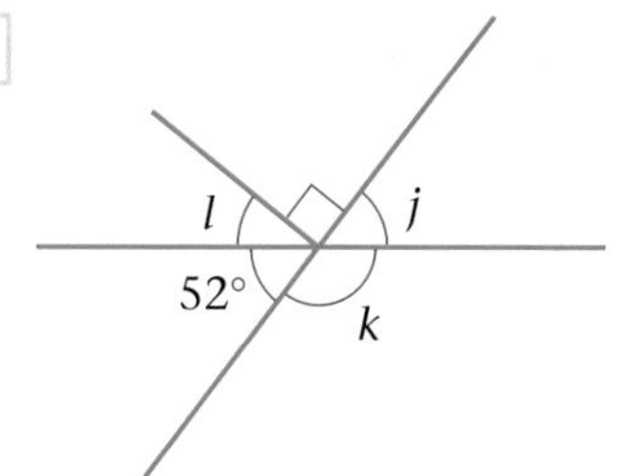

8

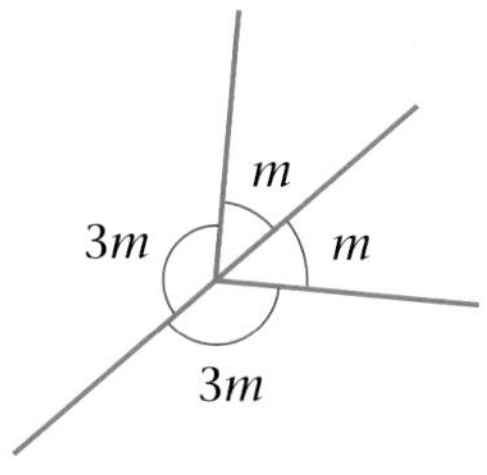

9

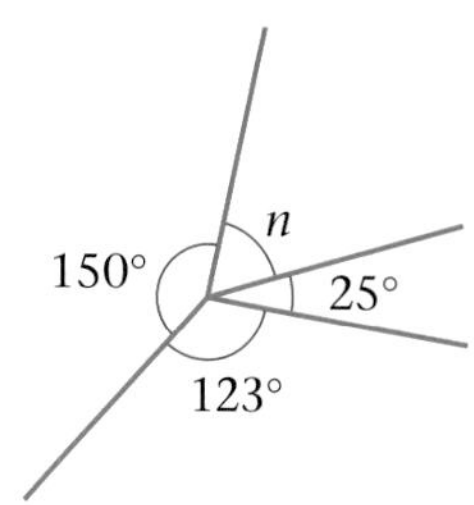

10

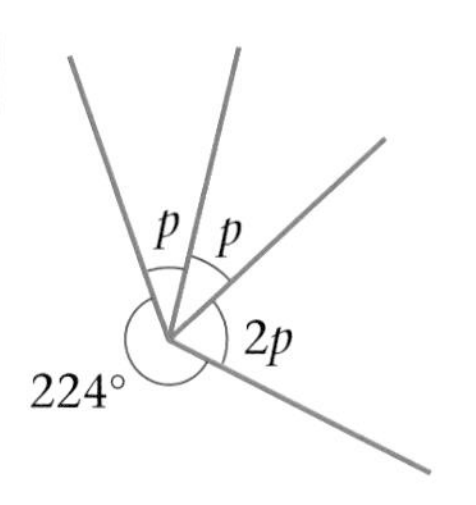

11

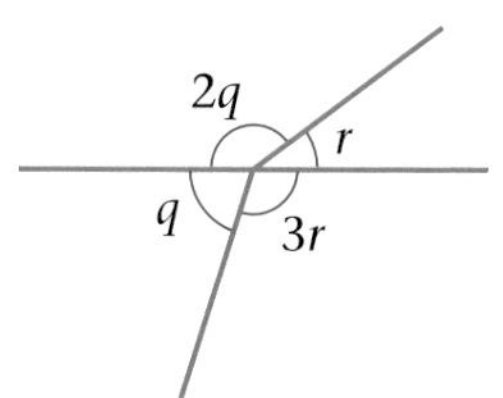

12

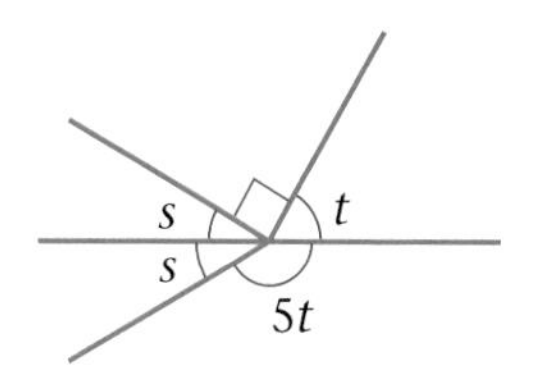

13.2 Parallel lines create alternate angles and corresponding angles.

Example 2

Write down the size of the named angles in the diagrams.

a

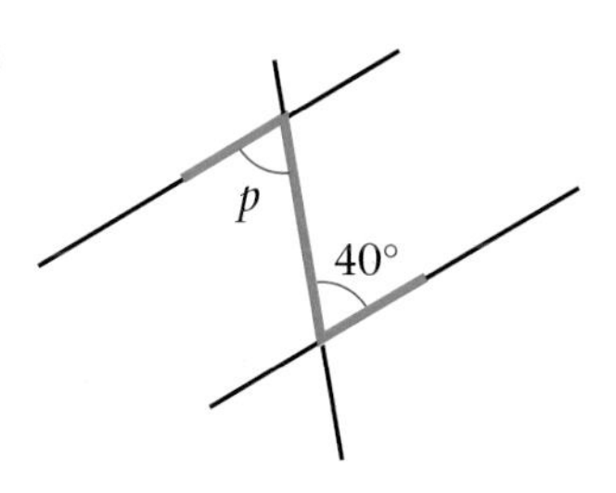

b

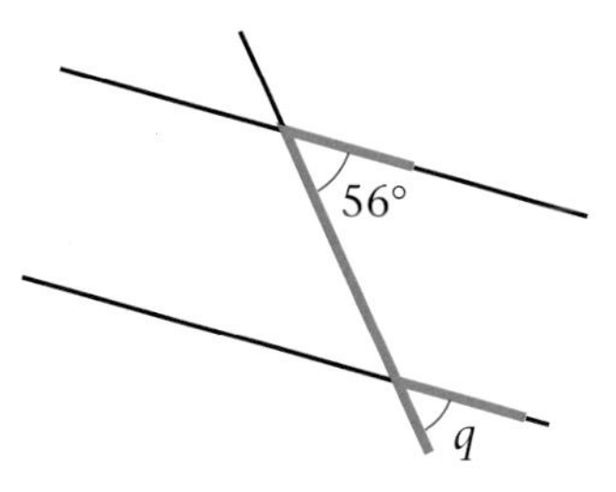

a $p = 40°$ — These are called **alternate** angles. Alternate angles are equal. Notice the 'Z' shape.

b $q = 56°$ — These are called **corresponding** angles. Corresponding angles are equal. Notice the 'F' shape.

Practice 13B

In questions 1–4 work out the value of the marked angle.
Give reasons for your answers.

> **Hint:**
> Remember: lines marked with equal numbers of arrowheads are parallel.

1

2

3

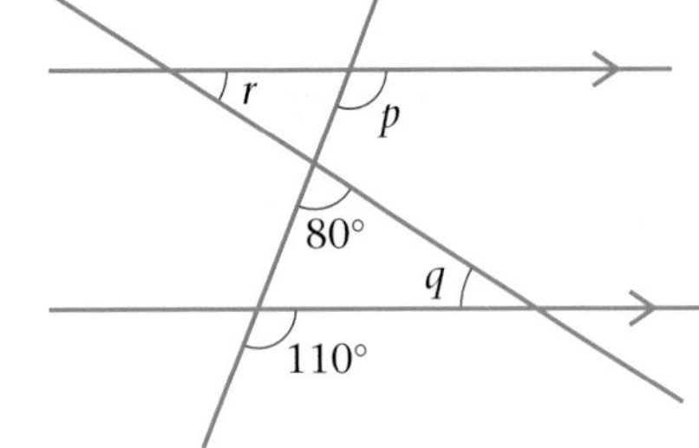

4

5

Explain why d is 73°.

6

Explain why e is 134°.

13.3 Angles in triangles add up to 180° and angles in quadrilaterals add up to 360°.

Example 3

Use the diagram to show that the angles in a triangle add up to 180°.

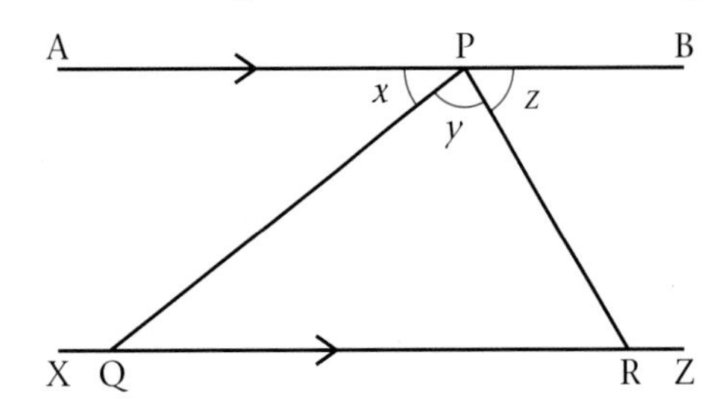

Angle PQR = $x°$ (alternate angles)
Angle PRQ = $z°$ (alternate angles)
But $x + y + z = 180°$
(angles on a straight line add up to 180°.)
As the angles of the triangle are x, y and z, they must also add up to 180°.

Giving reasons is expected.
Doing this line by line is a good way.

Example 4

Find the size of the interior and exterior angles of a regular pentagon.

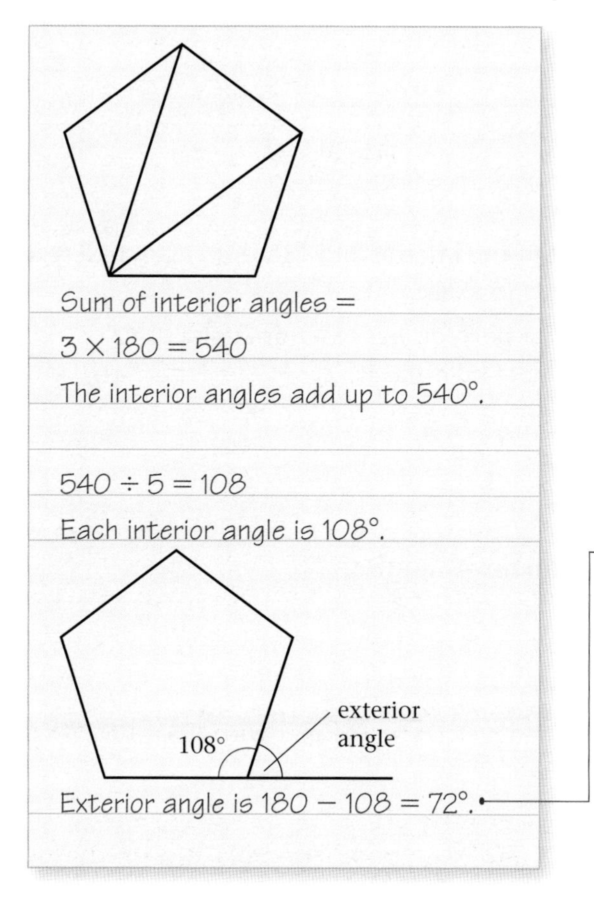

The pentagon can be divided into 3 triangles.
Each triangle has 180°

Interior with exterior angle makes a straight line.

The sum of the exterior angles of any polygon is one complete turn or 360°.
So $5 \times 72° = 360°$

Practice 13C

In questions 1–4 work out the value of the marked angles. Give reasons for your answers.

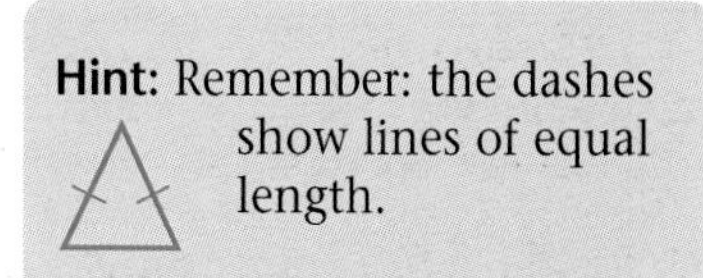

1 **a**

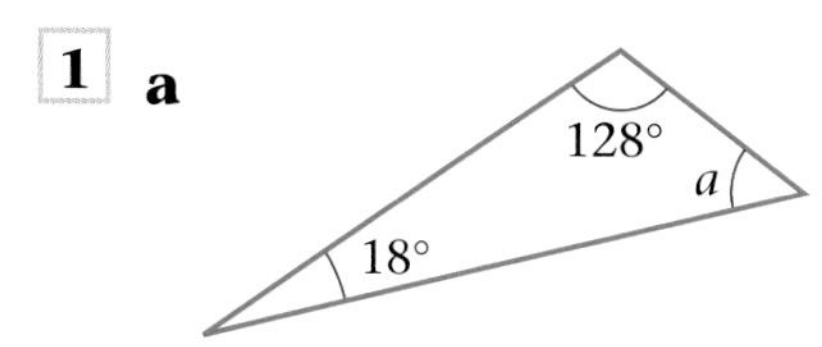

b (isosceles triangle with angles b and 62°)

c

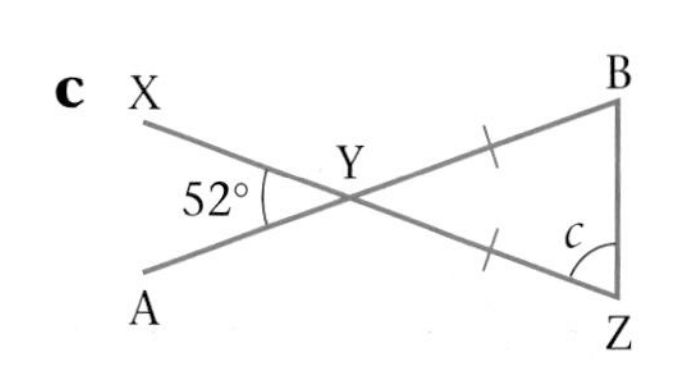

2 **a**

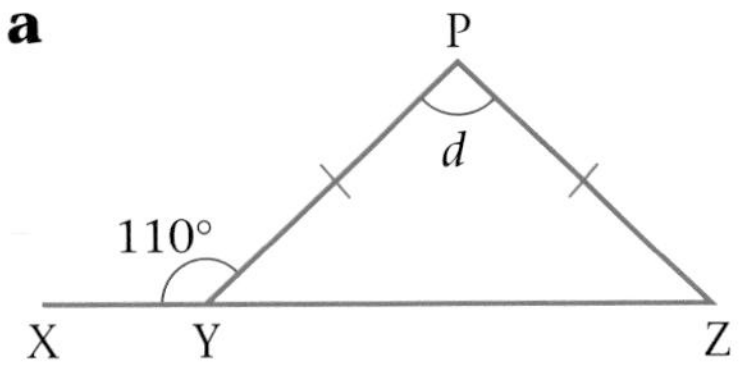

b

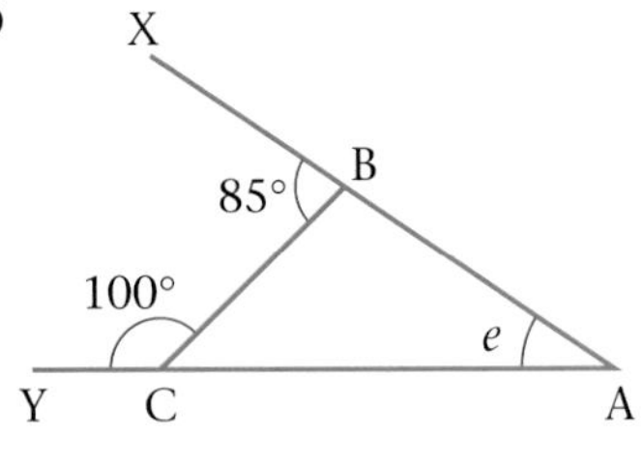

c

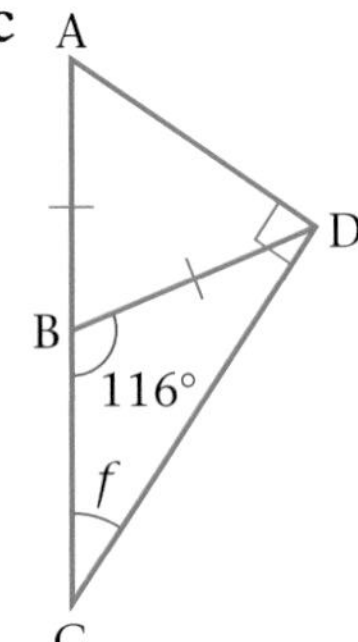

3 **a**

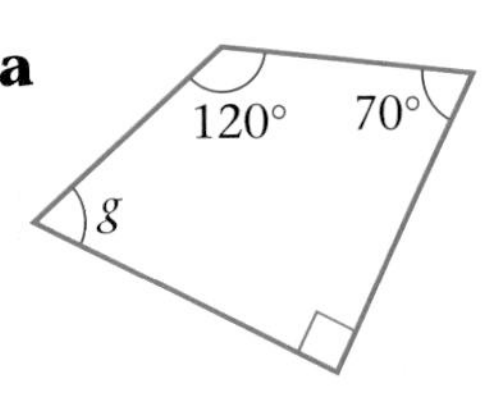

b

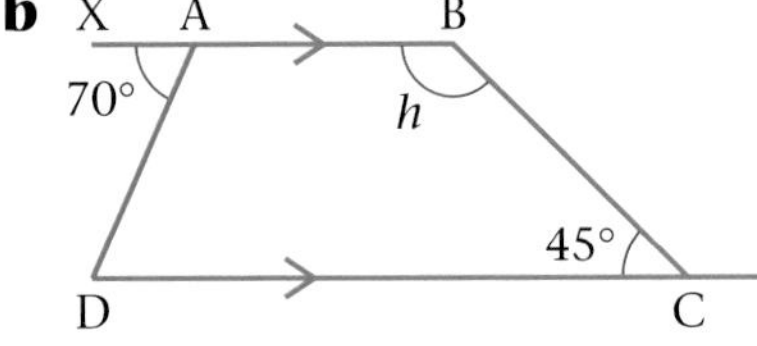

c

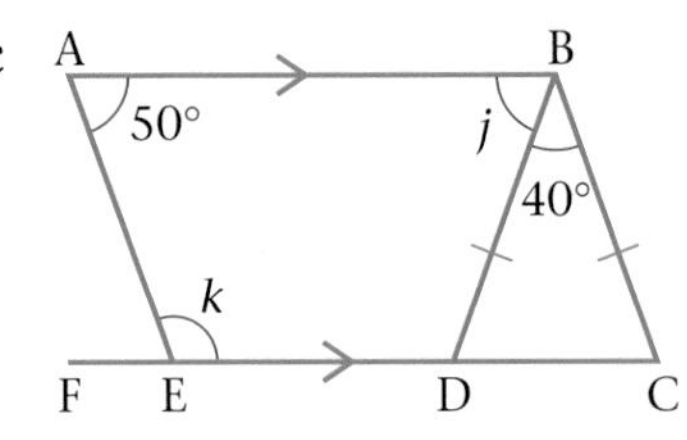

4

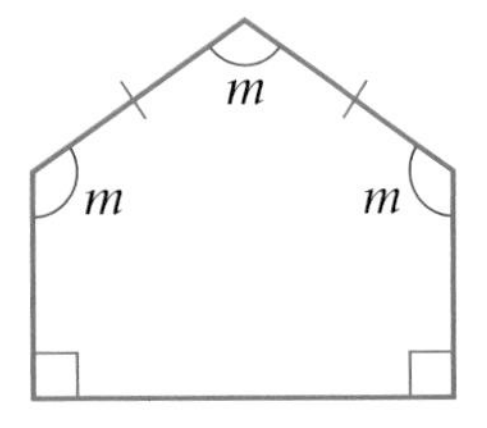

5 **a** Work out the interior and exterior angles of a regular nonagon (9 sides).

b A regular polygon has exterior angles of 12°. How many sides has it?

13.4 You can use angle facts to work out unknown angles.

Example 5

In the diagram, PQ is parallel to RS, and triangle CDS is isosceles with CD = CS. Angle PBA = 118°

Work out angle CSD, giving reasons for your answer.

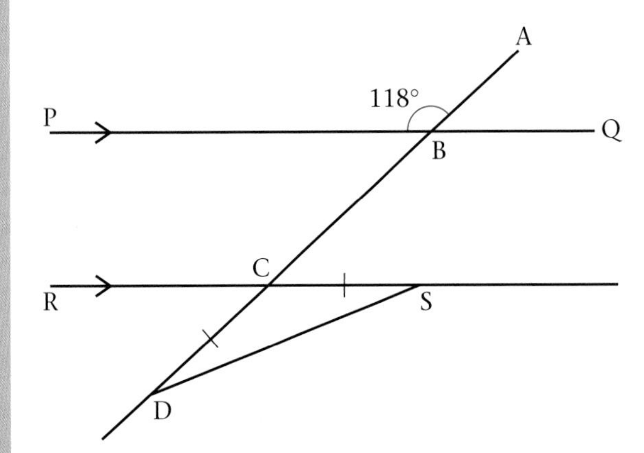

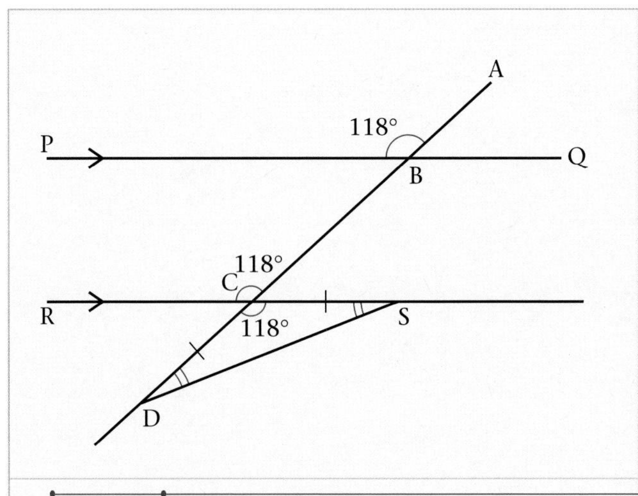

$\angle RCA = \angle PBA$ corresponding angles
$= 118°$

$\angle SCD = \angle RCA$ vertically opposite angles
$= 118°$

$\angle CDS = \angle CSD$ triangle CDS is isosceles

$\angle CDS + \angle CSD = 180° - 118°$ – angles in a
$= 62°$ triangle add up to 180°

So $\angle CSD = 31°$

The symbol ∠ means 'angle'.

A good clear way to show the answer is to present the statements line by line.

Each line has one step with the reason.

Practice 13D

1 CBA and DEA are straight lines.
CD is parallel to BEX.
BE = EA
Angle EAB = 25°.

Find angles DEB and DCB.
Give reasons for your answers.

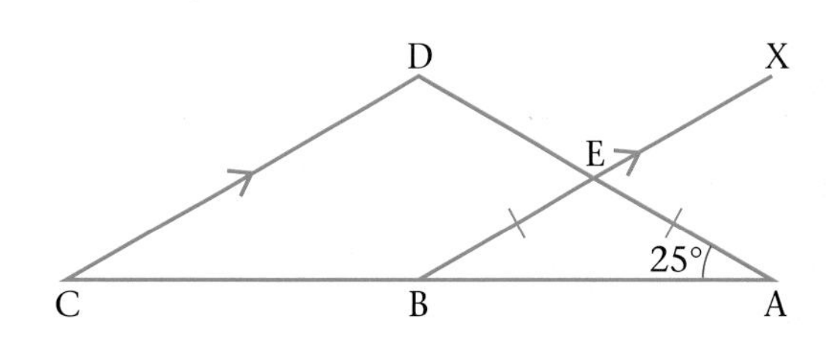

2 Triangles CYX, BXY and ABY are isosceles triangles.
CYA and CXB are straight lines.
Angle BYX = 35°.

Work out angle ABY.
Give reasons.

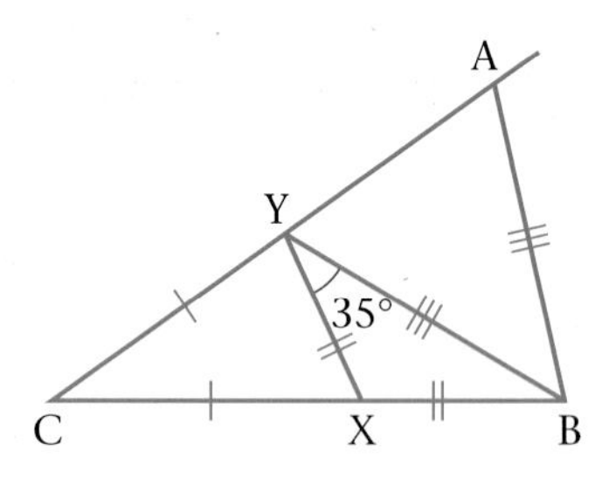

3 ABC is an isosceles triangle.
PQR is parallel to BC.
CR = CQ
Angle BCQ = 65°.

Work out angles ACB and APQ.
Give reasons for your answers.

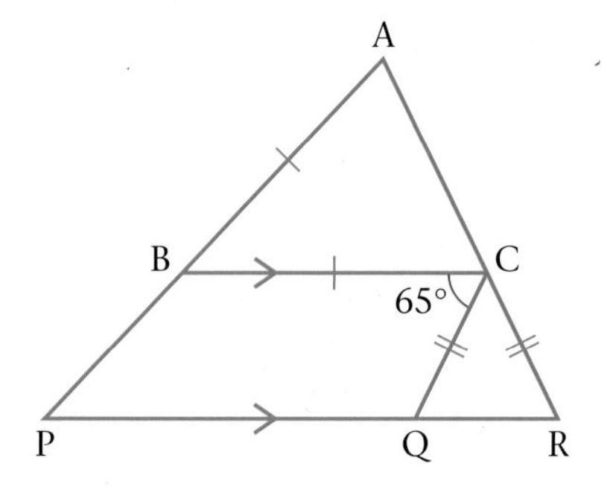

4 ABP and ACQ are straight lines.
BQ = CQ
BC is parallel to PQ.
Angle A = 12° and angle BPQ = 44°

Work out angles BCQ and BQP.
Give reasons.

Hint: Find the angles in triangle ABC first.

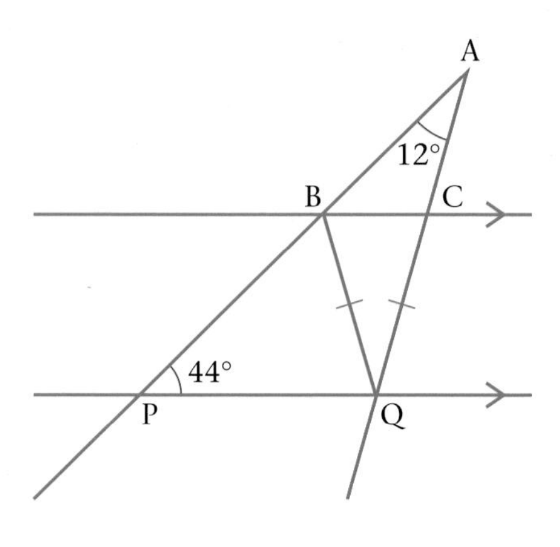

5 Explain why the exterior angles of a quadrilateral add up to 360°.

i.e. $a + b + c + d = 360°$

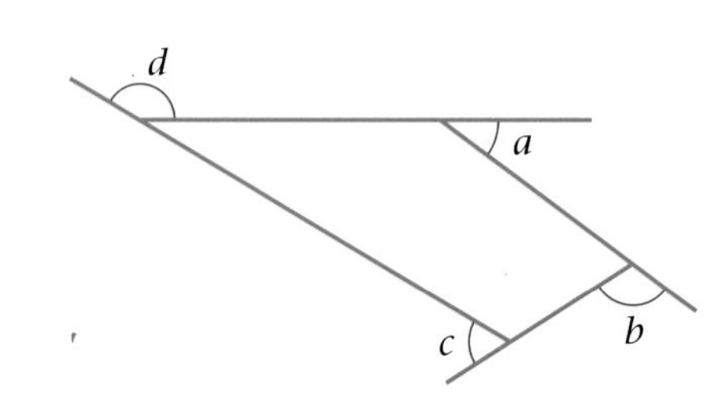

13.5 The angle subtended at the circumference by a semicircle is always 90°.

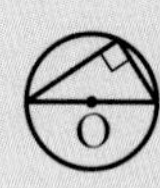

Example 6

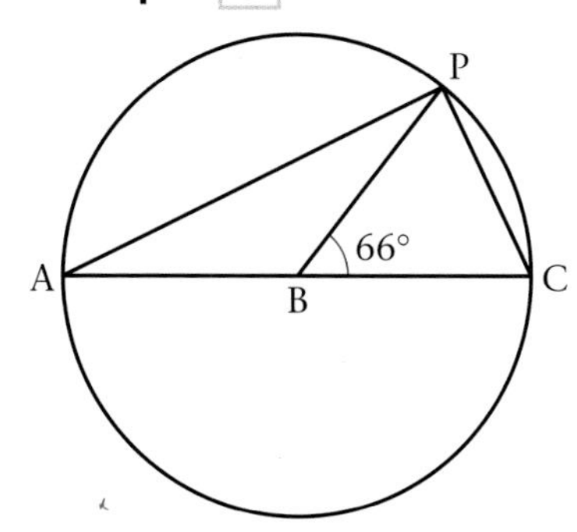

B is the centre of the circle.
Find angles BPC, BCP, ABP and PAB.

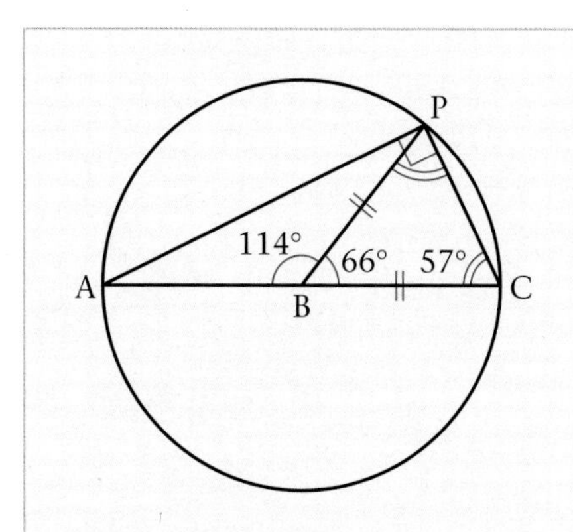

$\angle BPC + \angle BCP = 180 - 66 = 114°$
(angle sum of a triangle is 180°)
But $\angle BPC = \angle BCP = 57°$
(triangle BPC is isosceles) — BP and BC are radii.
$\angle ABP = 180 - 66 = 114°$
(angles on a straight line = 180°)
$\angle PAB = 180° - \angle APC - \angle ACP$ (angles in a triangle)
$= 180° - 90° - 57°$
$= 33°$

$\angle APC$ is an angle subtended at the circumference by a semicircle.

Practice 13E

1

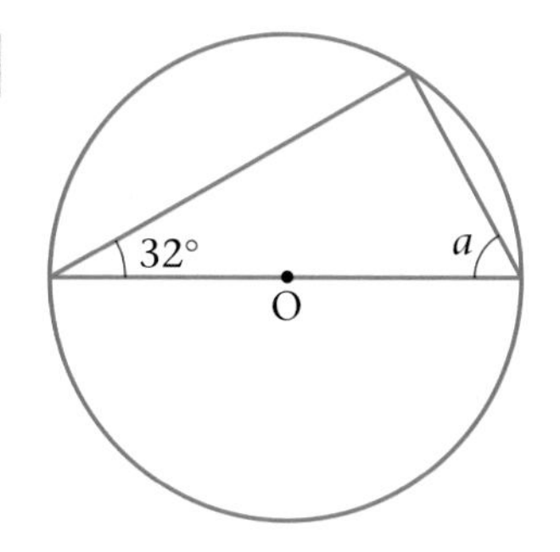

Calculate angle a.

13.6 Opposite angles in a cyclic quadrilateral sum to 180°.

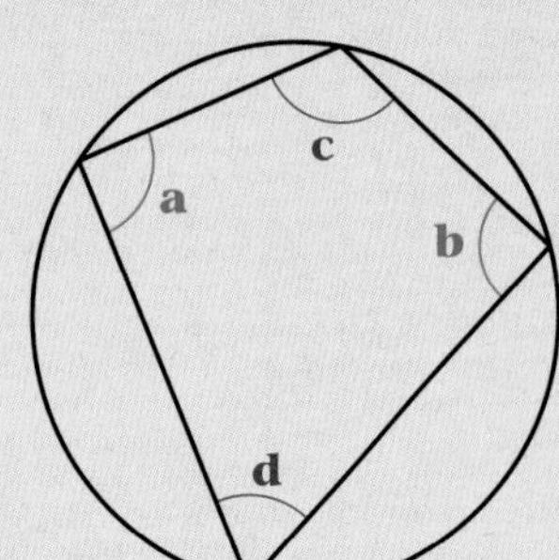

a + b = 180°

c + d = 180°

Example 7

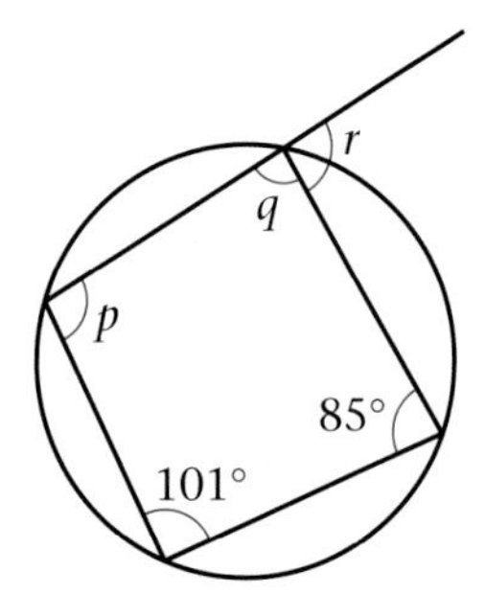

Calculate angles p, q and r.

$p = 180 - 85 = 95°$
(opposite angles in a cyclic quadrilateral)
$q = 180 - 101 = 79°$
(opposite angles in a cyclic quadrilateral)
$r = 180 - q$ (angles on a straight line)
$= 180 - 79$
$= 101°$

Practice 13F

Find the named angles, giving reasons for your answers.

1

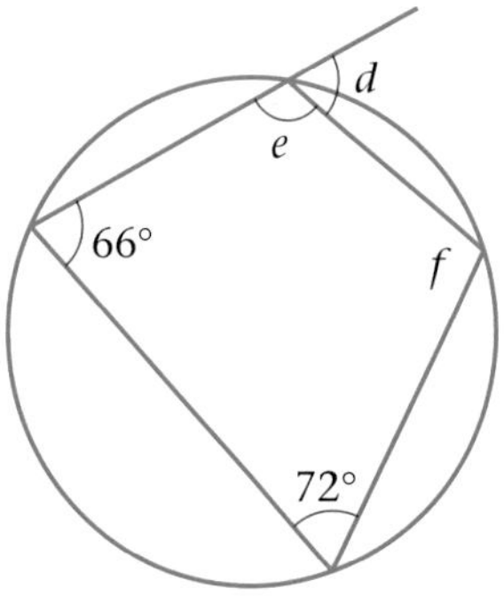

2

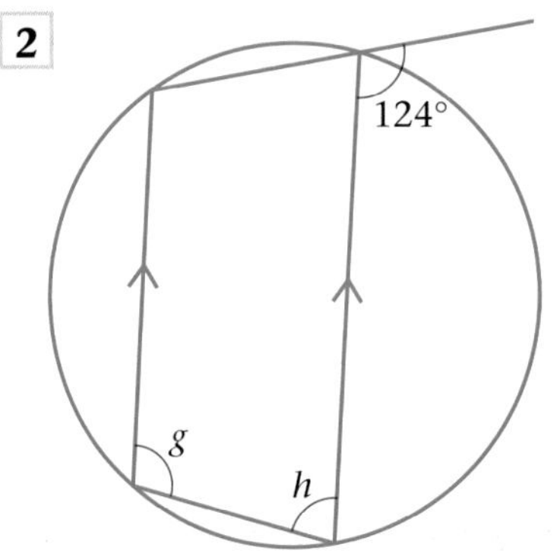

3

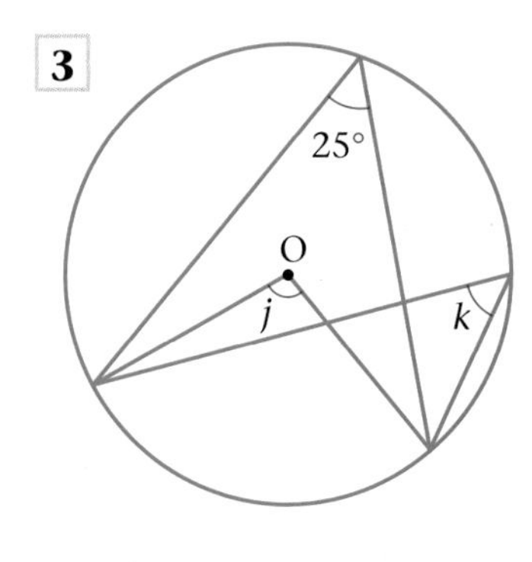

13.7 Angles subtended at the circumference by the same arc are equal.

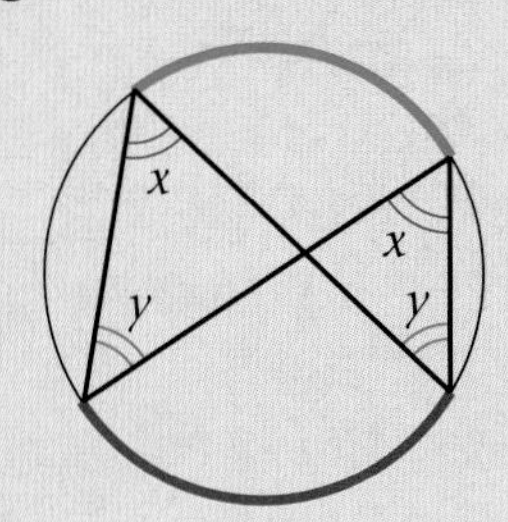

Example 8

Work out angles a, b, c and d.

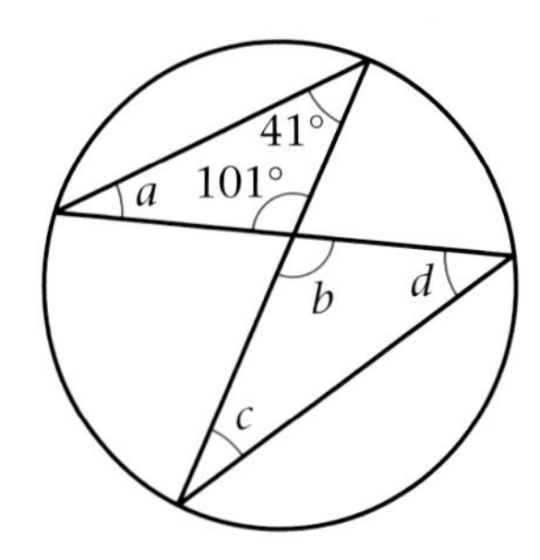

$a = 180° - 41° - 101° = 38°$ — Angles in a triangle add up to 180°.

$b = 101°$ — Vertically opposite angles.

$c = 38°$ — $c = a$; angles subtended by the same arc are equal.

$d = 41°$ — Angles subtended by the same arc are equal.

Practice 13G

1

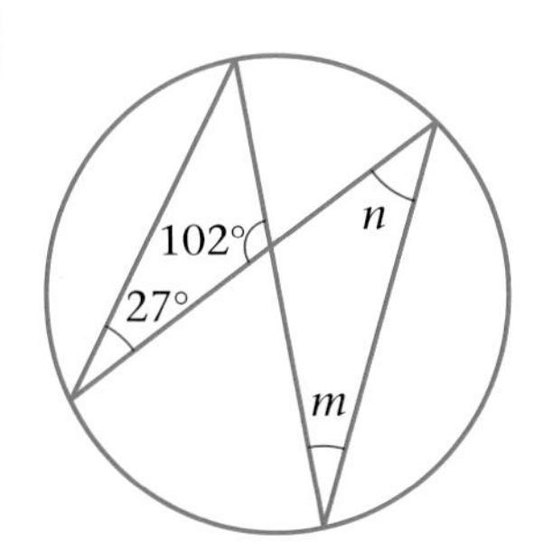

Find angles n and m, giving reasons for your answer.

13.8 An angle subtended by an arc at the centre of a circle is twice that subtended at the circumference.

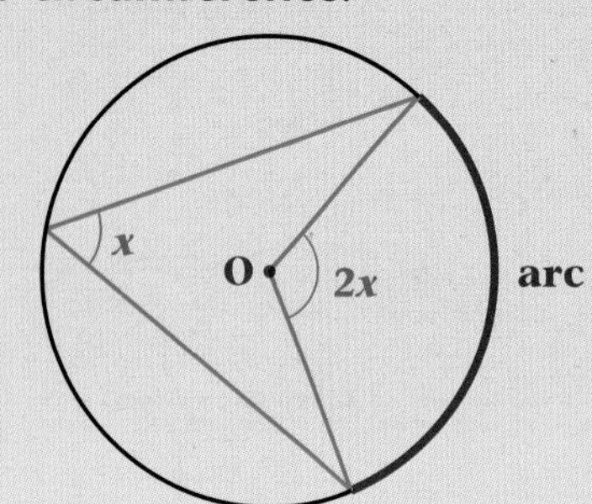

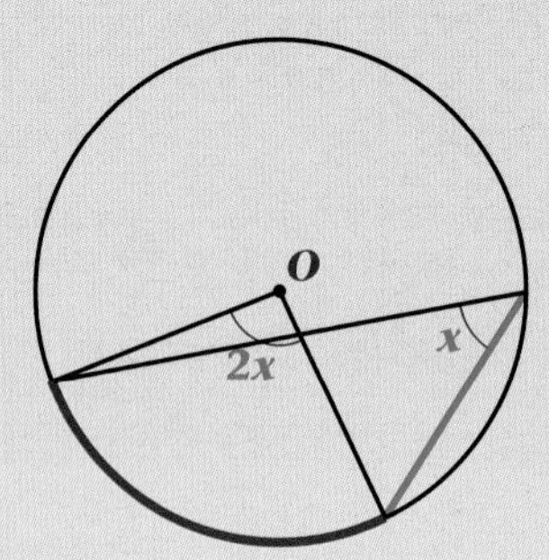

Example 9

Find angle m.

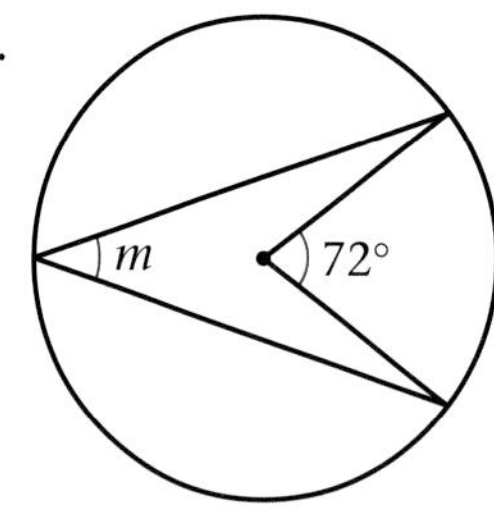

$2m = 72°$

$m = 36°$

The angle subtended by an arc at the centre is twice that subtended at the circumference.

Practice 13H

1

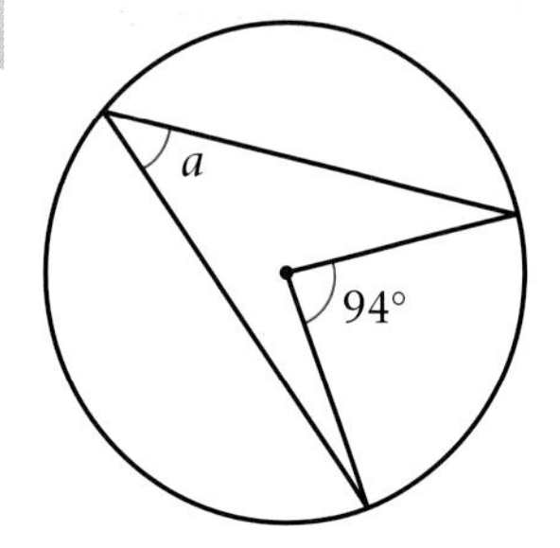

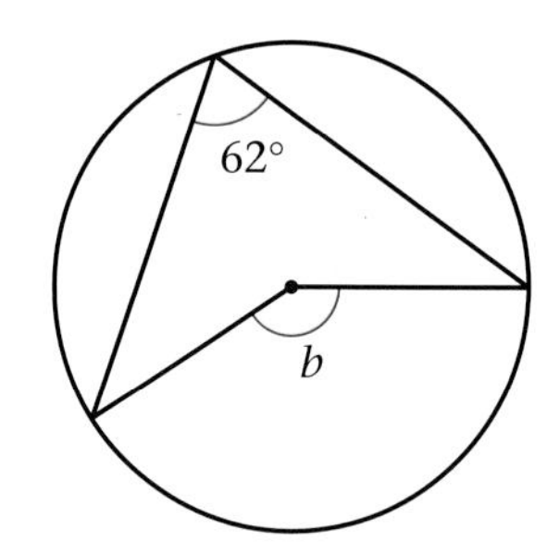

Find the named angles.

2

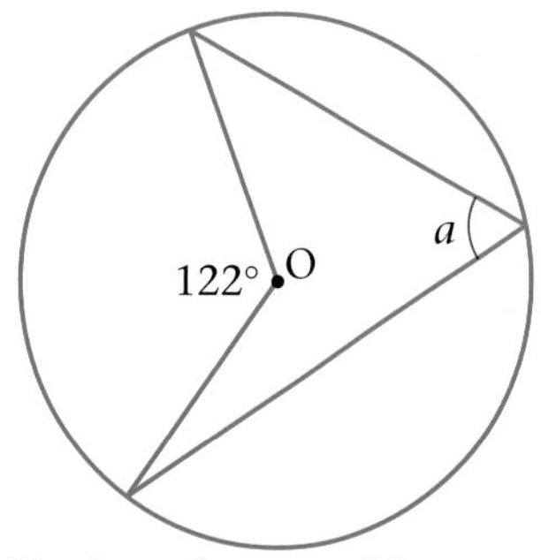

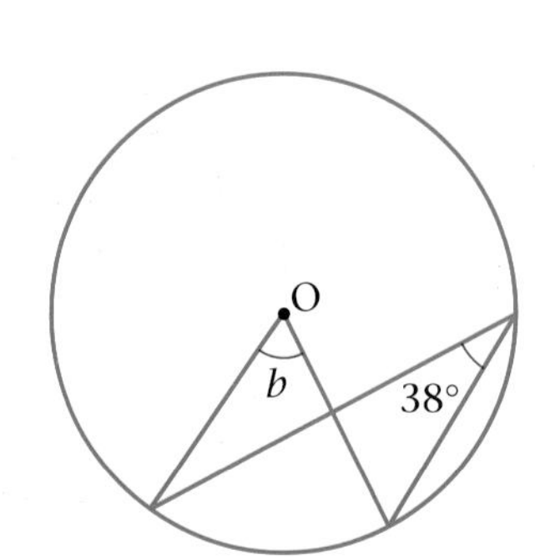

Find angles a and b.

13.9

A chord is a straight line between two points on the circumference, which does not pass through the centre. A line drawn from the centre of a circle to the middle point of a chord is perpendicular to the chord.

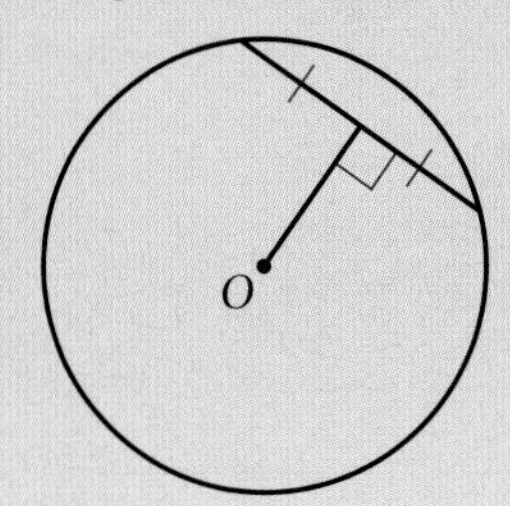

Example 10

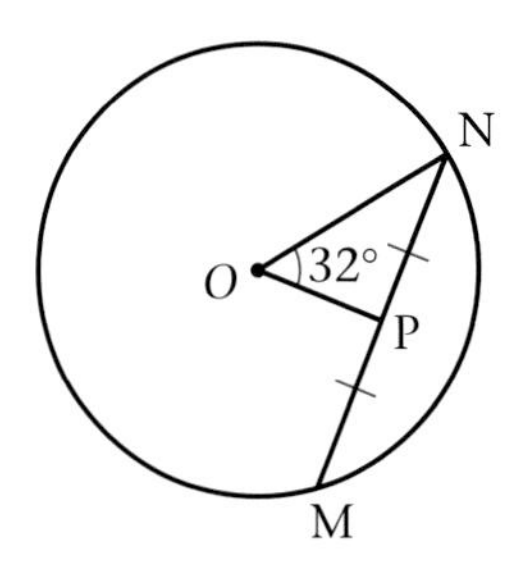

Find angle ONP.

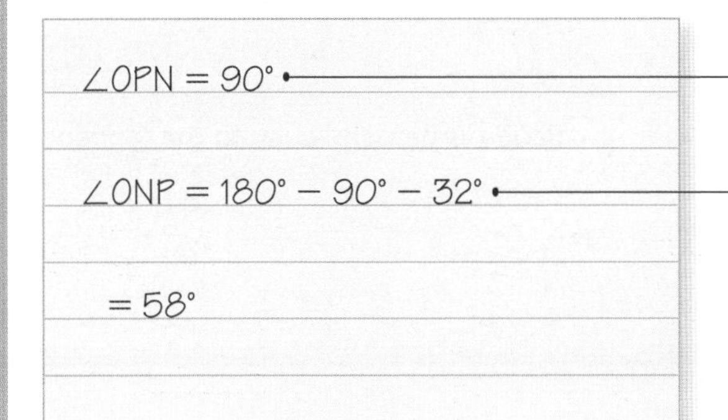

Line from centre to mid-point of a chord is perpendicular to the chord.

Angles in a triangle sum to 180°.

Practice 13I

1

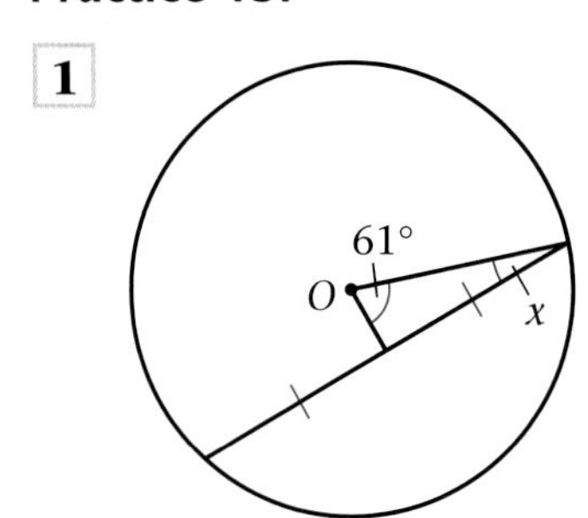

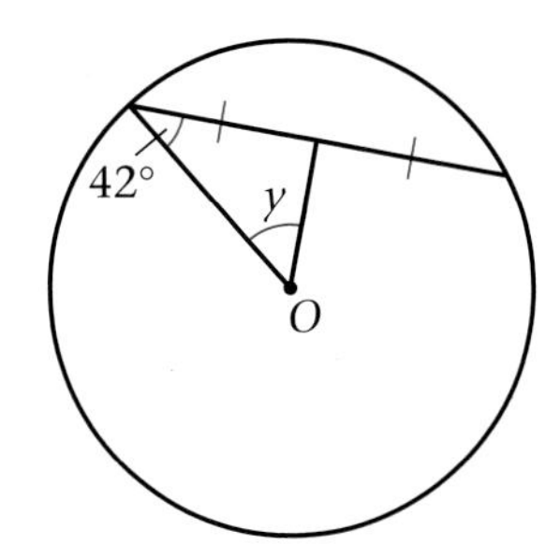

Find x and y.

13.10 A radius drawn from the point where a tangent touches a circle is perpendicular to the tangent.

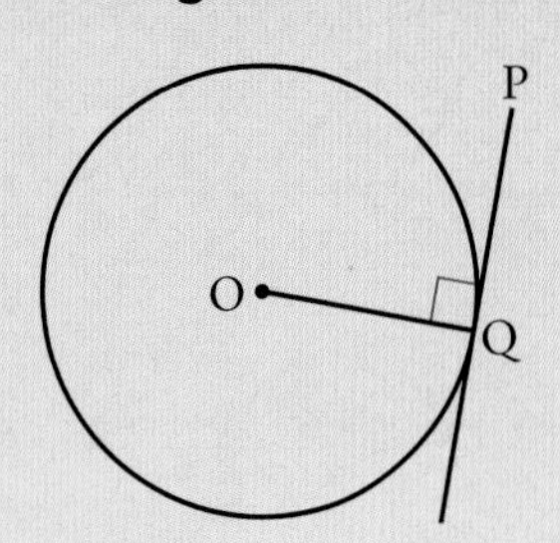

Angle OQ̂P = 90°

Example 11

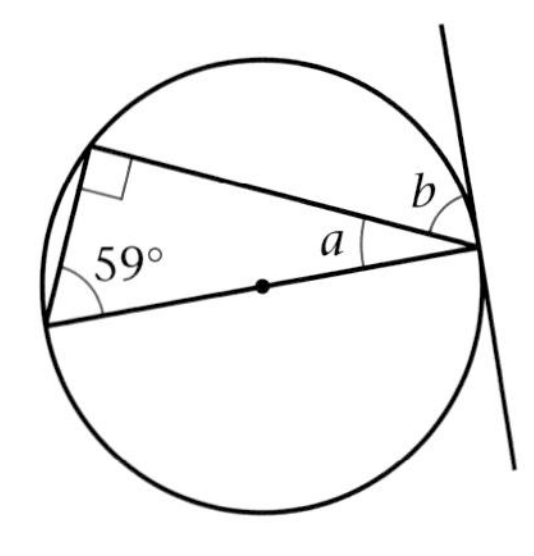

Work out angles *a* and *b*.

$a = 180° - 90° - 59°$ — Angles in a triangle sum to 180°.

$a = 31°$

Radius is perpendicular to the tangent.

$b = 90° - a$

$= 90° - 31°$

$= 59°$

Practice 13J

1

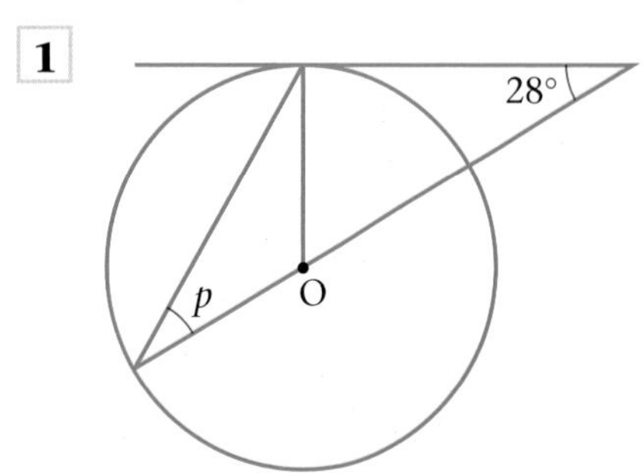

Find angle *p*.

13.11 Tangents drawn to a circle from a point outside the circle are equal in length.

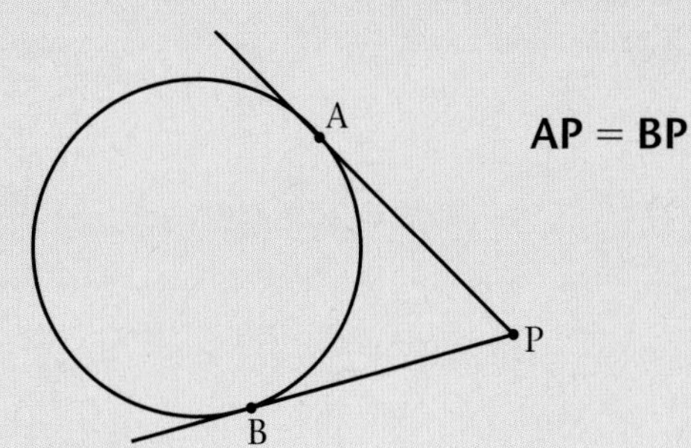

AP = BP

Example 12

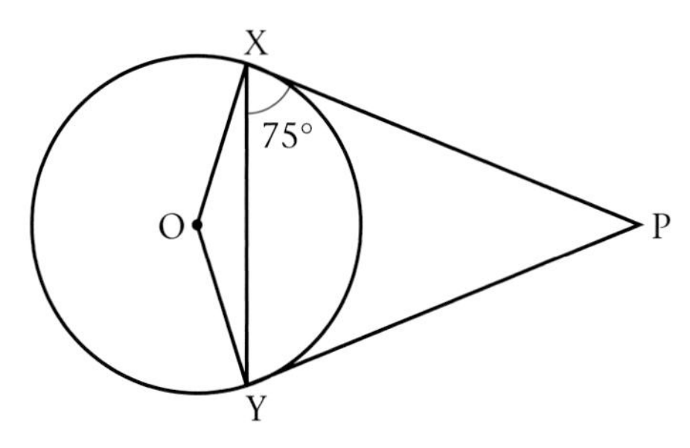

Calculate angles XPY and OXY.

PX = PY — Tangents from P are equal.

∠XYP = ∠PXY = 75° — Base angles in an isosceles triangle are equal.

∠XPY = 180° − 75° − 75°
= 180° − 150°
= 30° — Angles in a triangle sum to 180°.

∠OXY = 90° − 75°
= 15° — Radius is perpendicular to tangent.

Practice 13K

1

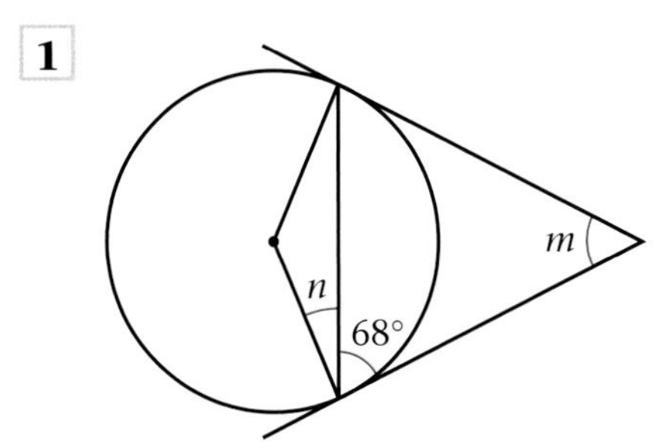

Find angles m and n, giving reasons.

2

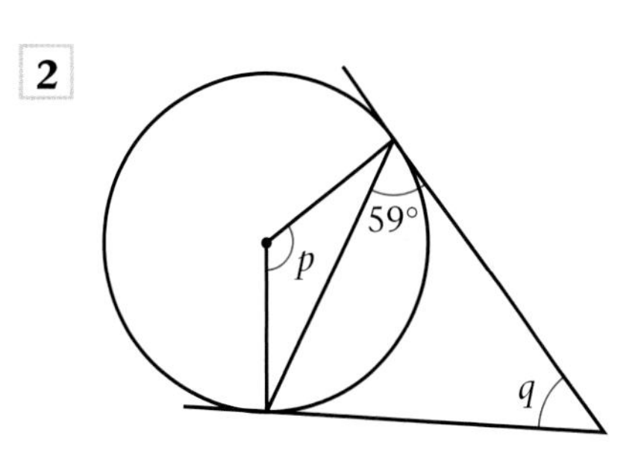

Find angles p and q, giving reasons.

13.12 To solve problems involving angles, you often need to use more than one rule. Solve the problem step by step, noting down your reasons.

Example 13

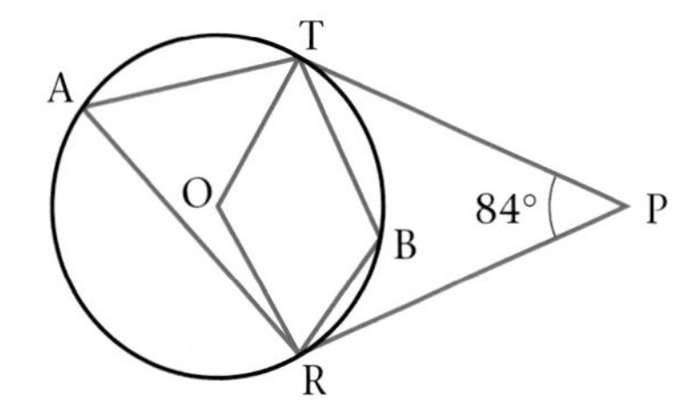

Work out ∠TAR and ∠TBR, giving reasons.

Angles PTO and PRO are 90°
(Angle between tangent and radius at point of contact is 90°.)

∠TOR = 360 − 90 − 90 − 84 = 96°
(Angles in a quadrilateral add up to 360°.)

∠TAR = 48°
(Angle at circumference = $\frac{1}{2}$ angle at centre.)

∠TBR = 180 − 48 = 132°
(Opposite angles in a cyclic quadrilateral = 180°.)

Practice 13L

1

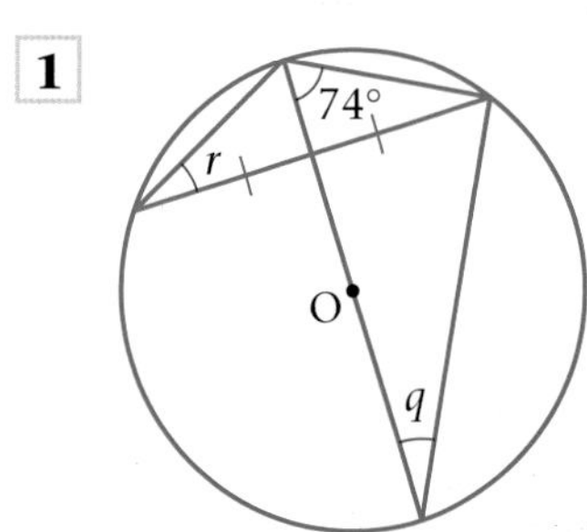

Work out angles q and r, giving reasons.

2 XTY is a tangent to the circle, centre O.
P and Q are points on the circumference.
OQ is parallel to PT.
Angle QOT = 37°

Work out angle OPT and angle PTY.
Give reasons for your answers.

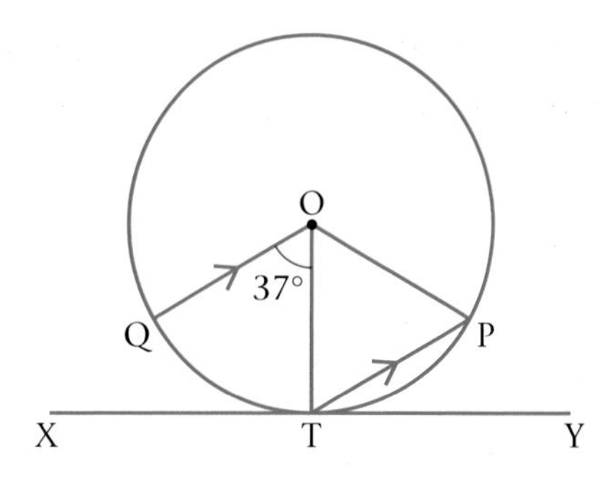

3 PTR is a tangent to the circle, centre O.
The chord AB is parallel to PR.
X is a point on the circumference.
Angle ORT = 18°

Work out angle AXB.
Give reasons for your answer.

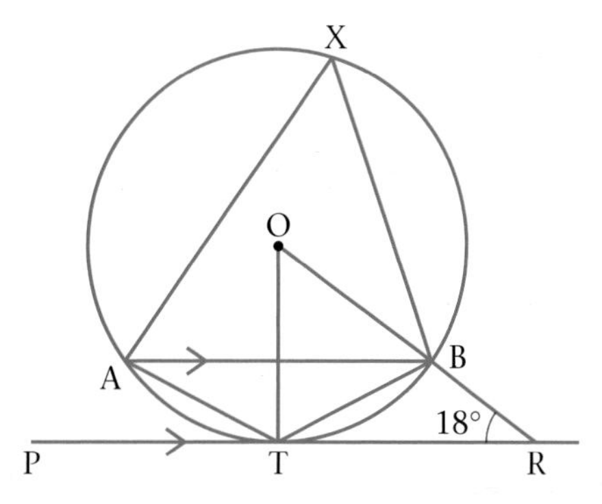

4 PA and PB are tangents to the circle, centre O.
X and Y are points on the circumference.
Angle APB = 75°.

Work out angle AXB and angle AYB.
Give reasons for your answers.

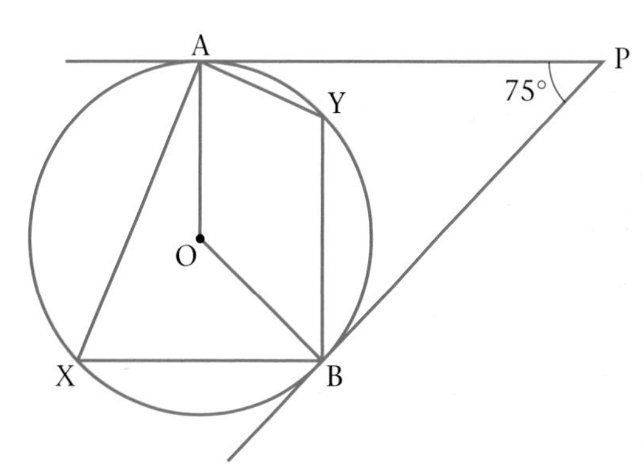

13.13 Direction between places on a map is given as a bearing.

Example 14

a What is the bearing of A from B?
b What is the bearing of B from A?

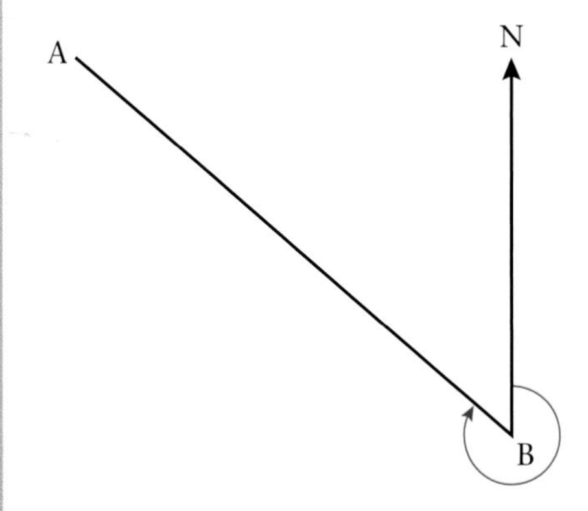

Hint: Bearings are always measured clockwise from North and written with three figures.

Hint: You are expected to be accurate to ±2°

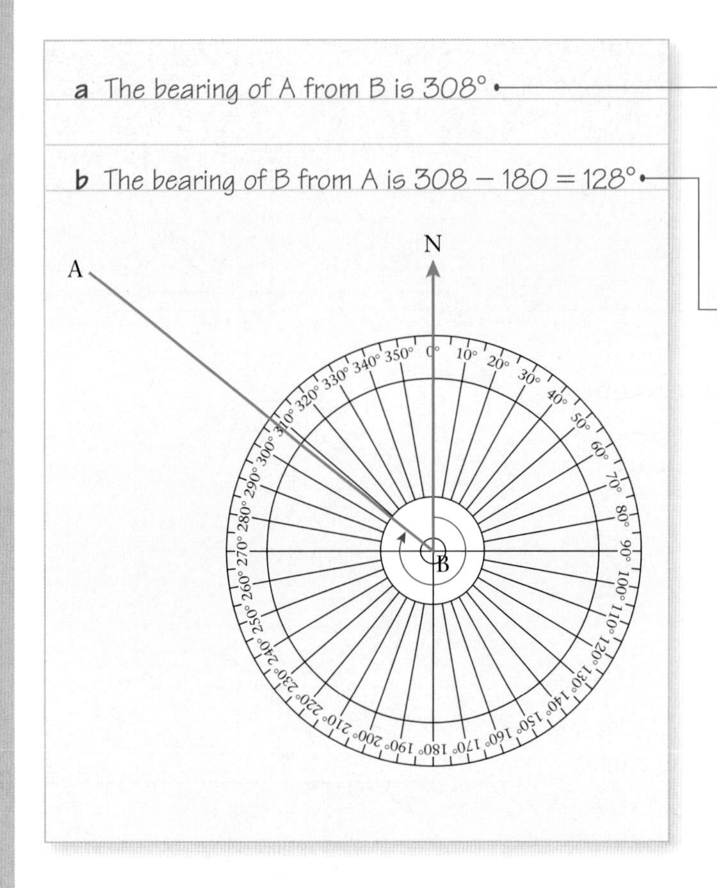

To obtain the bearing of A from B imagine you are standing at B and facing North.

Turning clockwise, measure the angle with a protractor.

To obtain the bearing of B from A involves a half turn.

Add or take away 180° for the half turn.

Practice 13M

1 Using the map of Cornwall find the bearing from Launceston of

a Tiverton
b Exeter
c Plymouth
d Bodmin
e Bude.

Also find the bearing of

f Penzance from Tavistock
g Ilfracombe from Newquay
h Truro from Bodmin.

2 P, Q and R are three ships at sea. The bearing of Q from P is 080°.
The bearing of Q from R is 155°. The bearing of R from P is 030°.
Draw a sketch to show the relative positions of the three ships.

Hint: you may find it helpful to pretend that the distance of Q from P is 80 km, to be represented by 8 cm. Because your diagram only shows relative positions the actual distance does not matter.

3 The bearing of Blackburn from Rochdale is 303°.
What is the bearing of Rochdale from Blackburn?

4 The bearing of Hull from York is 115°.
What is the bearing of York from Hull?

5 Measure and write down the bearing of
a A from B
a A from C
a B from C
a C from A

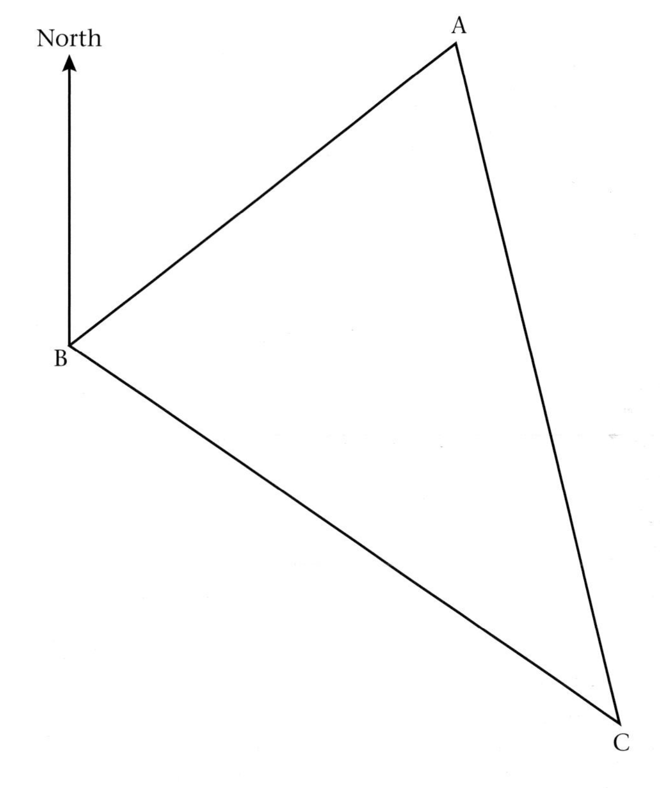

14 Shapes

This chapter shows you how to represent shapes on a coordinate grid and how to make accurate constructions. Symmetry of 2D and 3D shapes, and the use of loci are covered in the second half of the chapter. **(S9)**

14.1 You can investigate the properties of shapes using grids.

Example 1

$A(1, 4)$, $B(5, 5)$ and $C(6, 2)$ are three points.
Give the coordinates of the point that will complete the parallelogram $ABCD$.

There is more than one way to complete a parallelogram, with these points. Follow the clockwise order, to find point D.

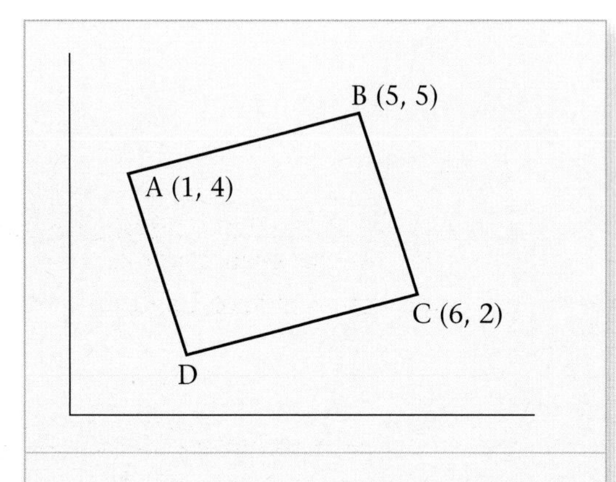

Plot the points A, B, C and sketch the parallelogram.

To get to C from B, go 1 unit across and 3 units down.

So to get to D from A, go 1 unit across and 3 units down.

$A(1, 4) \rightarrow D(1 + 1, 4 - 3)$
$\rightarrow D(2, 1)$

Opposite sides of a parallelogram are equal and parallel.

Check that the coordinates look right for the point that you have sketched.

Example 2

The mid-point of the line segment joining $A(3, -8)$ to $B(7, y)$ is $M(x, 1)$.

a Find the values of x and y.

b Calculate the length of AB.

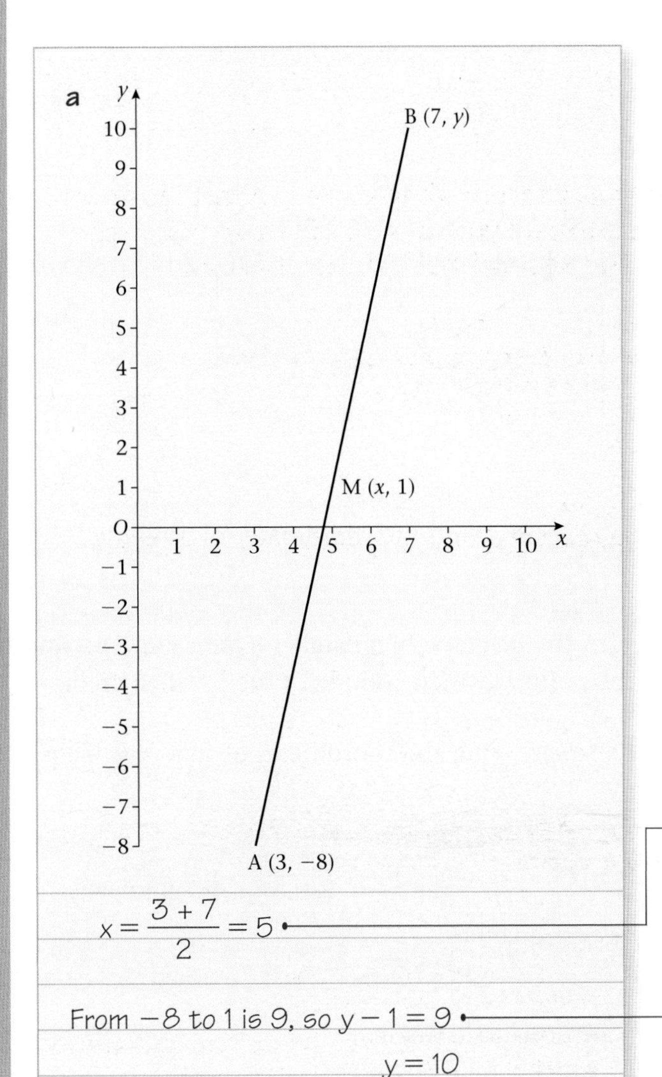

$x = \frac{3 + 7}{2} = 5$

From −8 to 1 is 9, so $y - 1 = 9$

$y = 10$

A sketch can help.

You know M is above the x-axis.

You know B is 4 units right of A and above M.

As M is the mid-point of AB, x must be half way between 3 and 7.

It is the average of 3 and 7.

Also, in the y direction, y must be as far from 1 as −8 is from 1.

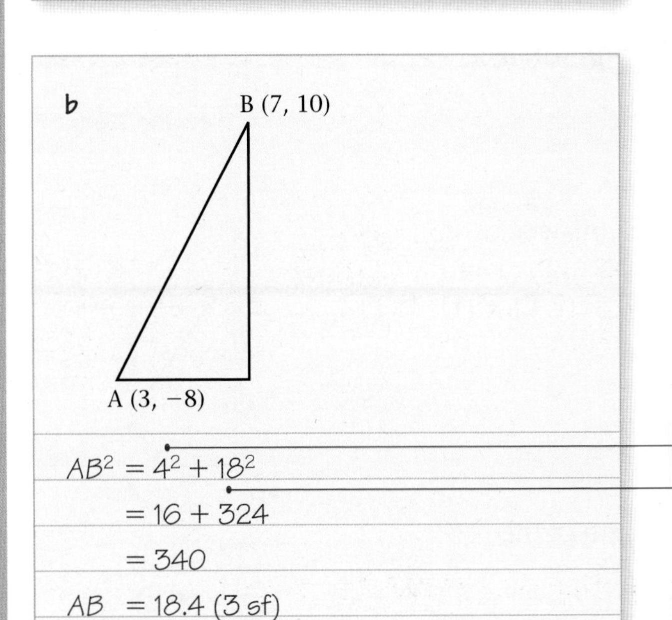

$AB^2 = 4^2 + 18^2$

$= 16 + 324$

$= 340$

$AB = 18.4$ (3 sf)

$7 - 3 = 4$

$10 - -8 = 18$

See Pythagoras' theorem, Section 17.1.

Practice 14A

1 Copy and complete the parallelograms $ABCD$.
Write down the coordinates of D.

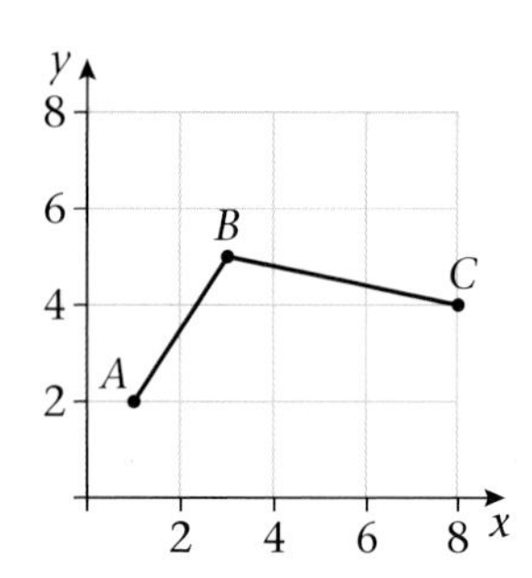

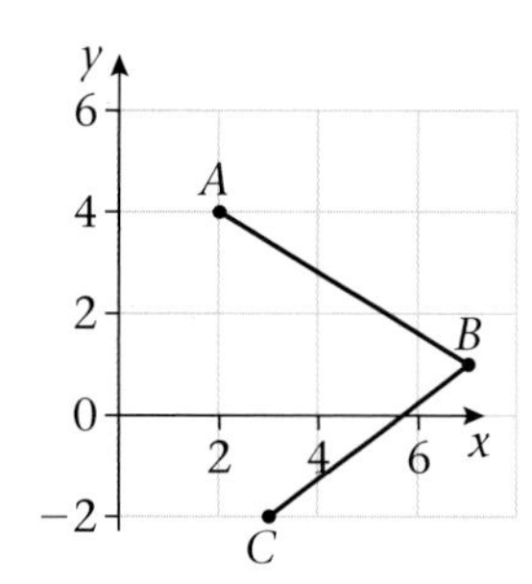

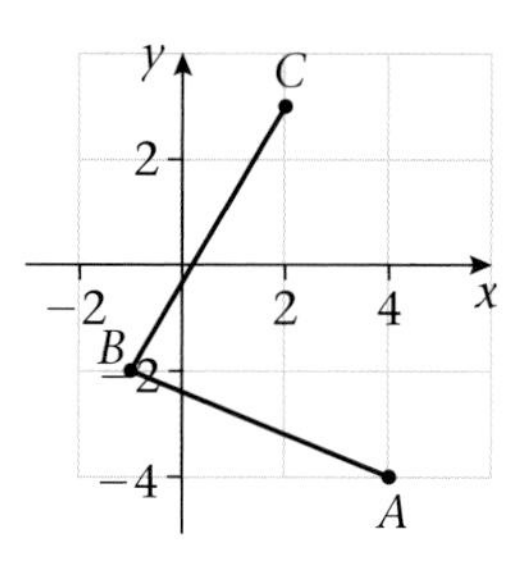

2 $A(2, 5)$, $B(3, 7)$ and $C(7, 4)$ are three points. Find the coordinates of 3 points D which could complete a parallelogram.

3 $A(3, 4)$ and $C(9, -6)$ are the ends of the diagonal of a parallelogram. $B(6, 1)$ is another vertex. Find the coordinates of the point D which completes the parallelogram.

4 $A(3, 2)$ and $B(3, 4)$ is one side of a square. Find the coordinates of two possible points which would complete the square.

5 $A(1, 5)$ and $C(2, 2)$ is one side of a square.
Find the coordinates of the points C and D which complete the square in the first quadrant.

Hint: The first quadrant is the upper right hand quadrant.

6 In the diagram, AB is one side of an isosceles triangle.

a Find the coordinates of a point which would complete the isosceles triangle if AB is the unequal side.

b Describe the locus of the points which could complete the isosceles triangle in part **a**.

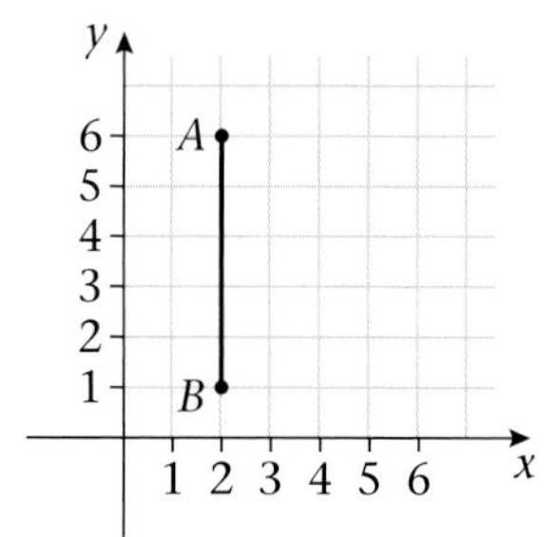

7 The diagram shows part of a quadrilateral.
Write down the shape that is made if the fourth point is **a** (3, 5), **b** (5, 4), **c** (7, 3), **d** (9, 2), **e** (11, 1).

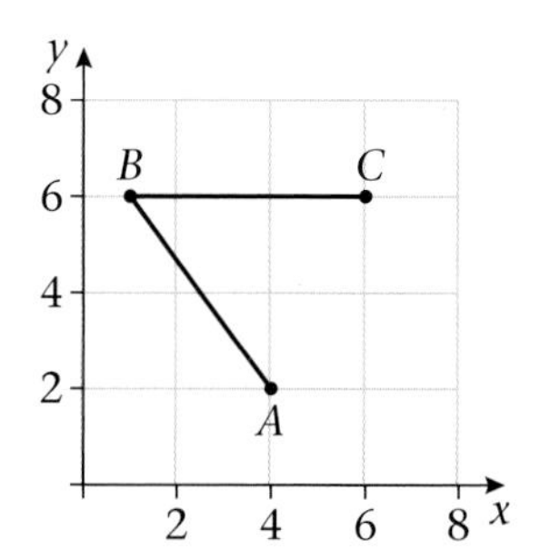

8 Work out the mid-point and length of the line segment joining $A(5, 1)$ to $B(3, 7)$.

9 Work out the mid-point and length of the line segment joining $C(-3, 4)$ to $D(-6, -6)$.

10 $M(3, y)$ is the mid-point of the line segment joining $L(x, -2)$ to $N(5, -6)$.
Work out x and y.
Work out the lengths of LM and LN.

11 $Q(-1, -4)$ is the mid-point of the line segment joining $P(3, 1)$ to $R(x, y)$. Find x and y.

14.2 You can use straight edge and compasses to make standard constructions.

(S10)

Example 3

Construct a triangle with sides of length 5 cm, 6 cm and 7 cm.

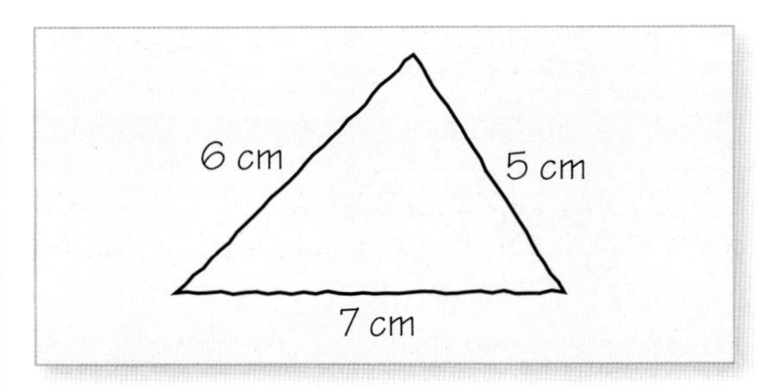

Making a rough sketch first is a good idea. It improves the chances of keeping the construction on the paper.

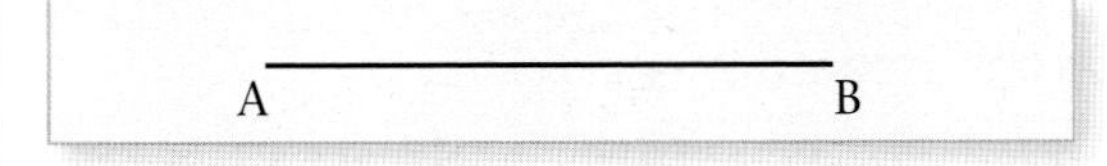

Draw one of the sides. The longest is best. In this case, 7 cm.

This is *AB*.

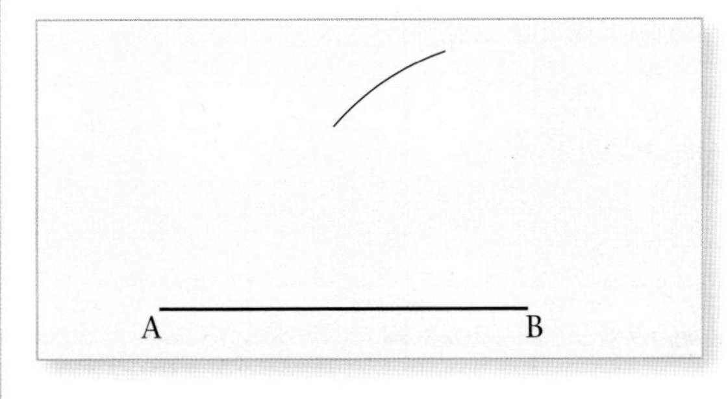

Open the compasses to 5 cm.

Use one end of the line (B) as centre.

Draw an arc near where you expect C to be.

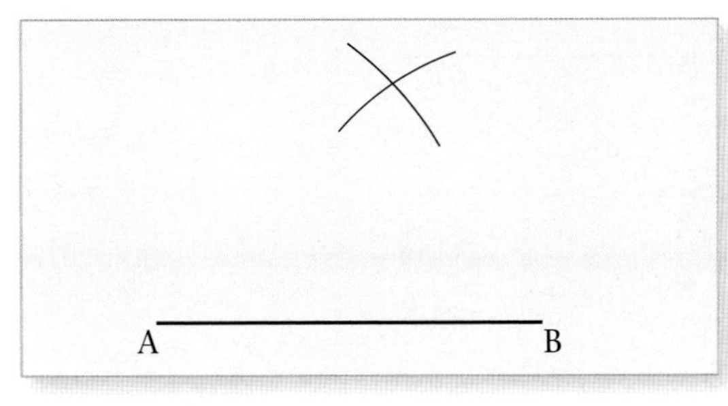

Open the compasses to 6 cm.

Use the other end of the line (A) as centre and draw an arc to intersect the arc previously drawn.

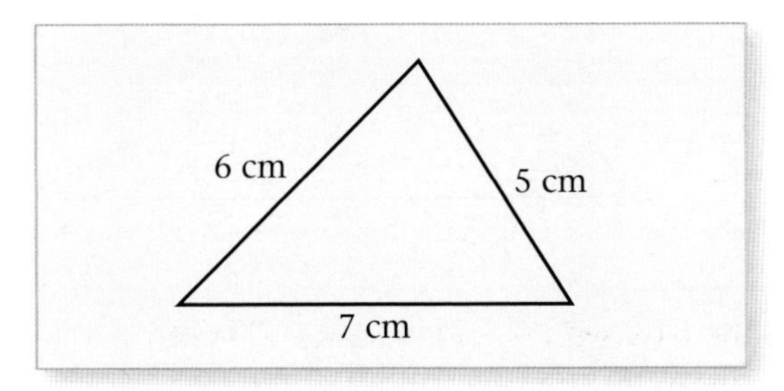

Complete the triangle.

Example 4

Draw a net to construct a cuboid of sides 3 cm × 4 cm × 5 cm.

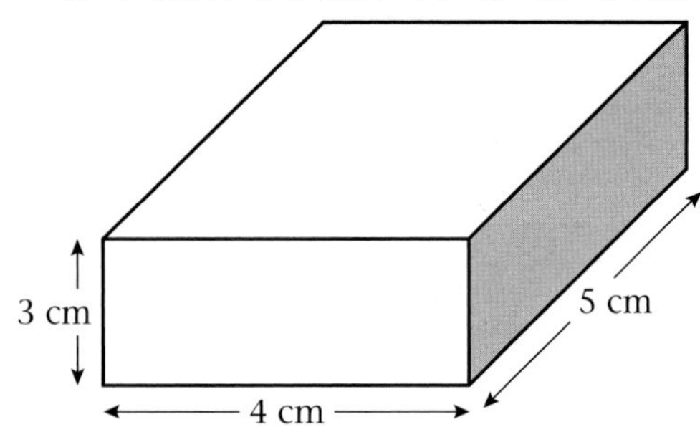

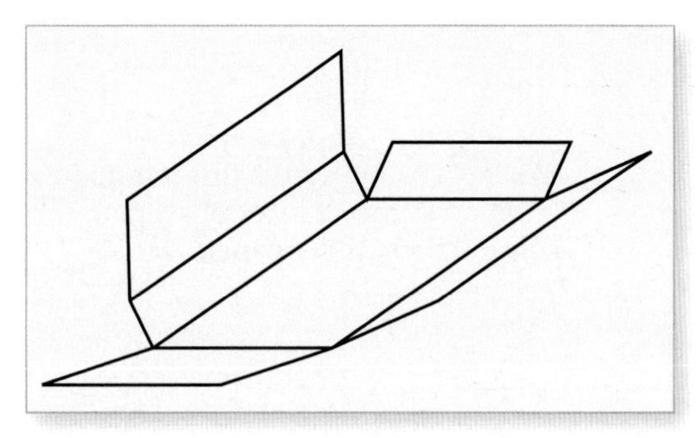

A net is the flat shape which folds up to make a solid shape.

To obtain a net, imagine the shape being unfolded.

There may be more than one way of doing this. Each will be a net.

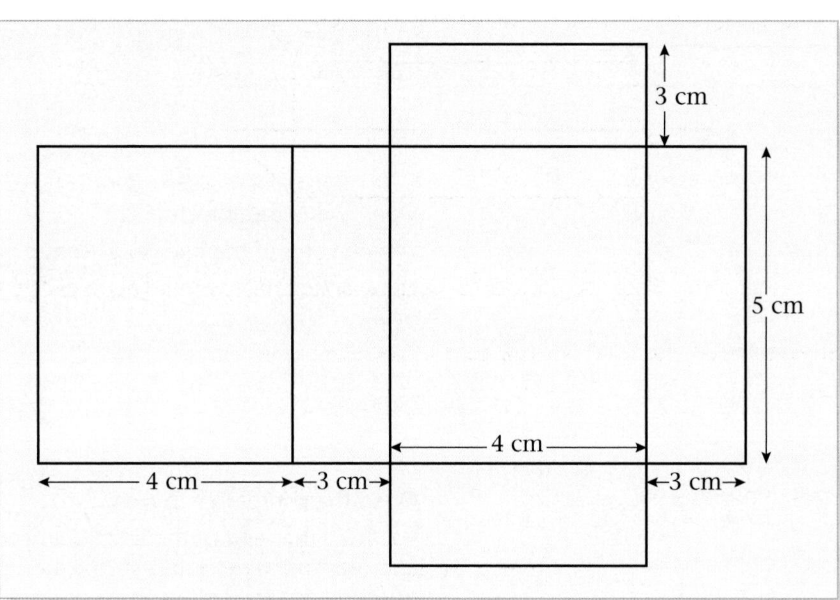

Practice 14B

1 Here is a sketch of a triangle.
The lengths of the sides of the triangle are 8 m, 9 m and 12 m.
Use a scale of 1 cm to 2 m to make an accurate drawing of the triangle.

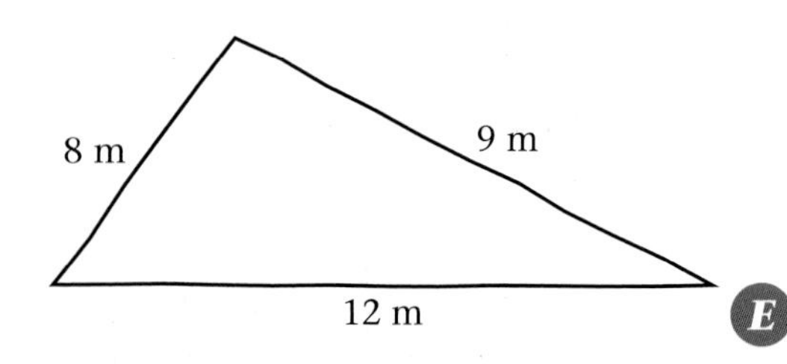

E

2 $ABCD$ is a quadrilateral. $AB = 4$ cm. $AC = 6$ cm. $BC = 3.3$ cm. $AD = 2.4$ cm.
Angle $BAD = 66°$. Make an accurate drawing of the quadrilateral $ABCD$.

E

3 In the triangle ABC, $AB = 6.3$ cm, $BC = 7$ cm, angle $ABC = 103°$.
Make an accurate drawing of the triangle ABC.

4 Draw an accurate net for the cuboid shown.

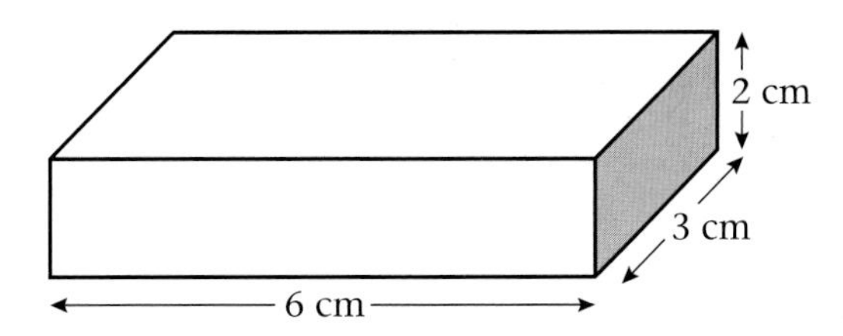

5 Sketch a net for the solid shown.

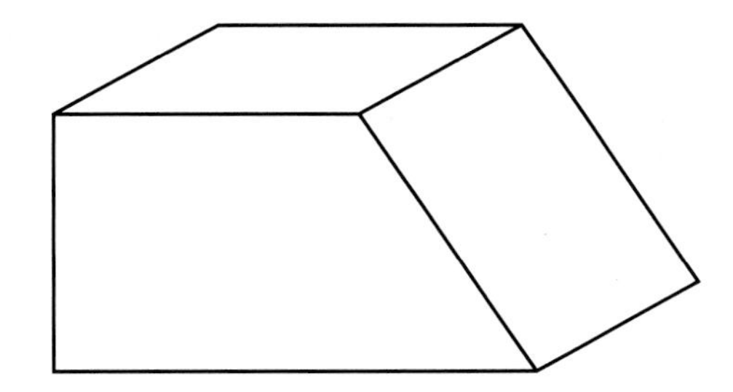

6 Sketch the solid that can be made from the net shown.

7 Which of these nets could produce the dice shown?

a

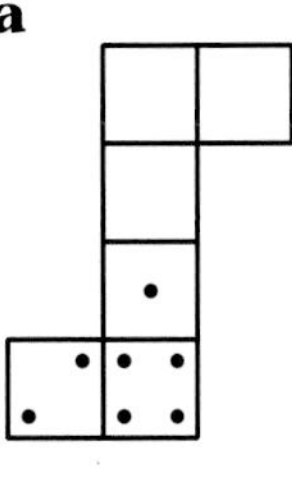

b

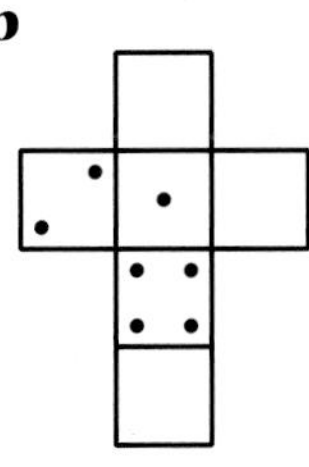

c

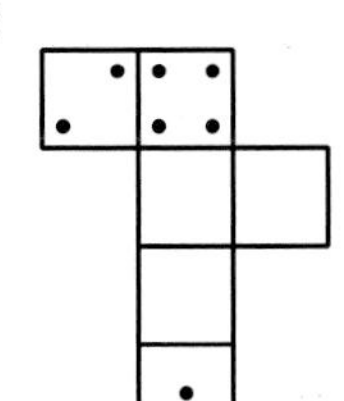

d

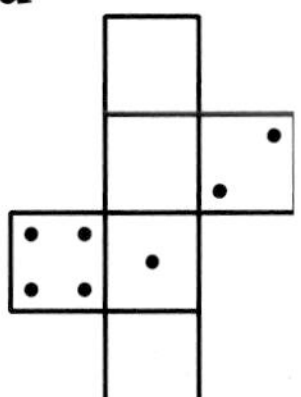

e

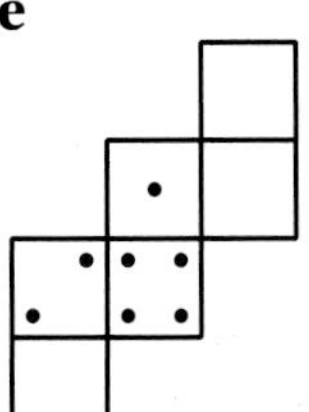

f

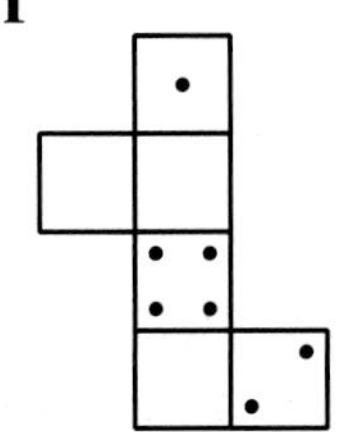

14.3

A 2-D shape has a line of symmetry if there is a line which divides the shape into two halves and one half is the mirror image of the other half.
A 2-D shape has rotational symmetry if it fits onto itself two or more times in one turn. The order of rotational symmetry is the number of times a shape fits onto itself in one turn.

Example 5

a Copy and draw on the lines of symmetry for the regular pentagon.
b What is the order of symmetry of the regular pentagon?

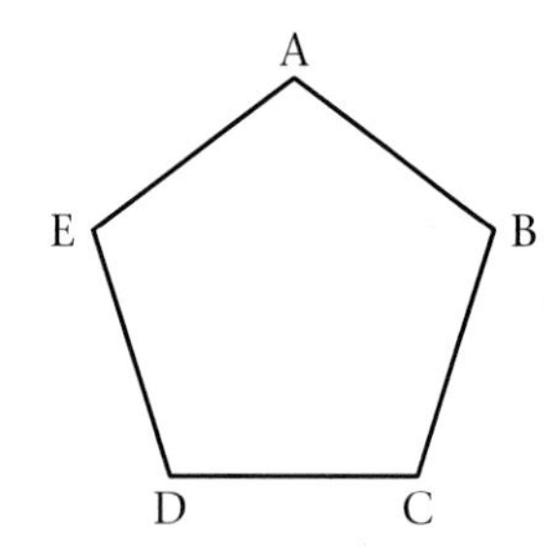

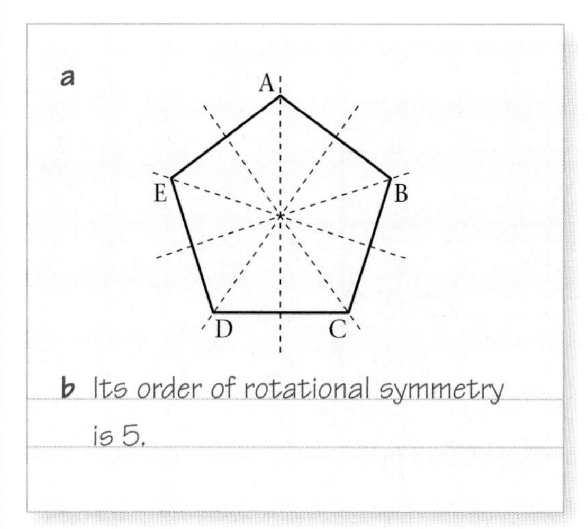

b Its order of rotational symmetry is 5.

As it turns, AB goes to BC,
to CD,
to DE,
to EA,
and returns to AB.

The pentagon fits onto itself 5 times in one complete turn.

Practice 14C

1 Copy the shapes shown and draw in all the lines of symmetry.

a

b

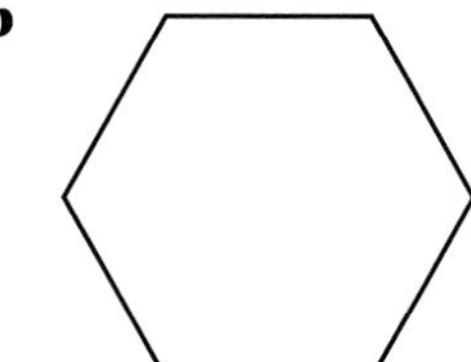

c

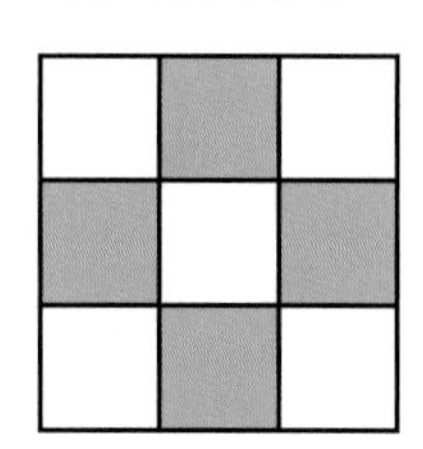

d

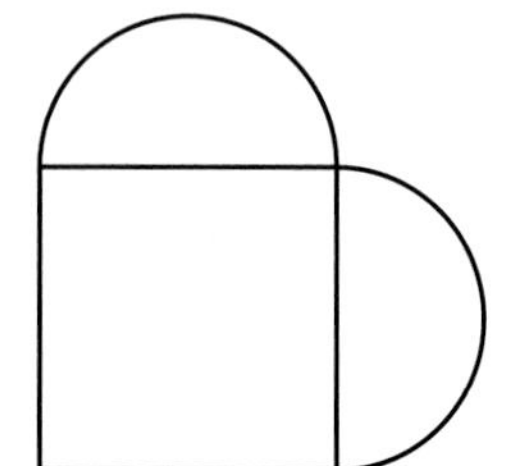

2 Write down the order of rotational symmetry for each of the shapes shown.

a

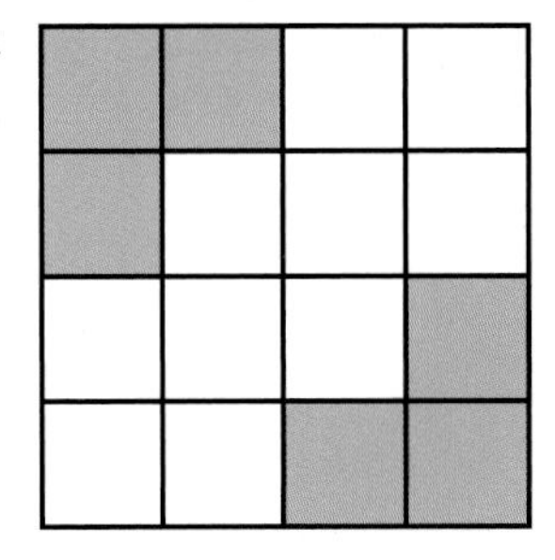

b

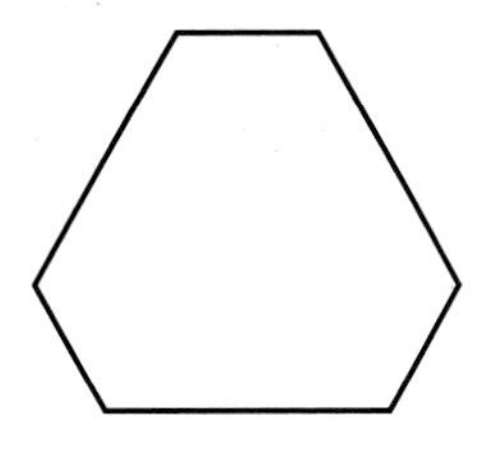

c

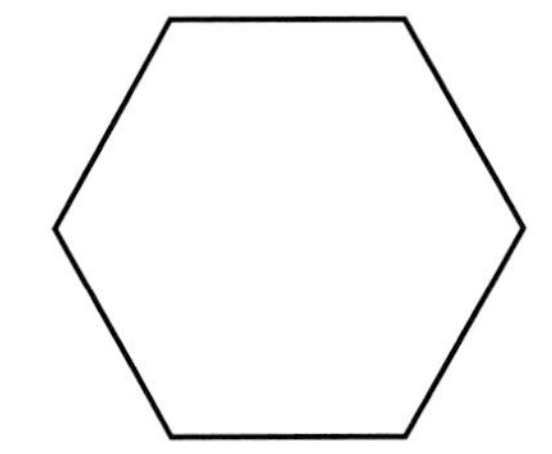

d

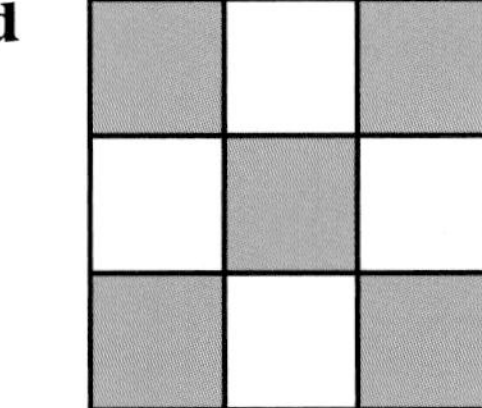

3 Copy and complete the diagram so that the final shape has rotational symmetry, order 4.

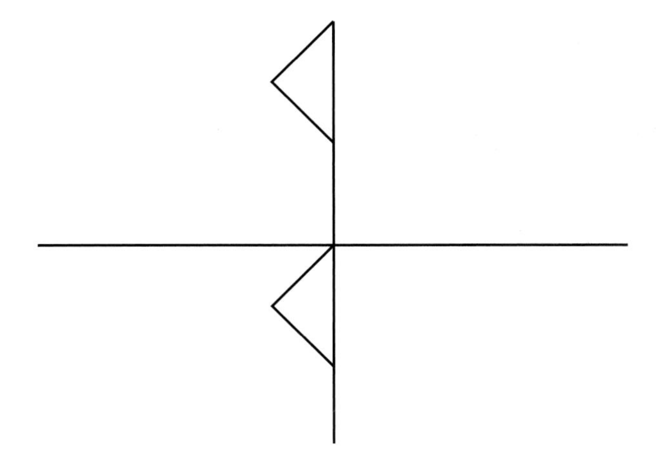

4 Copy the diagram.

a Tick two squares which would give the diagram one line of symmetry.

b Shade as few squares as possible to give the shape rotational symmetry. State the order of rotational symmetry for your diagram.

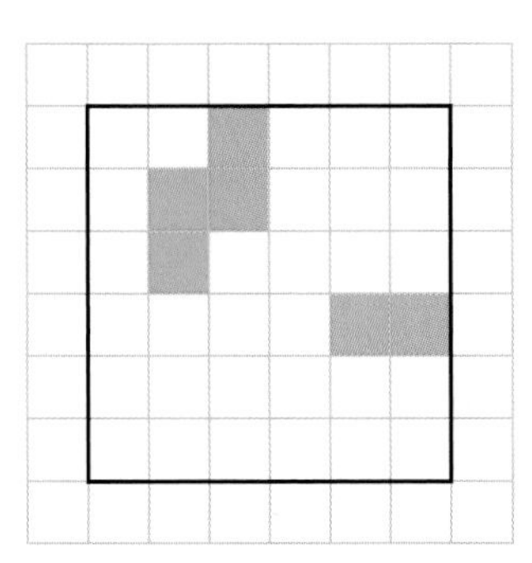

14.4 A plane of symmetry separates a solid into two halves which mirror each other.

Example 6

Sketch the planes of symmetry of the shape shown.

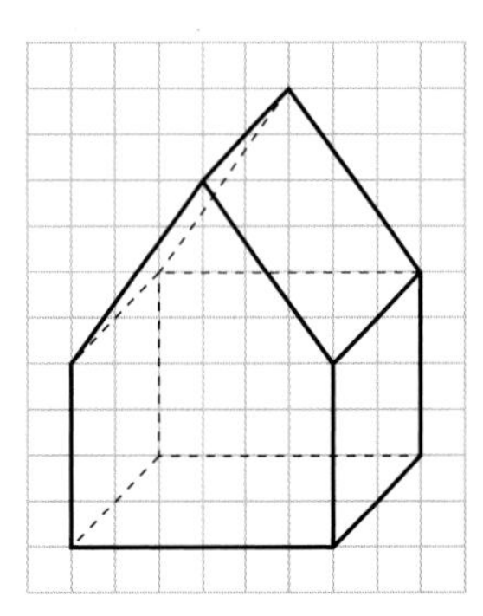

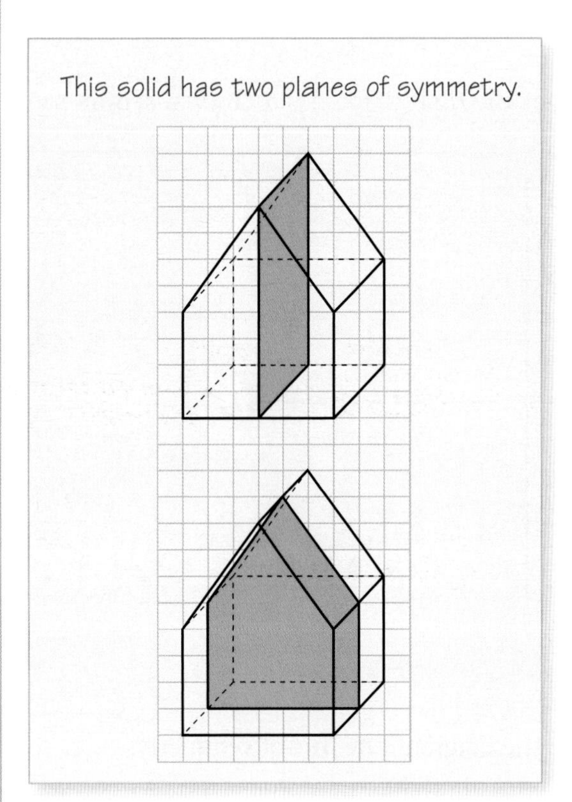

These are hard to draw. An acceptable way is to shade the solid where the plane coincides with it.

Where this shaded part of the plane is the same as the ends that it is parallel to, then it is a cross-section.

Practice 14D

1 Copy the shapes and draw on one plane of symmetry.

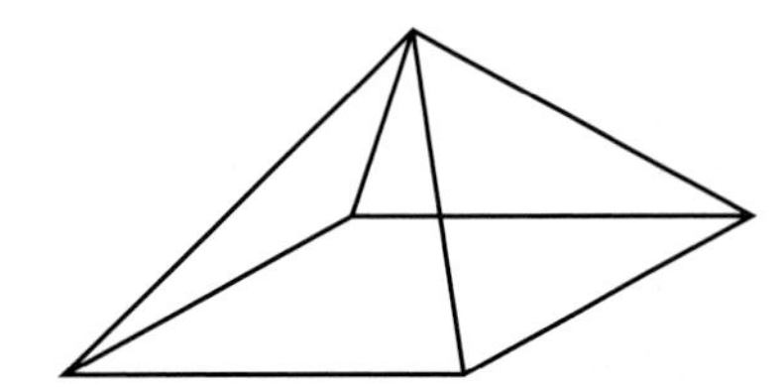

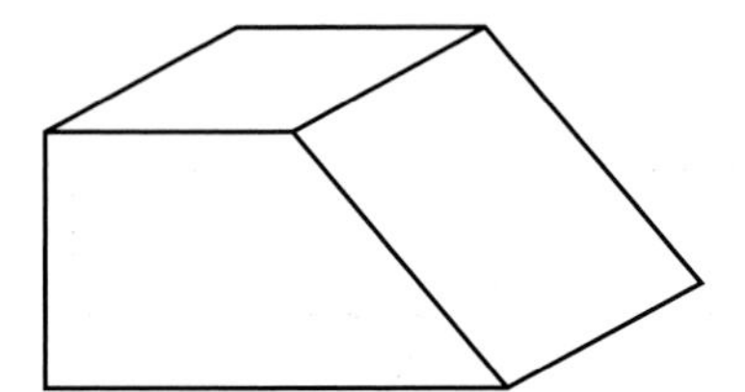

14.5 **The view from the front and the side of a solid are the elevations. The plan is the view from above.**

Example 7

Draw the views of the solid shown when seen from the front, the side and above.

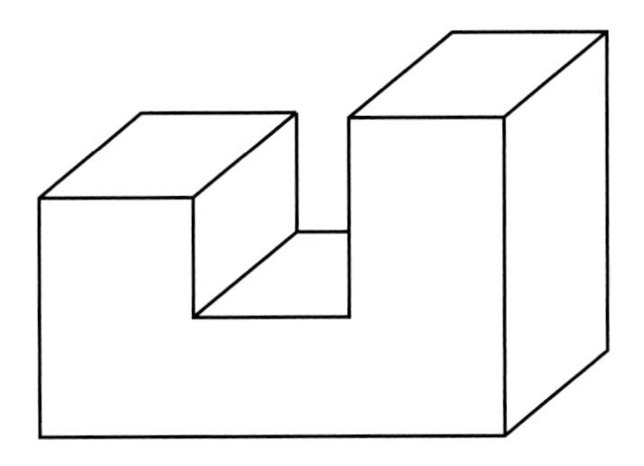

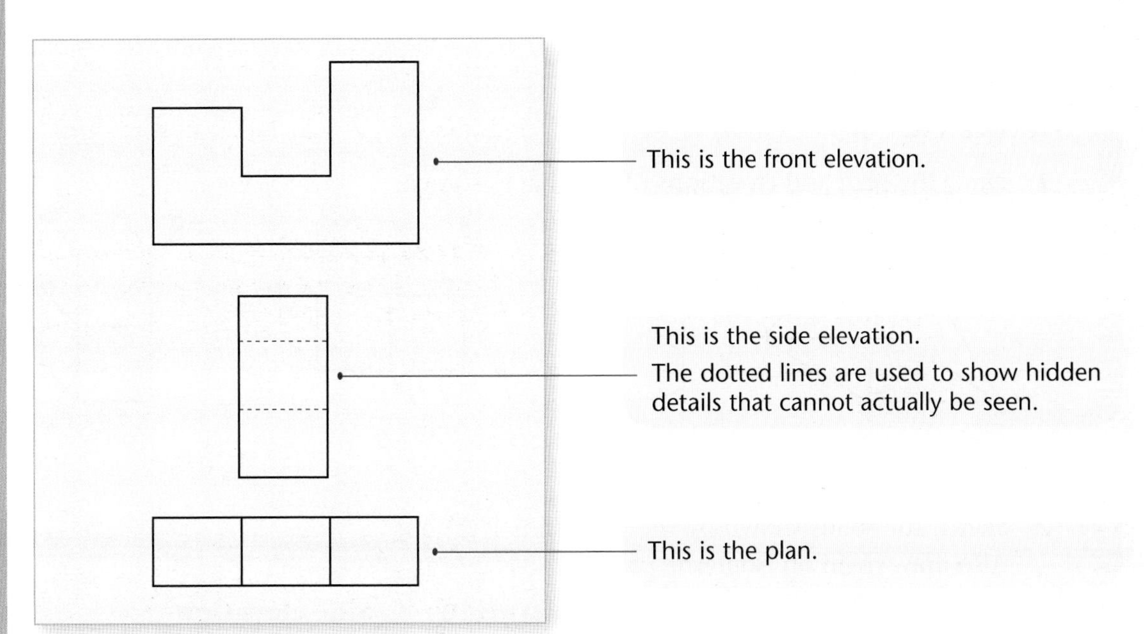

These three views are enough to determine the exact shape of the solid.

Example 8

Use these plan and elevations to sketch the solid.

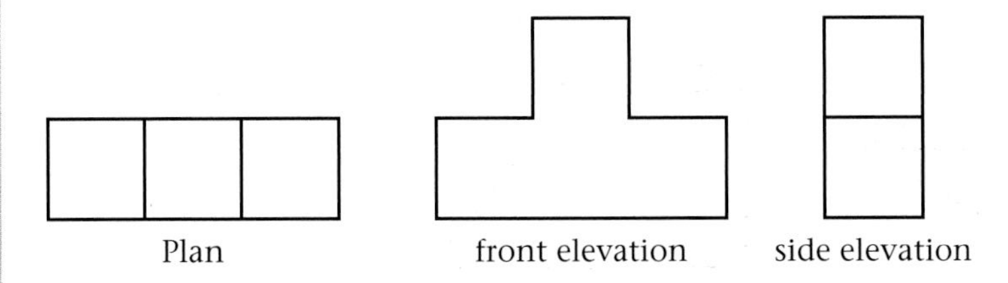

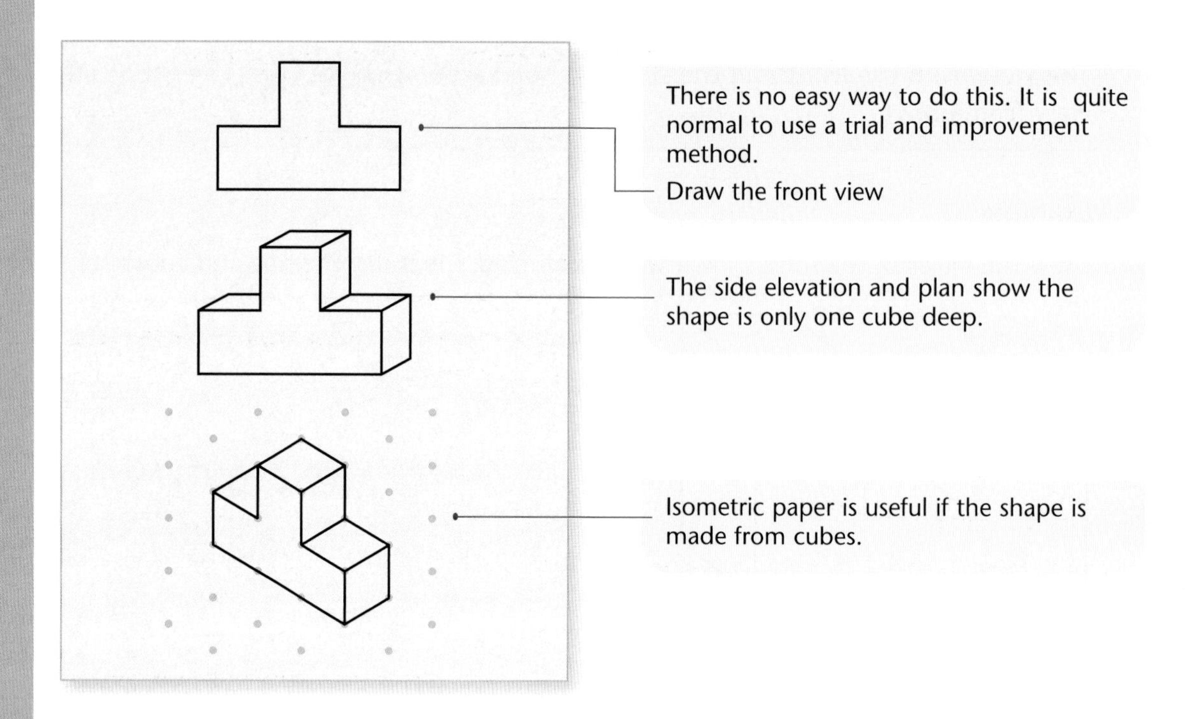

Practice 14E

1 Draw the plan and elevations of the shapes shown.

a front

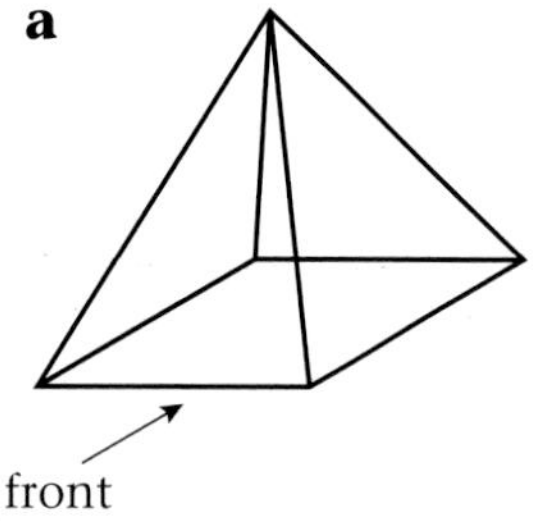

b

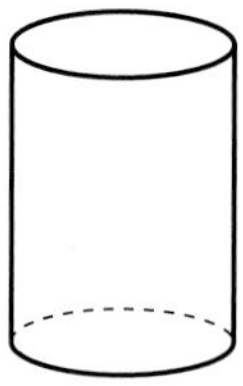

c

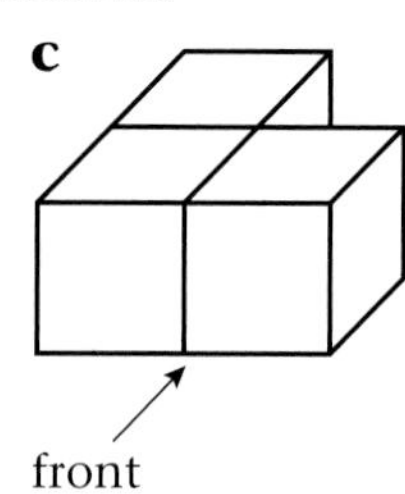

d

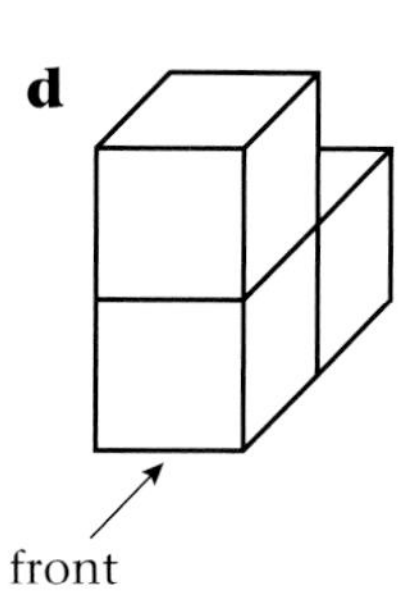

2 Sketch the solids shown by the plan and elevations.
Also draw them on isometric paper.

Plan **Front elevation** **Side elevation**

a

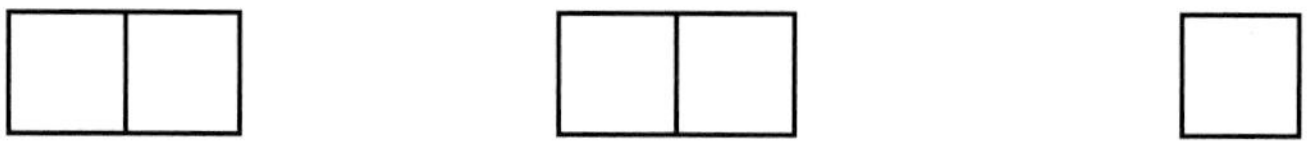

b

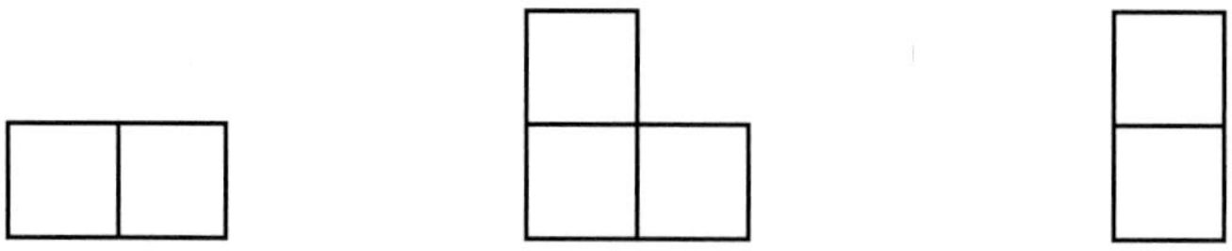

c

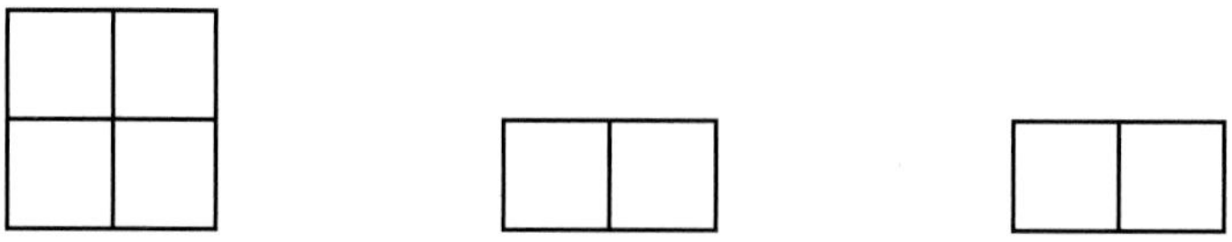

d

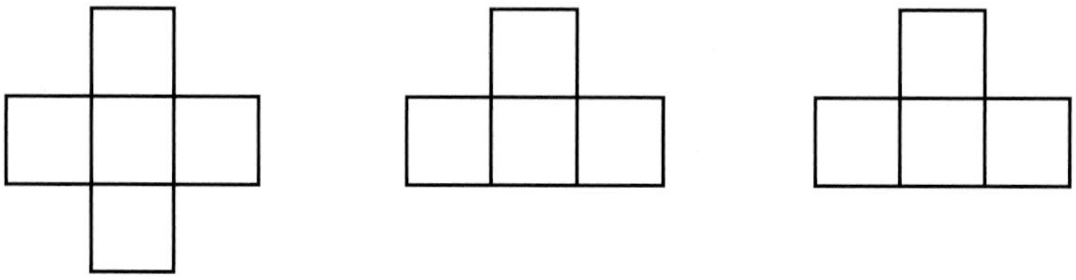

e

Plan	Front elevation	Side elevation

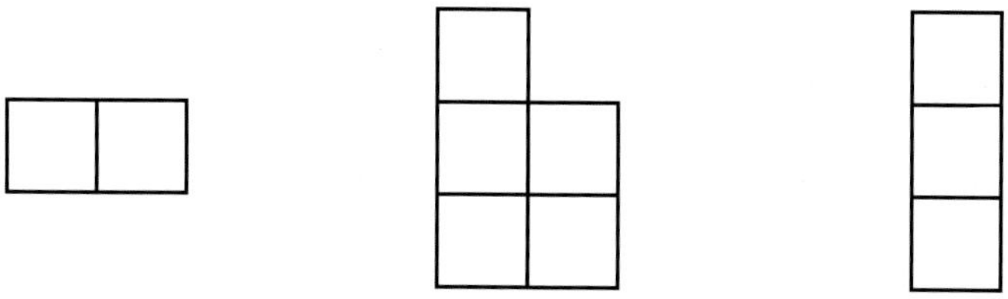

14.6 **A locus is a path traced out by a moving point. The moving point is often obeying a rule, such as: the path is equidistant from two points.**

Example 9

On the diagram, *AB* and *AC* represent two sides of a field. A drainage pipe is installed which is equidistant from the sides of the field. Construct the line of the pipe on the diagram. You must show all your construction lines.

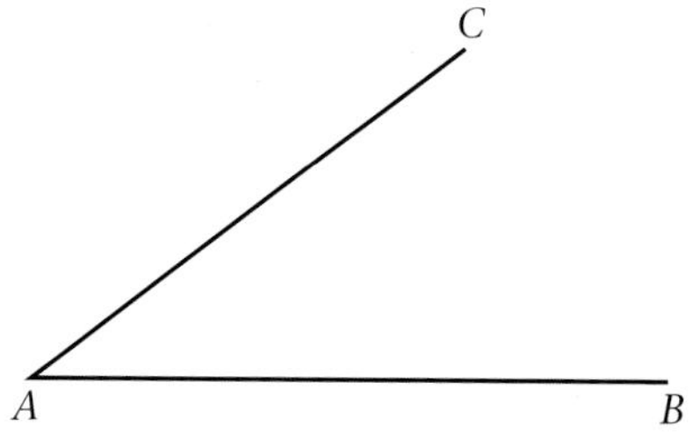

C
A B

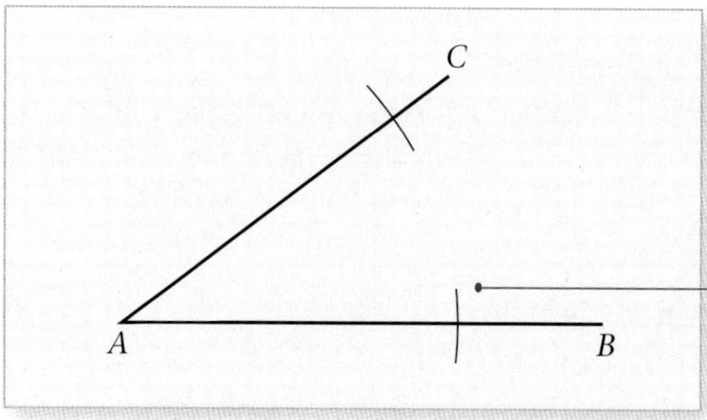

The line required bisects angle *BCA*.

C
A B

Open your compasses about 5 cm and with centre *A* mark *AB* and *AC* with arcs.

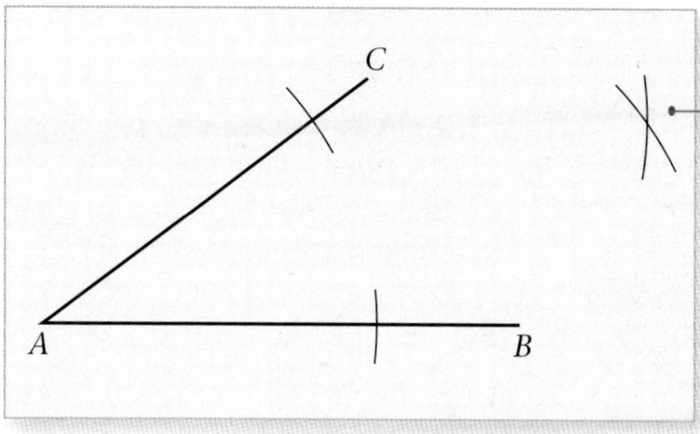

C
A B

Using each point of intersection in turn as centre, draw two intersecting arcs.

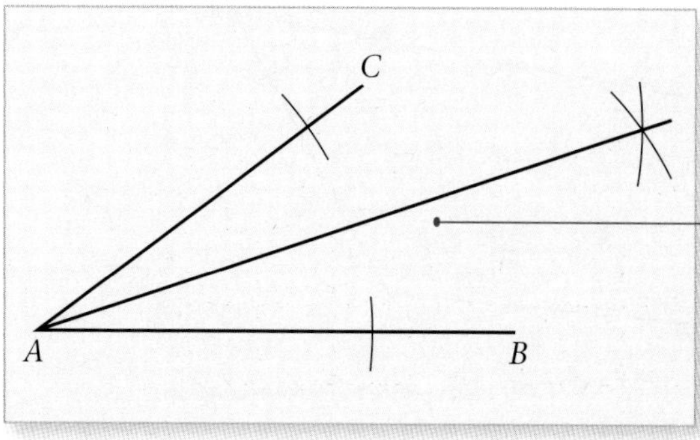

C
A B

Join *A* to the intersection to complete the bisector.

Example 10

The diagram is a plan of a field drawn to a scale of 1 cm to 20 cm.

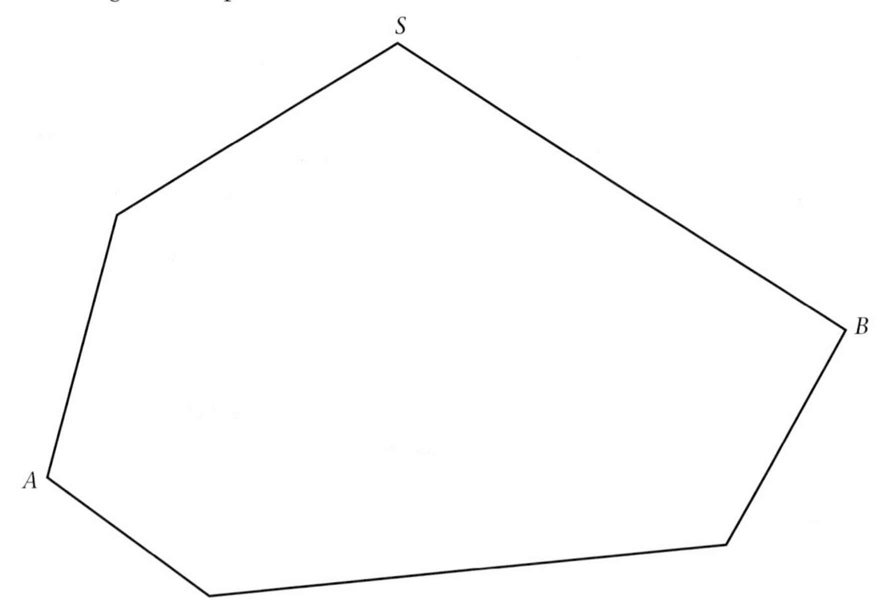

There is a water sprinkler at *S*.
The sprinkler can water that region of the field which is 60 metres or less from the sprinkler.

When points obey a rule containing an inequality, the locus is a region, not a line.

a Shade, on the diagram, the region of the field which is 60 metres or less from the sprinkler.

b A farmer is going to lay a pipe to help water the field.
A and *B* are posts which mark the widest part of the field.
The pipe will cross the field so that it is always the same distance from *A* as it is from *B*.
On the diagram, construct a line to show exactly where the pipe should be laid. **E**

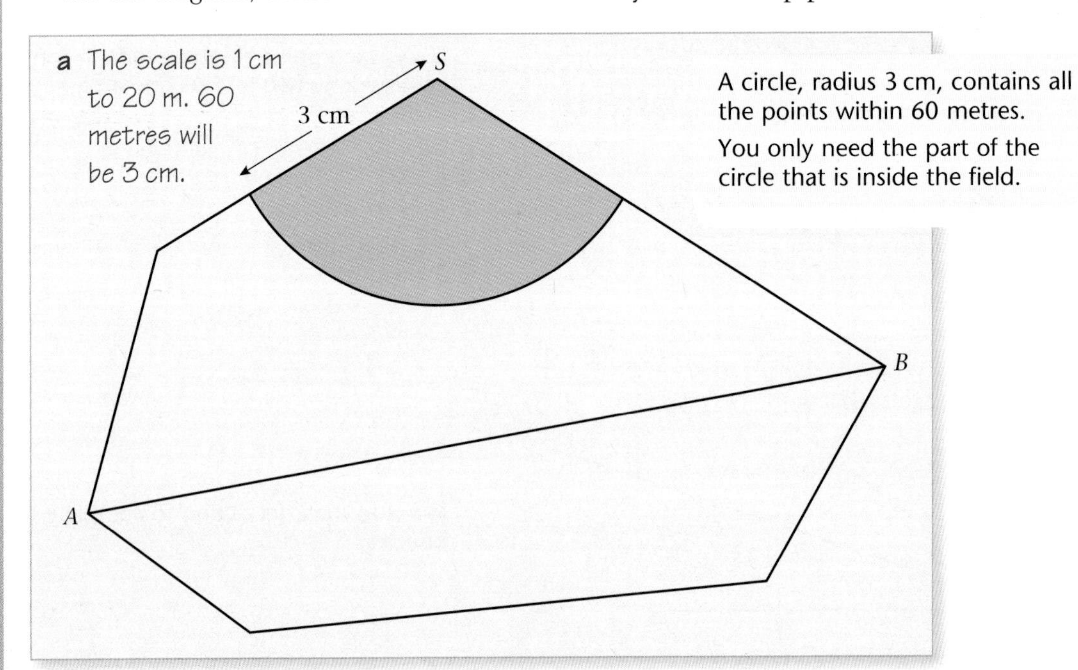

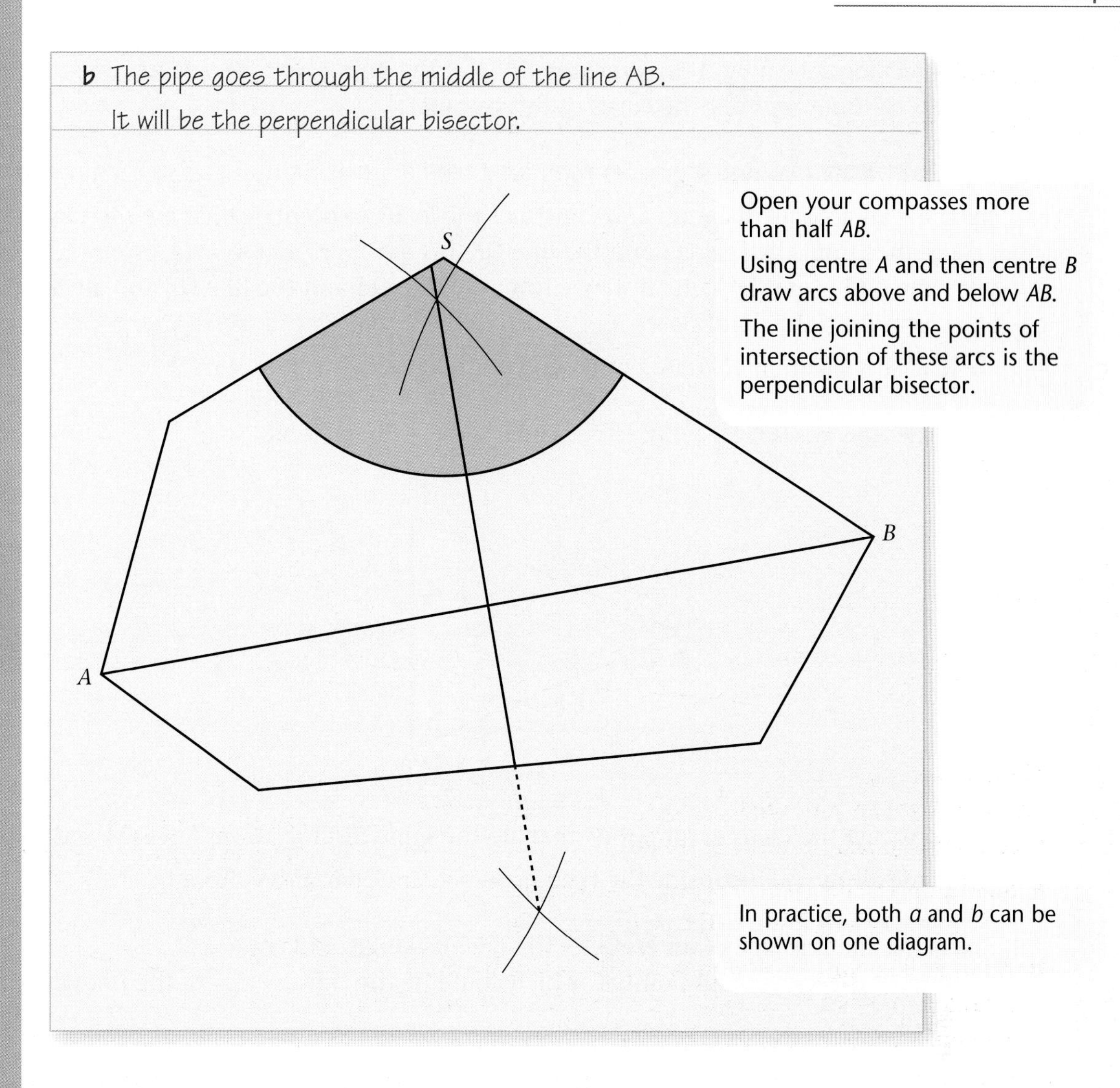

Practice 14F

1 Copy the diagram below.
Shade in the region within triangle ABC that satisfies all three of these conditions:

a closer to A than to B,

b closer to the line AC than to line AB,

c more than 2 cm from A.

×C

A×　　　　　　　　　　×B

E

2 On a rectangle 3 cm by 2 cm, draw the locus of the points, *outside the rectangle*, that are 1.5 centimetres from the edges of this rectangle.

3 *OA* and *OB* are two line segments 6 cm long which are at right angles to each other.

a Draw an accurate diagram and construct the locus of points which are the same distance from the line *OA* and the line *OB*.
Some points are the same distance from the line *OA* and the line *OB* and are also 4.5 cm from the point *B*.

b Mark the position of these points on your diagram.

4 Triangle *ABC* is isosceles with *AB* = 6 cm and *AC* = *BC* = 10 cm.

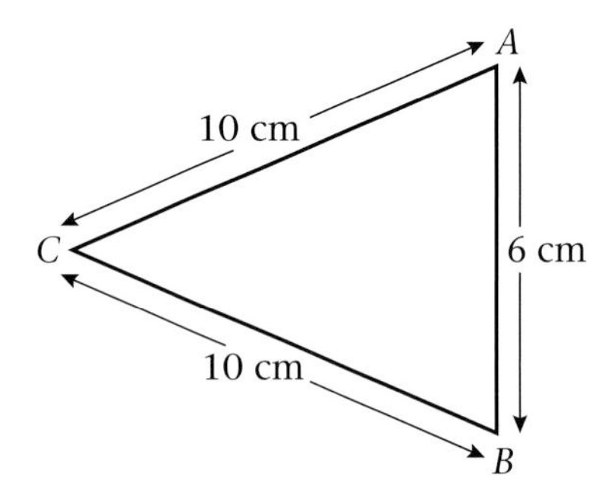

a Construct the locus of all points that are the same distance from lines *CA* and *CB*.

b Shade all the points inside the triangle which are 4 cm or less from *C*.

5 *AB* is the shortest side of an isosceles triangle. Its length is 3 cm.
Construct the locus of all points *C* which could be the third vertex of the triangle.

6 *ABCD* is a rectangle which measures 4 cm by 3 cm.

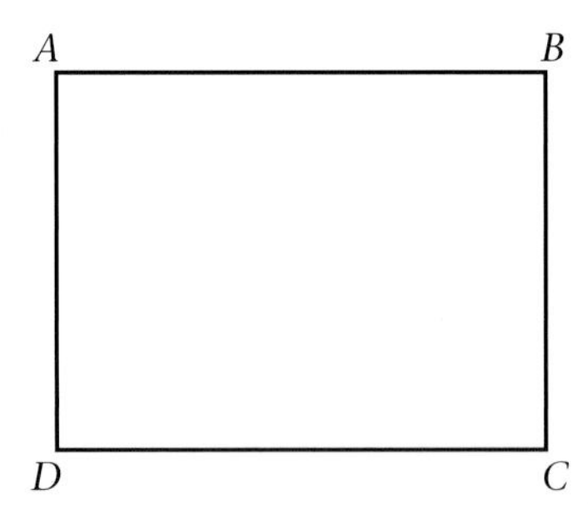

Shade the area defined by the following conditions:

a points nearer *A* than *B*

b points within 2.5 cm of *D*

c points nearer *D* than *A*.

You must show all your construction lines.

15 Transformations

Transformations are about changing the position or size of a shape. This may involve sliding, turning, reflecting or scaling the size.

15.1

A reflection in a line produces a mirror image. To describe a reflection fully you need to give the line of reflection.

Example 1

Reflect the shape in the mirror line.

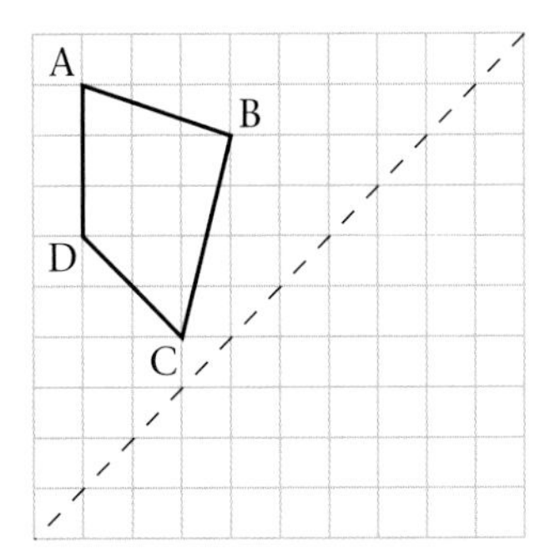

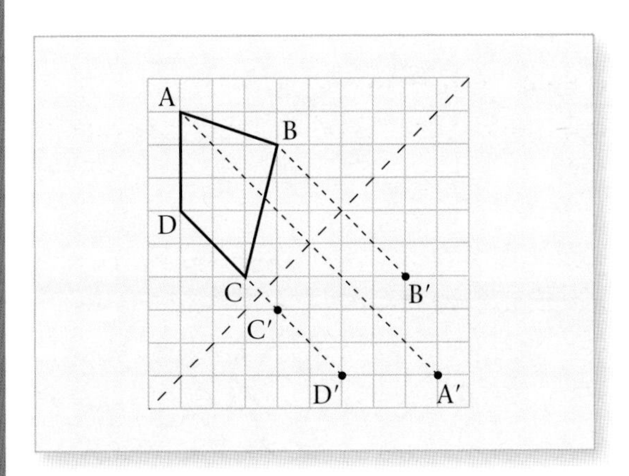

The way to do this is to find the images of points on the shape. The image of A is A′, the image of B is B′, etc.

Each image point is the same distance behind the mirror line as its object point is in front. A line joining object to image is perpendicular to the mirror line.

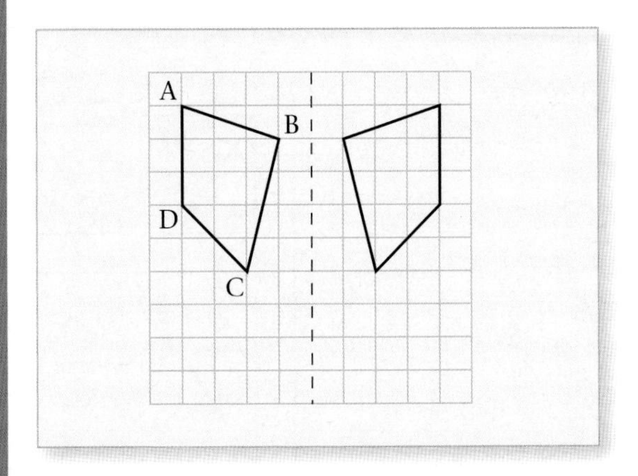

It helps to turn the page so that the mirror line is vertical.

Example 2

Describe fully the transformation which maps shape **A** onto shape **B**.

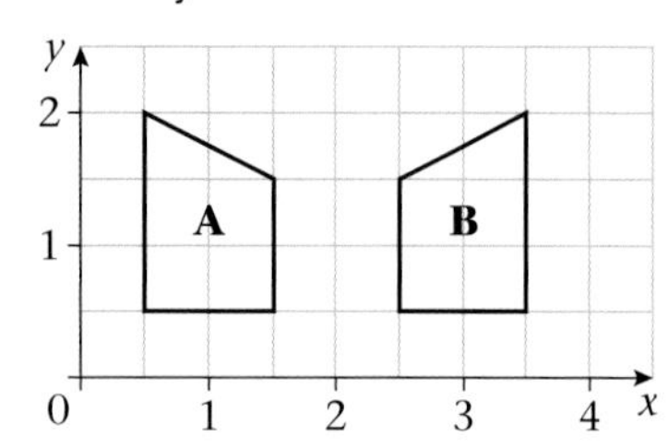

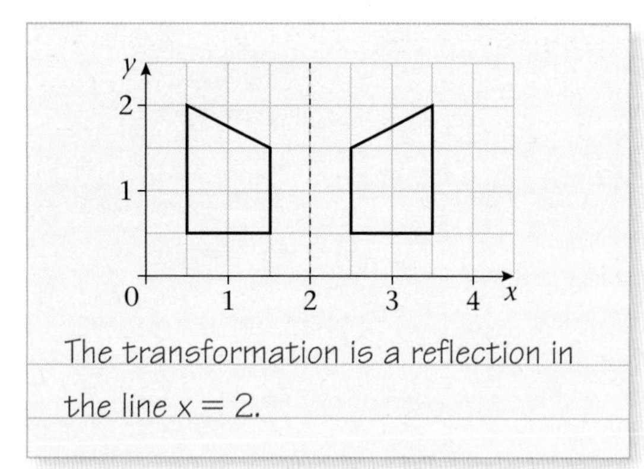

The transformation is a reflection in the line $x = 2$.

You can see the mirror line that goes down the middle. Its equation is $x = 2$.

Practice 15A

1 Copy the diagrams and draw the reflection using the dotted line as the line of reflection.

a

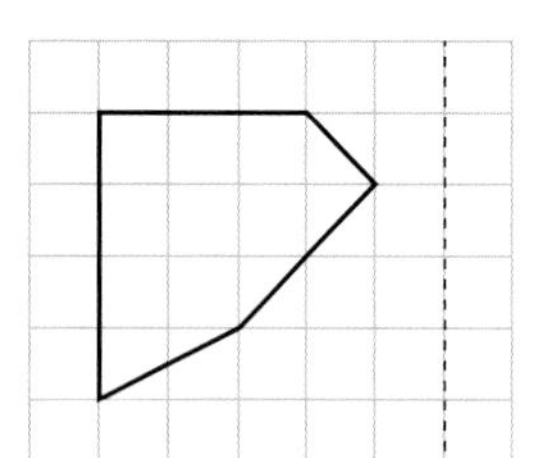

b

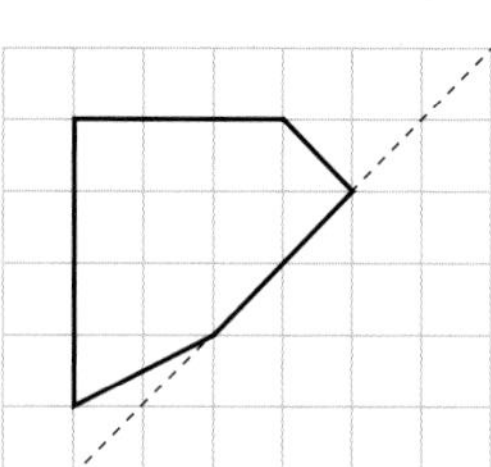

c

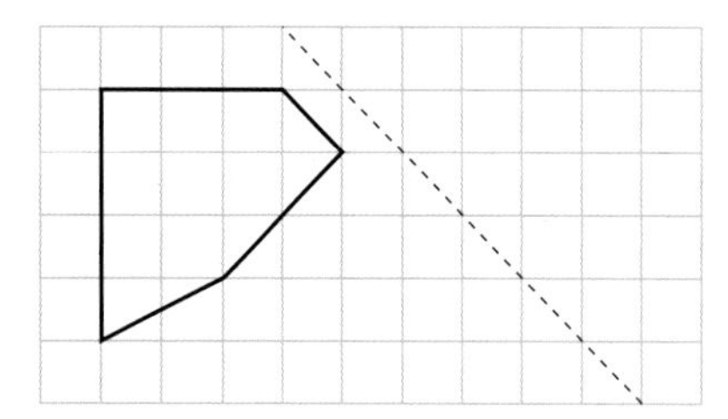

d

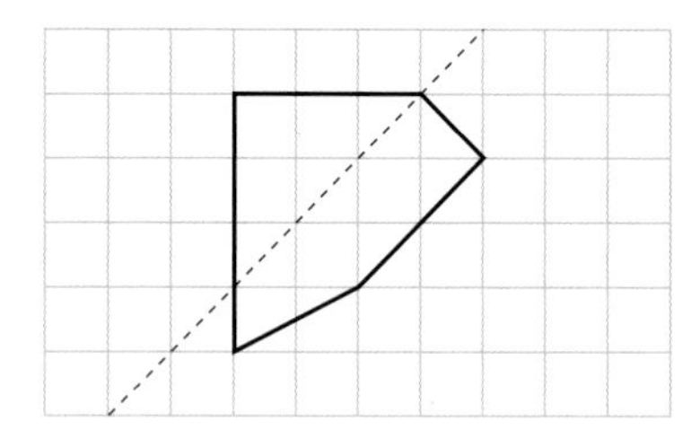

2 Fully describe the transformation which maps:

a **A** onto **B**

b **A** onto **C**

c **A** onto **D**.

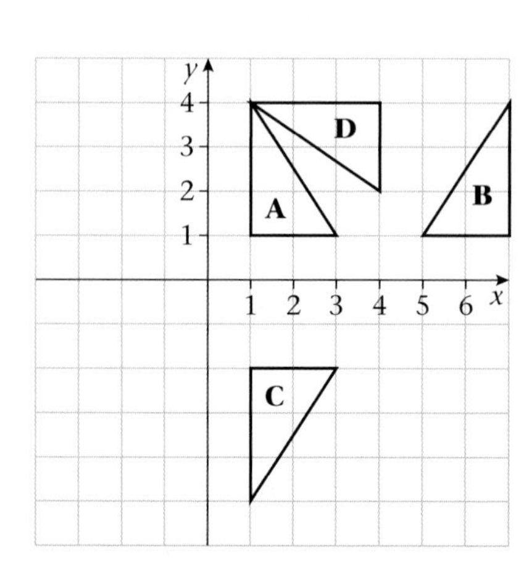

3 A shape is defined by the points $A(-1, 2)$, $B(0, 4)$, $C(2, 5)$ and $D(2, 2)$.
Draw the shape and its reflection in the lines:

a $x = 3$

b $x = y$

15.2

A rotation turns a shape through an angle about a fixed point.
To describe a rotation fully you need to give the centre of the rotation, angle of turn and direction of turn.

Example 3

Rotate the shape a quarter turn clockwise. Use X as the centre.

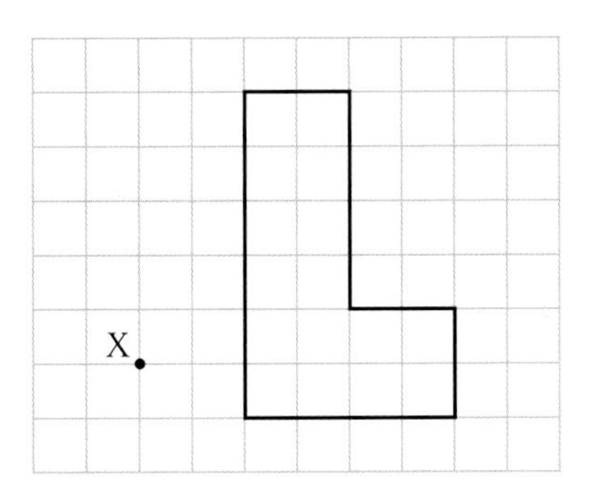

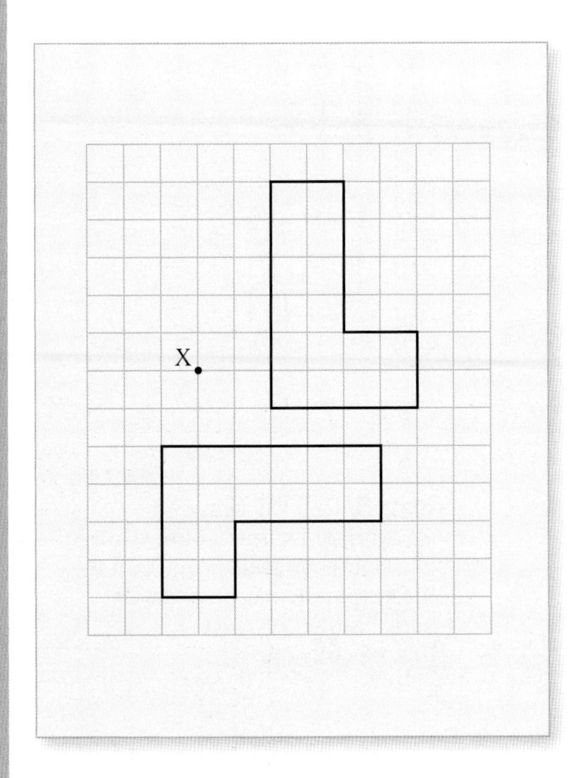

You can use tracing paper for this.
Trace the shape and include cross lines so that you can tell the angle turned.

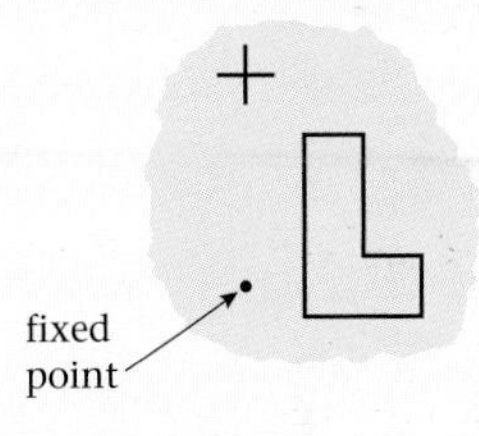

Use a pencil to fix the tracing paper at the centre of rotation.

Example 4

Describe fully the rotation which maps shape **P** onto shape **Q**.

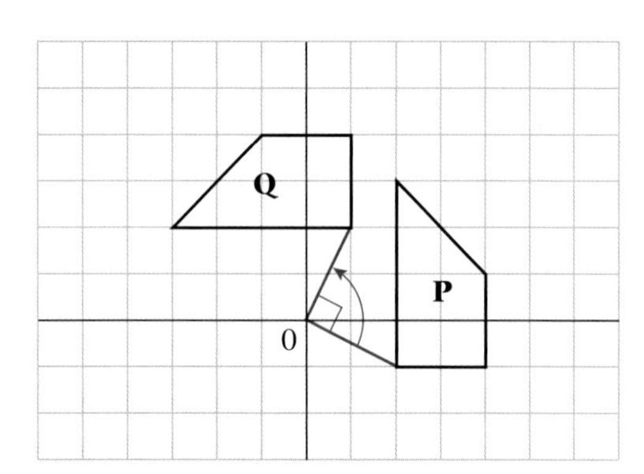

The transformation is a rotation with the origin as the centre.

It is a quarter turn anti-clockwise.

Use tracing paper to help you describe the transformation.

Each image point is the same distance from the centre of rotation as its original point. In this case, the centre is the origin O.

Each side of Q is at right angles to the corresponding side of P, so P has made a quarter turn anticlockwise to get to Q.

Practice 15B

1 In each case, describe fully the transformation which maps shape **A** onto shape **B**.

a

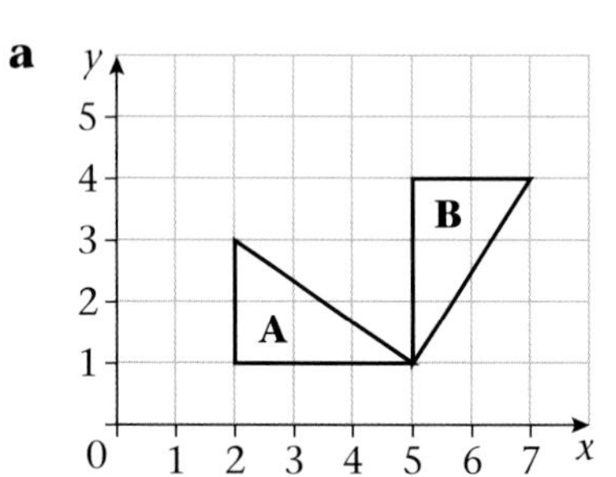

b

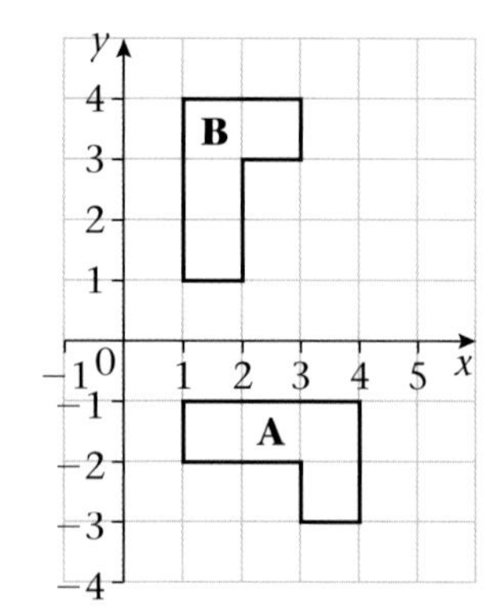

c

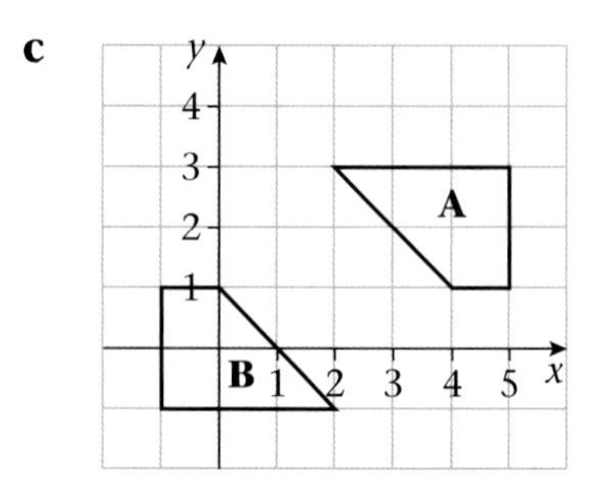

Hint: To find the centre of rotation, draw lines connecting two points to their images. Draw the perpendicular bisectors of both lines. The centre is where the bisectors cross.

2 Rotate the triangle through 90° **clockwise** about the point (0, 0).
Draw the triangle in its new position.

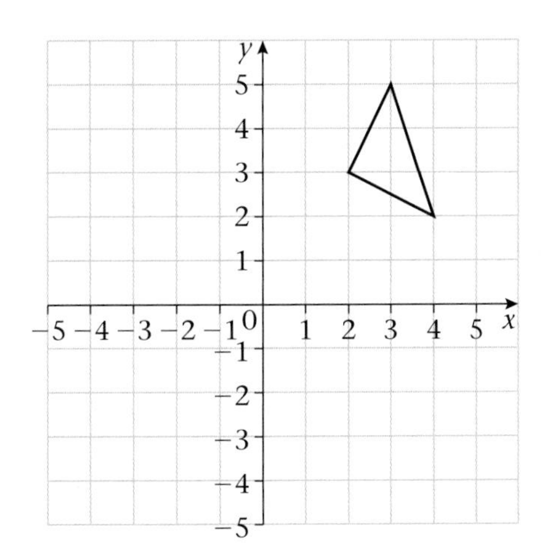

3 **a** Describe fully the single transformation that maps shape **P** onto shape **Q**.
b Rotate shape **P** 90° anticlockwise about the point $A(1, 1)$.

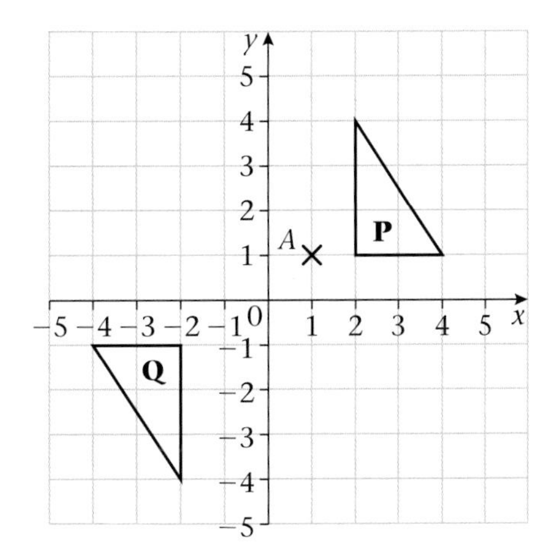

4 Copy the diagram. Draw the image after a rotation of 90° clockwise, about each of the points A, B, C and D.

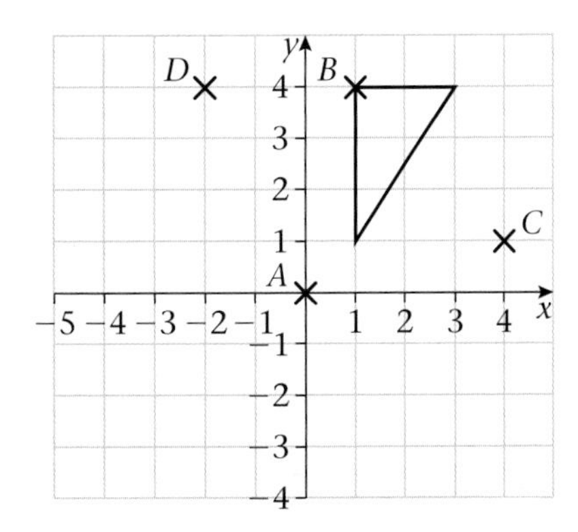

(E13)

15.3

A translation moves every point on a shape the same distance in the same direction. To describe a translation fully you need to give the distance moved and the direction of movement. You can also do this by writing down the column vector of the translation.

Example 5

Translate the shape **R** by the vector $\begin{pmatrix} 5 \\ 2 \end{pmatrix}$.

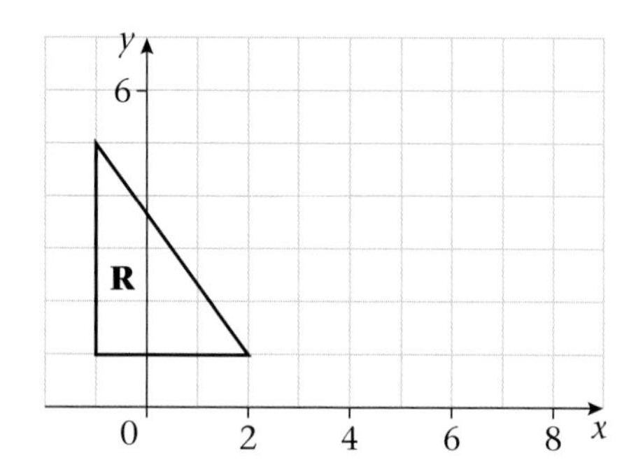

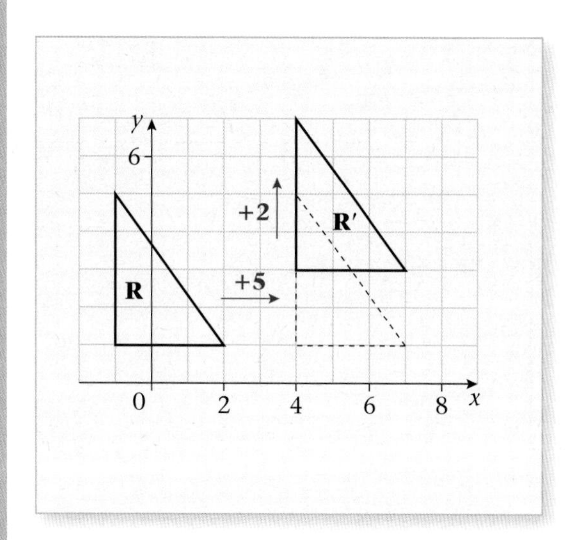

A column vector is a way of describing a translation.

In this case, the vector $\begin{pmatrix} 5 \\ 2 \end{pmatrix}$ means that a point moves +5 units in the x direction and +2 units in the y direction (or 5 units to the right and 2 units up).

Example 6

Describe fully the transformation which maps shape **C** onto shape **D**.

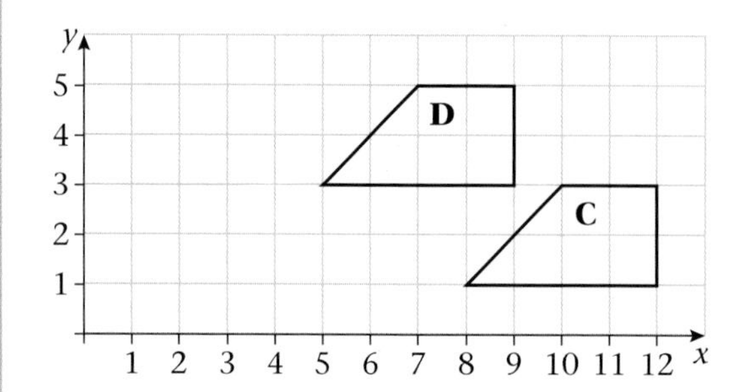

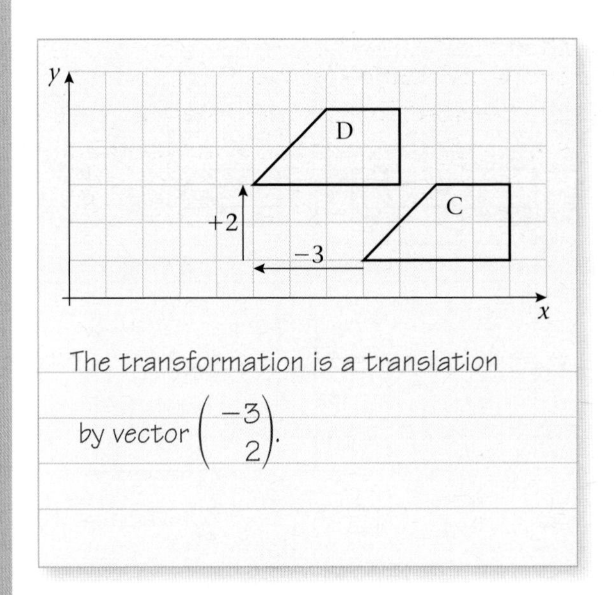

Look to see how far one of the points has moved.

Here the bottom left hand corner is the easiest.

It has moved backwards 3 and up 2.

The shape moves −3 in the x direction and 2 in the y direction.

Practice 15C

1 Describe fully the transformation which maps shape **A** onto shape **B**.

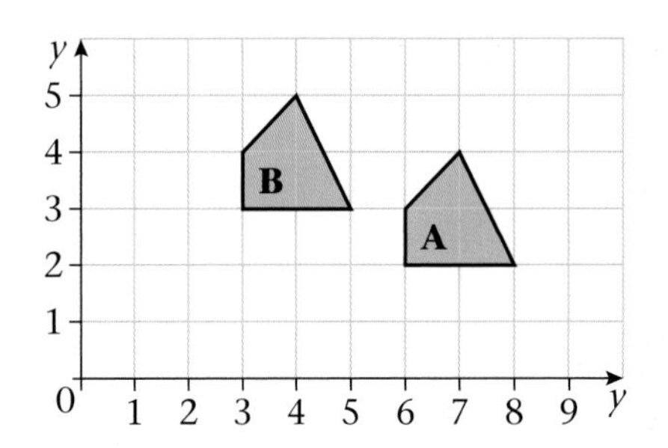

2 Copy the shape onto squared paper.
Draw the images:
X formed by translating 3 units in the x direction,
Y formed by translating 5 units in the x direction and 3 units in the y direction,
Z formed by translating −4 units in the x direction and −1 unit in the y direction.

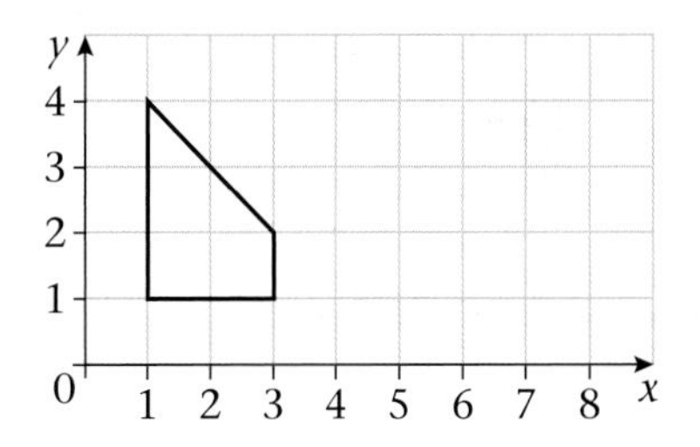

3 Copy the diagram and draw the images after the vector translations:

a 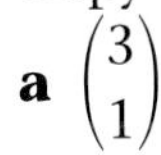 $\begin{pmatrix} 3 \\ 1 \end{pmatrix}$ **b** 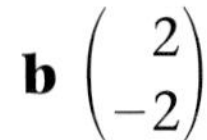 $\begin{pmatrix} 2 \\ -2 \end{pmatrix}$

c $\begin{pmatrix} -3 \\ 4 \end{pmatrix}$ **d** 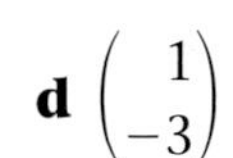 $\begin{pmatrix} 1 \\ -3 \end{pmatrix}$

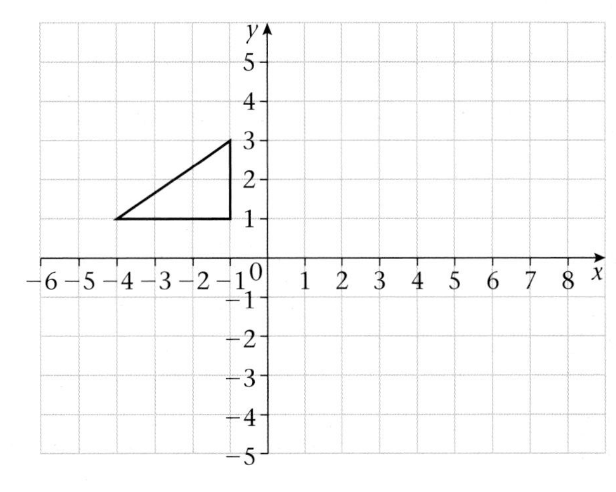

4 Use column vectors to describe the translations which map

a **A** onto **B** **b** **A** onto **E**

c **C** onto **D** **d** **E** onto **B**

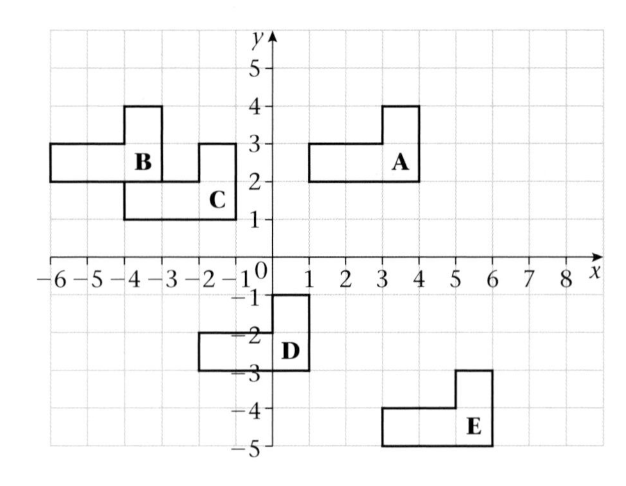

15.4

An enlargement changes the size but not the shape of an object. The scale factor of the enlargement is the number of times the lengths of the original object have been enlarged.

In an enlargement, angles are unchanged, corresponding lengths have the same scale factor, and corresponding sides are parallel.

Example 7

Enlarge the shape ABCDE by a scale factor of $1\frac{1}{2}$ using the centre P shown.

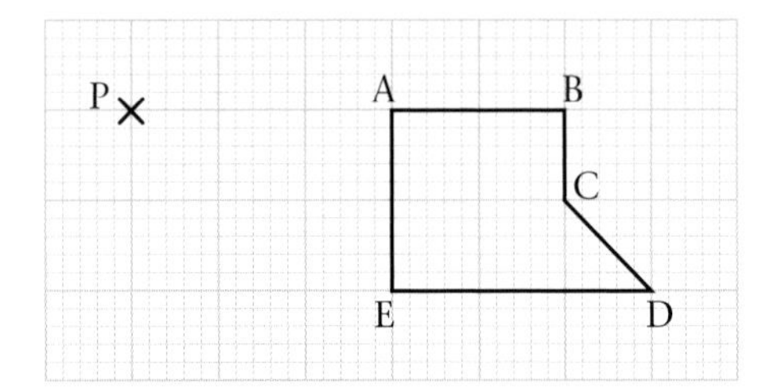

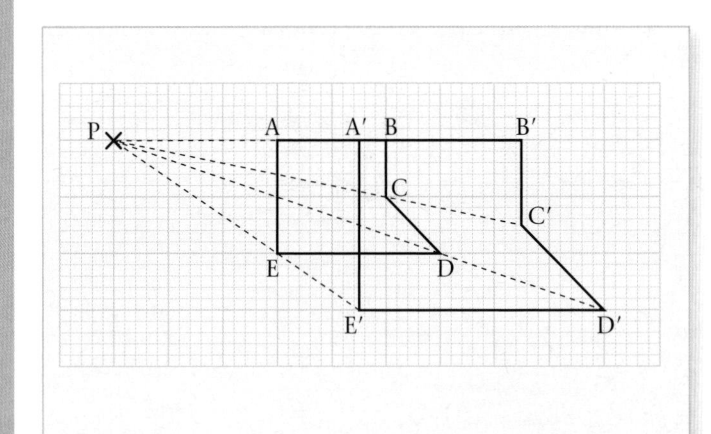

The image of a point will be $1\frac{1}{2}$ times the distance of that point from the centre of the enlargement.

So PA′ = $1\frac{1}{2}$PA
PB′ = $1\frac{1}{2}$PB
PC′ = $1\frac{1}{2}$PC
PD′ = $1\frac{1}{2}$PD
PE′ = $1\frac{1}{2}$PE

Example 8

Describe the transformations which map
a shape **A** onto shape **B** and
b shape **C** onto shape **D**.

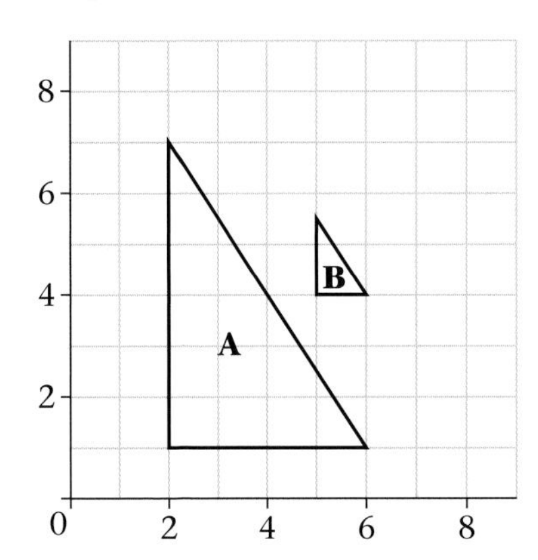

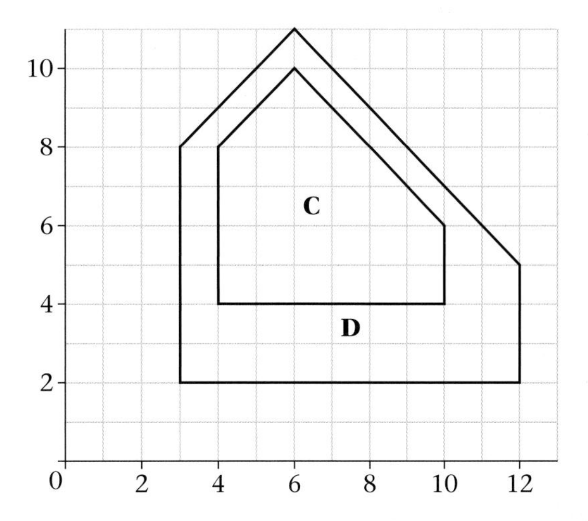

a

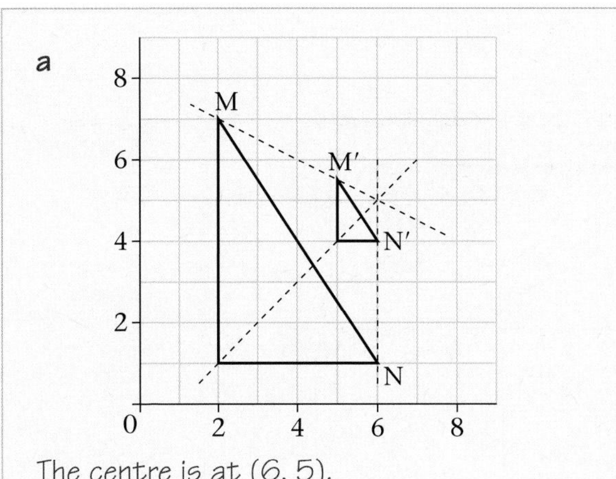

The centre is at (6, 5).

The scale factor is $\frac{1}{4}$.

Draw lines joining 2 sets of corresponding points, eg M and M′, N and N′. The centre of enlargement is where the 2 lines meet. You also need a scale factor. Write two corresponding sides as a fraction. If the image is smaller, the scale factor is less than one. It is still called an enlargement.

b

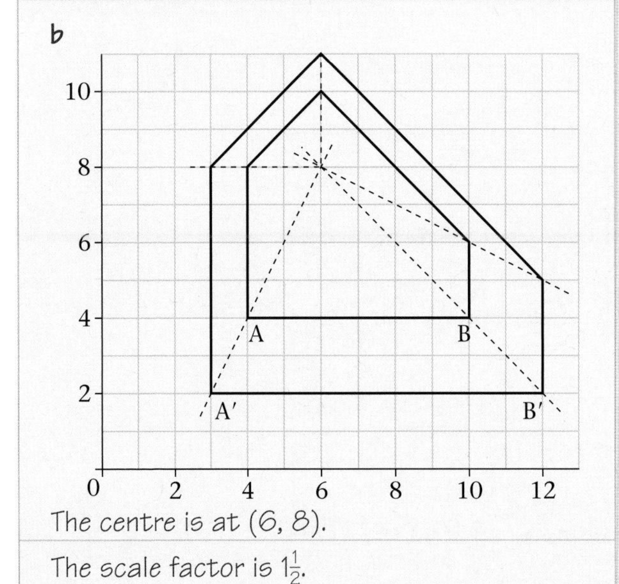

The centre is at (6, 8).

The scale factor is $1\frac{1}{2}$.

The centre is inside the shape.
The scale factor is

$$\frac{\text{Length A}'\text{B}'}{\text{length AB}} = \frac{9}{6} = \frac{3}{2} = 1\tfrac{1}{2}$$

Practice 15D

1 Copy shape **A** onto squared paper.

Using centre O, enlarge the shape by a scale factor of **a** 2 and **b** $\frac{1}{2}$.

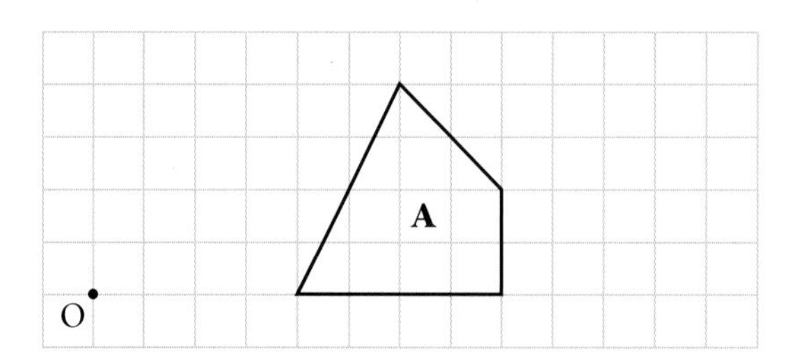

2 Describe fully the transformations which map the shaded triangle **A** onto **B** and **C**.

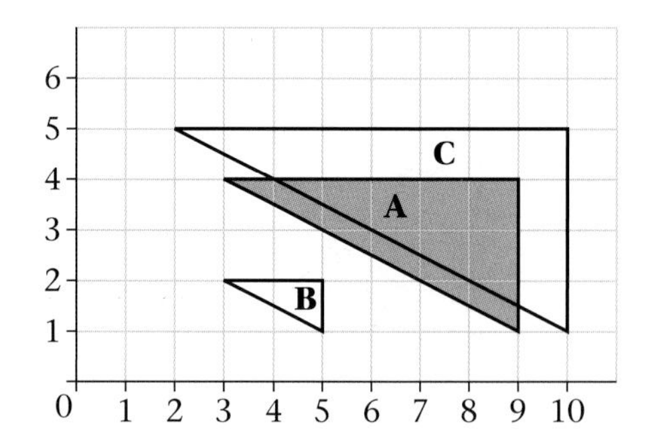

3 Copy the shape shown onto squared paper.

Enlarge the shape, using each of the centres marked, by a scale factor of $1\frac{1}{2}$.

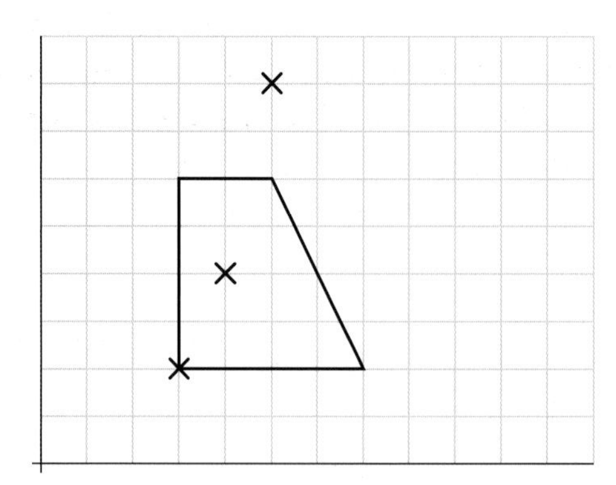

4 Triangle **A** is defined by its coordinates $(-4, 7)$, $(-1, 1)$ and $(-1, 7)$.

Draw the triangle **A**.

Enlarge triangle **A** by the scale factor 1/3 with centre the point P $(-7, 7)$.

E

5 Shape **A** is shown on the grid.
Shape **A** is enlarged, centre (0, 0), to obtain shape **B**.
One side of shape **B** is shown on the diagram.

a Write down the scale factor of the enlargement.

b On a copy of the grid, complete shape **B**.

The shape **A** is enlarged by a scale factor $\frac{1}{2}$, centre (5, 16) to give the shape **C**.

c On your grid, draw shape **C**.

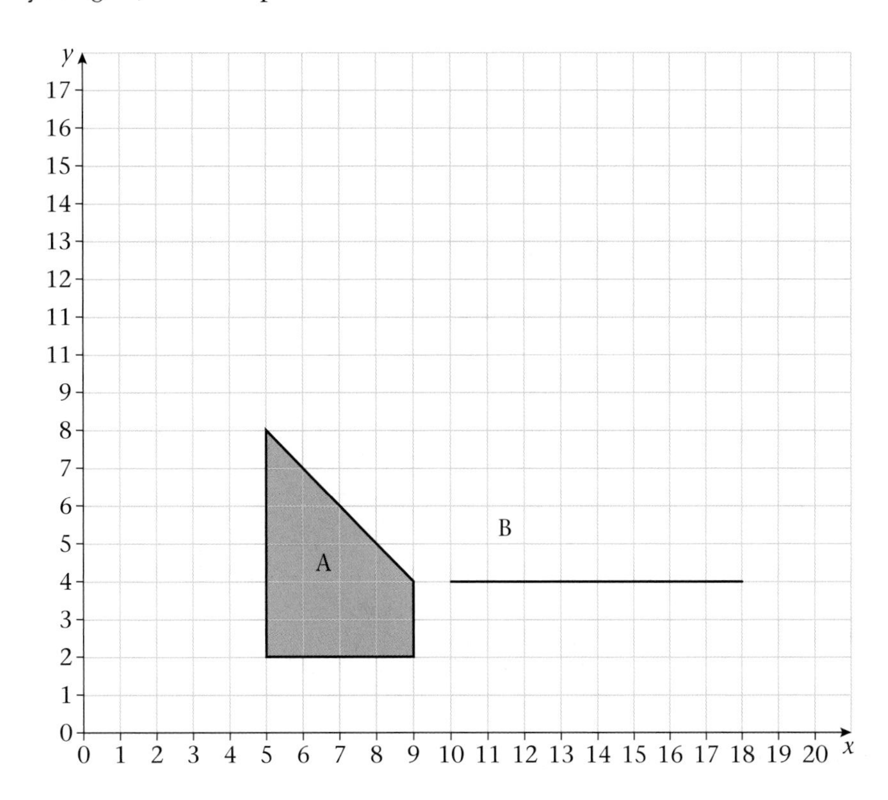

15.5

A shape and its image after enlargement are said to be mathematically similar. Similar shapes have corresponding sides in proportion and corresponding angles equal. Triangles are similar if their angles are the same. All squares are similar to each other. All circles are similar to each other.

Example 8

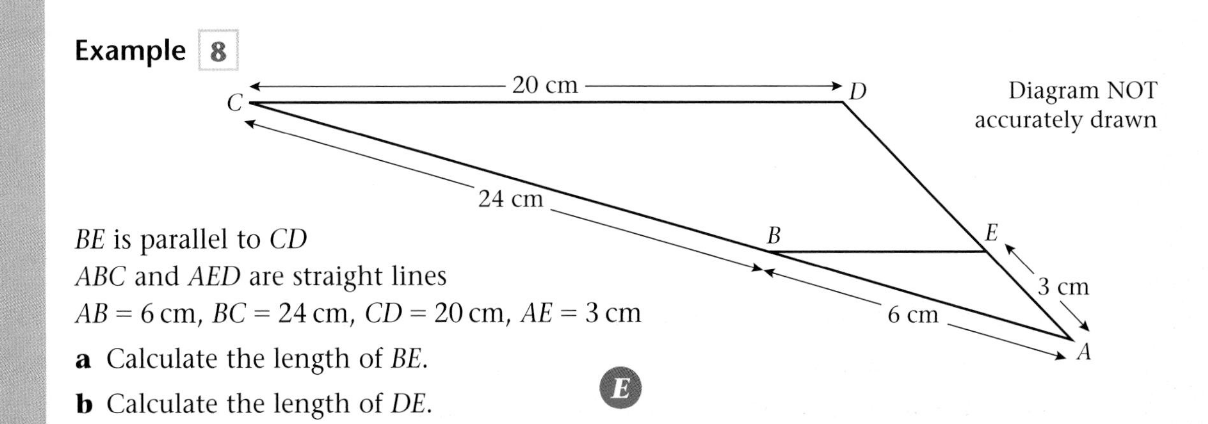

BE is parallel to *CD*
ABC and *AED* are straight lines
$AB = 6$ cm, $BC = 24$ cm, $CD = 20$ cm, $AE = 3$ cm

a Calculate the length of *BE*.

b Calculate the length of *DE*.

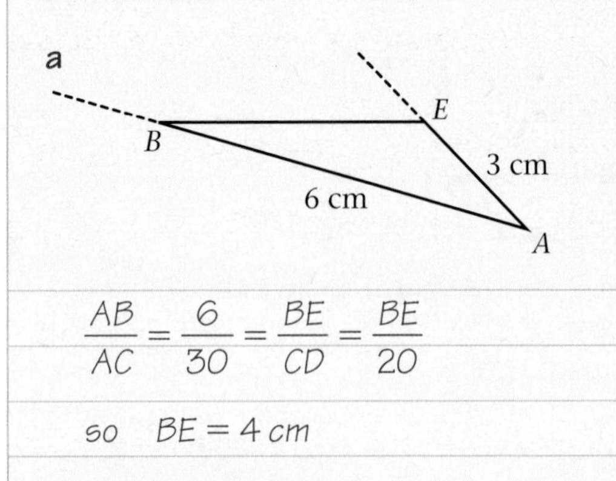

A separate diagram helps.

You can think of $\triangle AEB$ as an enlargement of $\triangle ADC$.

$\frac{AB}{AC} = \frac{6}{30} = \frac{BE}{CD} = \frac{BE}{20}$

Scale factor $\frac{AB}{AC} = \frac{6}{30} = \frac{1}{5}$.

so $BE = 4$ cm

b $\frac{AE}{AD} = \frac{AB}{AC} = \frac{3}{AD} = \frac{1}{5}$ — Using the scale factor.

$\therefore AD = 15$ cm — $AD = \frac{3 \times 5}{1}$

so $DE = 15 - 3 = 12$ cm

Practice 15E

1 In the diagram BC is parallel to DE.
$AB = 4.5$ cm, $AC = 6$ cm, $CE = 4$ cm, $DE = 5$ cm

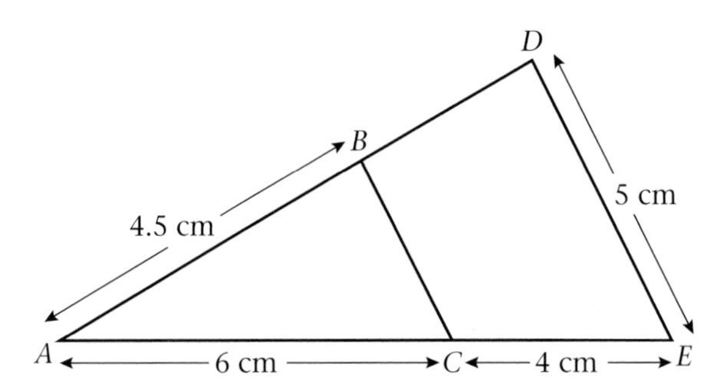

a Work out the length of AD.
b Work out the length of BC.

E

2

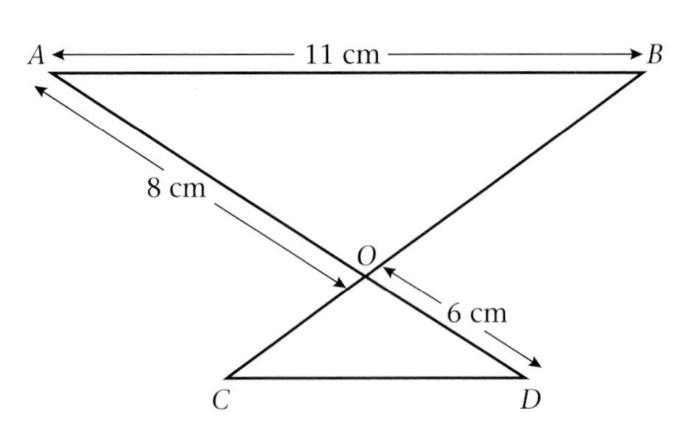

AB is parallel to CD.
The lines AD and BC intersect at point O.
$AB = 11$ cm, $AO = 8$ cm, $OD = 6$ cm.
Calculate the length of CD.

15.6 **A scale is a ratio which shows the relationship between a length on a drawing and the actual length of the real thing.**

The scale may be represented as a ratio (eg 1 : 25 000) or in the form of an equivalence (eg 1 cm = 2.5 km).

Example 10

The scale of a map is 1 : 50 000.

a What is the actual distance between two places which are 4.8 cm apart on the map?

b Two places are 12.2 km apart. How far apart are they on the map?

a Actual distance is $4.8 \times 50\,000$ cm
$= 240\,000$ cm
$= 2400$ m
$= 2.4$ km

Each cm on the map is 50 000 cm, so multiply by 50 000.

b 12.2 km = 12 200 m = 1 220 000 cm
$1\,220\,000 \div 50\,000 = 122 \div 5$
$= 24.4$ cm

Practice 15F

1 The scale of a map is 1 : 100 000.

a Two windmills are 4.2 cm apart on the map.
Find the actual distance between them.

b The distance between two junctions on a motorway is 7400 metres.
Work out the distance between these junctions on the map.

2

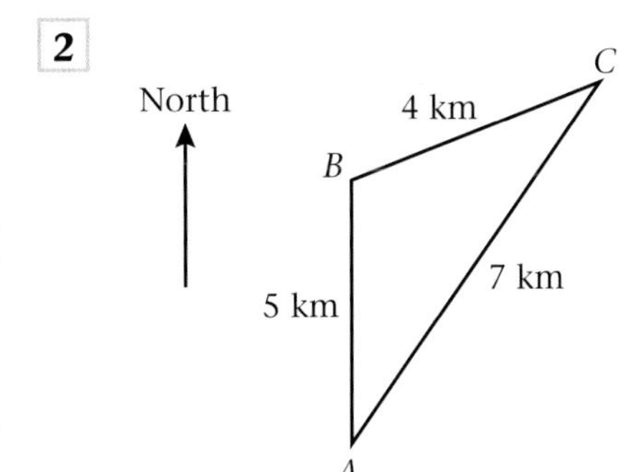

Diagram NOT accurately drawn

B is 5 km North of *A*
C is 4 km from *B*
C is 7 km from *A*

Hint: You have met bearings in Section 13.13.

a Make an accurate scale drawing of triangle *ABC*.
Use a scale of 1 cm to 1 km.

b From your accurate scale drawing, measure the bearing of *C* from *A*.

c Find the bearing of *A* from *C*.

E

3 Make a sketch and construct an accurate drawing for a journey which consists of 200 km on a bearing of 070° followed by 150 km on a bearing of 110°.
Use a scale of 1 cm = 50 km.

4 Make a sketch and construct an accurate drawing for a journey which consists of 30 km on a bearing of 220° followed by 50 km on a bearing of 330°.
Use a scale of 1 : 1 000 000.

16 Measure

This unit is about lengths, areas and volumes.

16.1 You can find areas of composite shapes by breaking them into parts.

Example 1

Find the area of this composite shape.

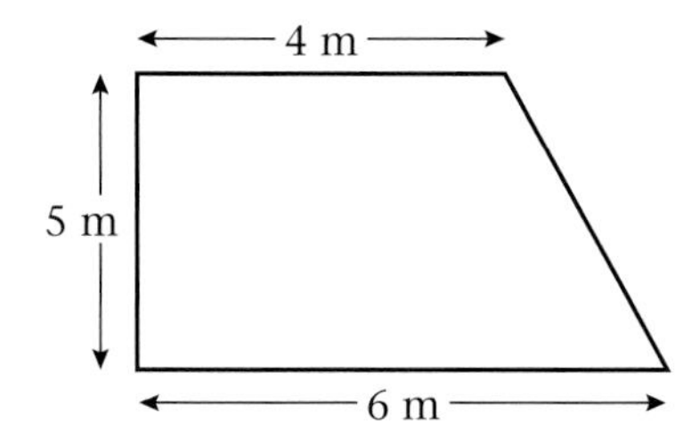

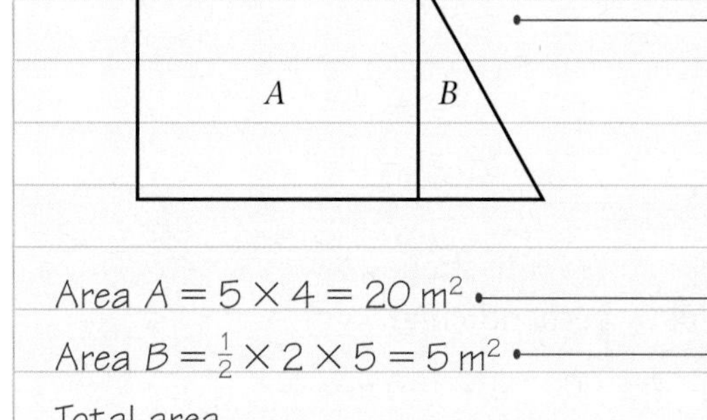

Split the shape into simple pieces, in this case a rectangle and a triangle.

Area $A = 5 \times 4 = 20\text{ m}^2$ — Area of a rectangle = length × width.

Area $B = \frac{1}{2} \times 2 \times 5 = 5\text{ m}^2$ — Area of triangle = $\frac{1}{2}$base × height.

Total area

$= \text{area } A + \text{area} B = 20 + 5 = 25\text{ m}^2$

Alternatively, you can use the formula for the area A of a trapezium: $A = \frac{1}{2}(a + b)h$.

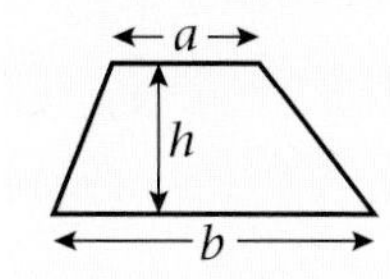

This formula will be provided on the exam paper.

Practice 16A

1 Work out the area of these shapes.

a

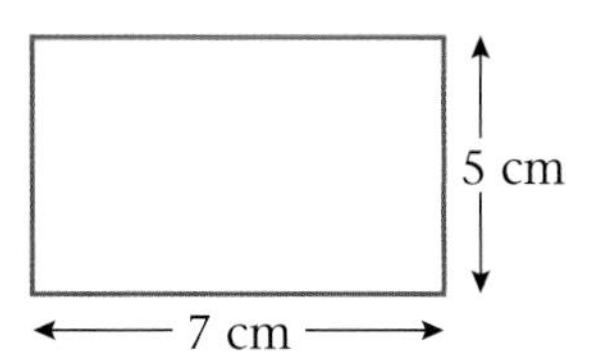

b

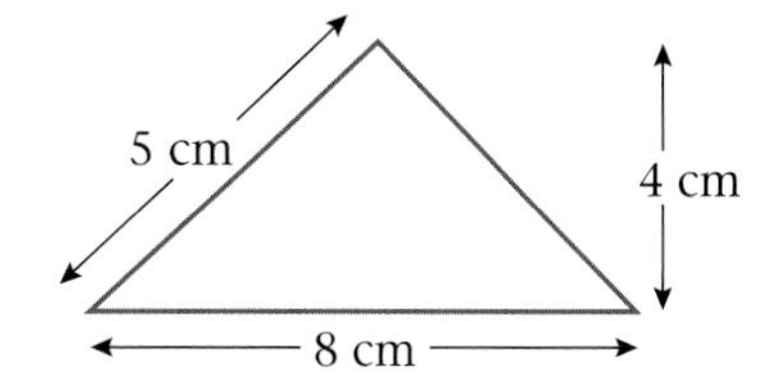

c

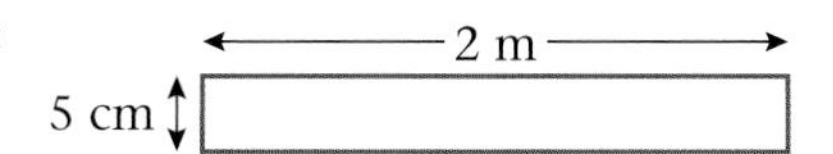

d

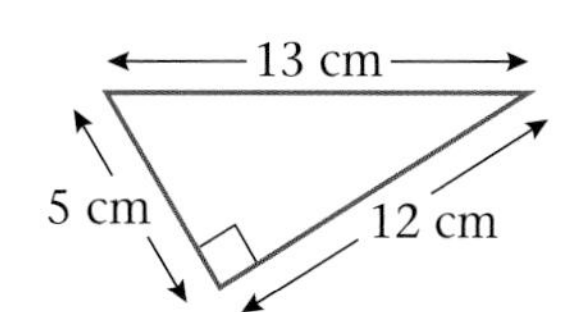

e

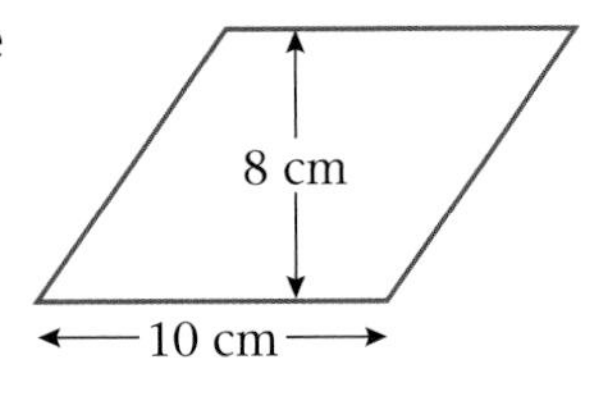

f

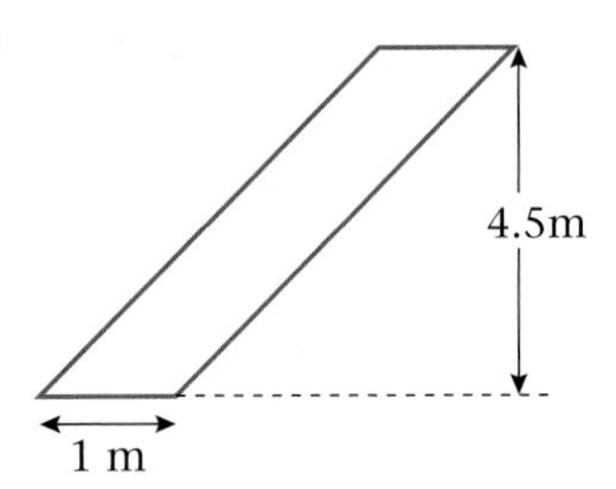

Hint: the area of a parallelogram = base × perpendicular height

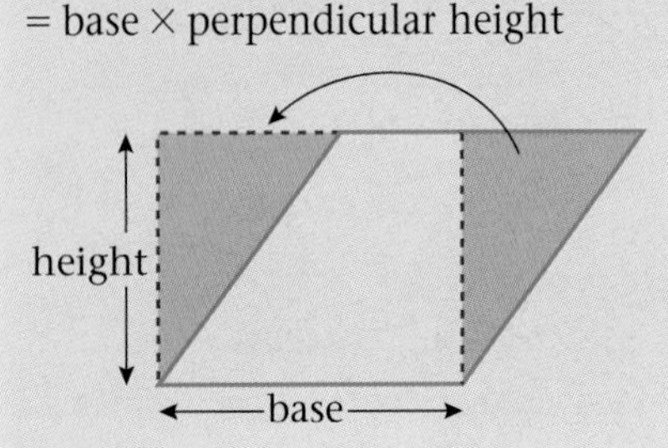

2

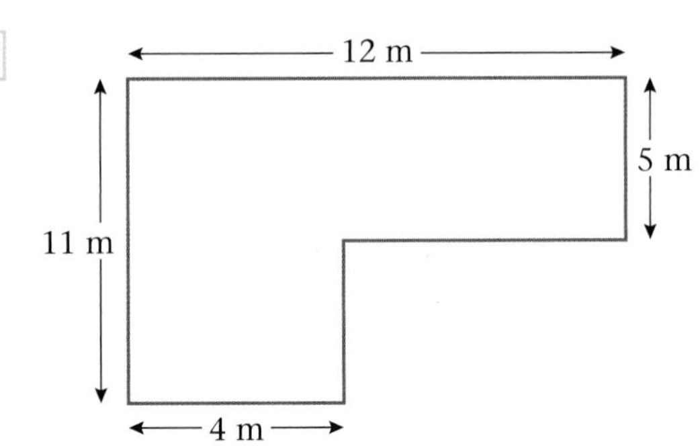

The diagram shows the plan of a floor.

a Work out the perimeter of the floor.

b Work out the area of the floor.

E

3 Work out the perimeter and area of these composite shapes.
In each shape the lines are either parallel or perpendicular.

a

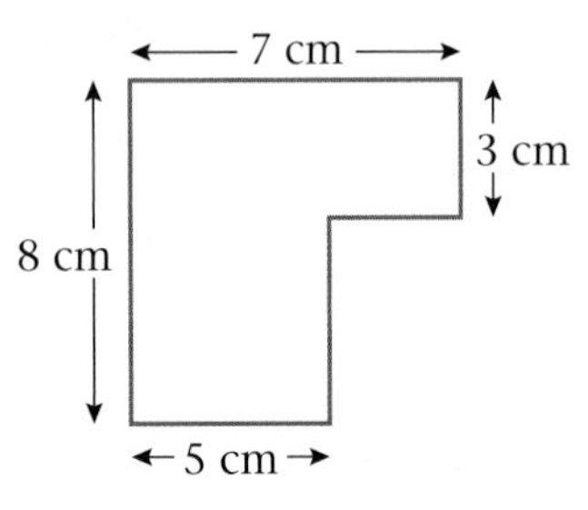

b

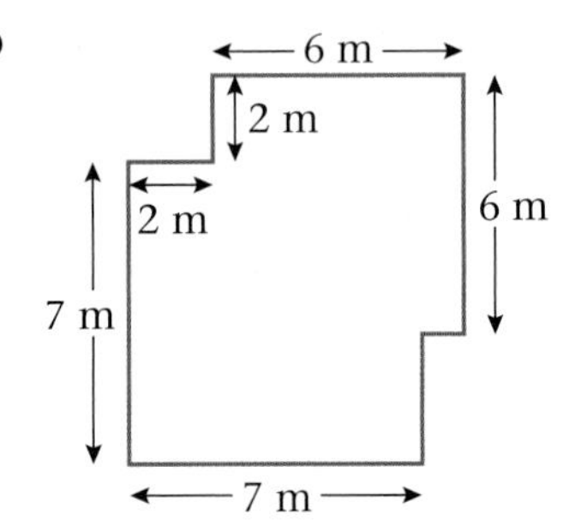

4 Work out the area and perimeter of these shapes.

a

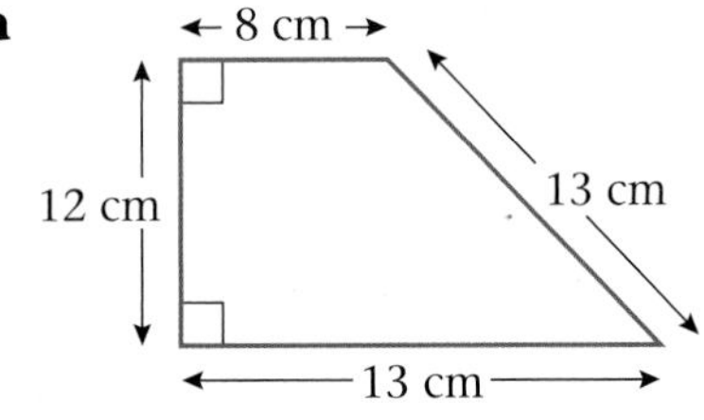

b

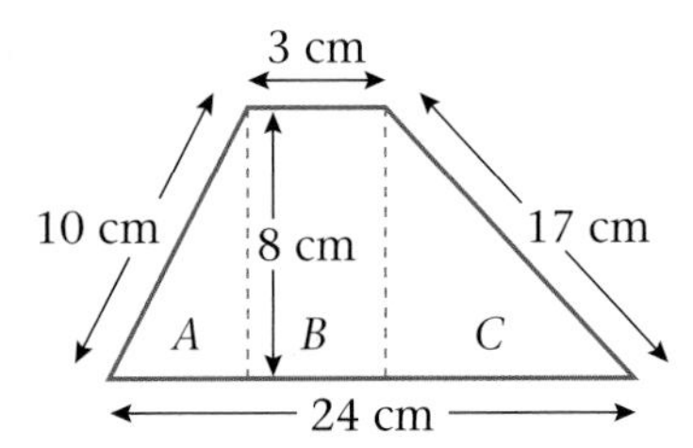

Hint: *A* and *C* can be treated as one triangle.

5 Work out the area of the kite shown.

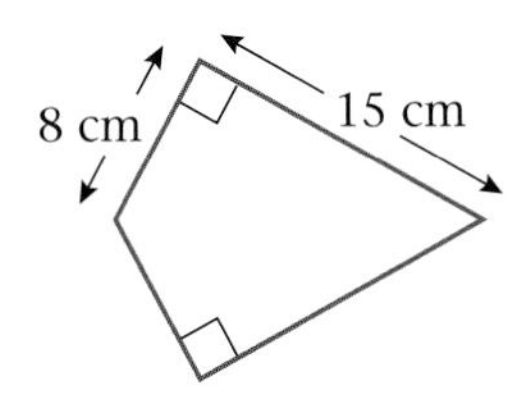

6 Work out the area of a rhombus with sides 10 cm if its diagonals have length 12 cm and 16 cm.

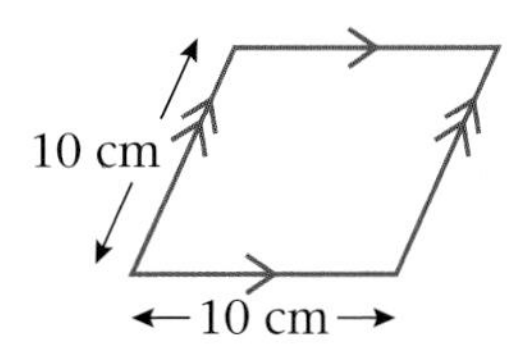

Hint: the diagonals of a rhombus meet at 90°.

7 The diagram shows a trapezium ABCD

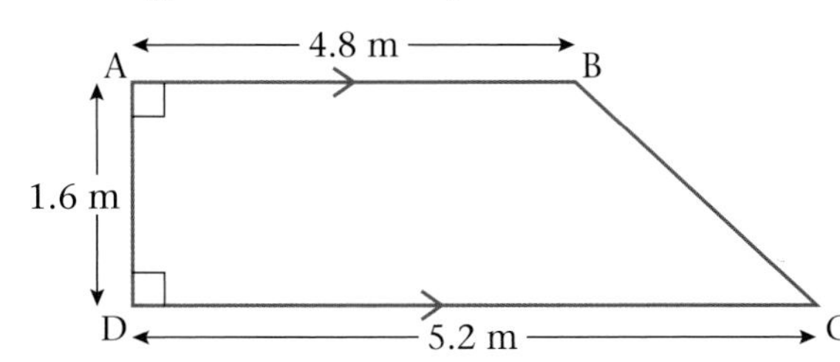

Diagram **NOT** accurately drawn

AB is parallel to DC
AB = 4.8 m, DC = 5.2 m, AD = 1.6 m, angle BAD = 90°, angle ADC = 90°

Calculate the area of trapezium ABCD.

16.2

You need to know the following formulae for working out the circumference and area of a circle:

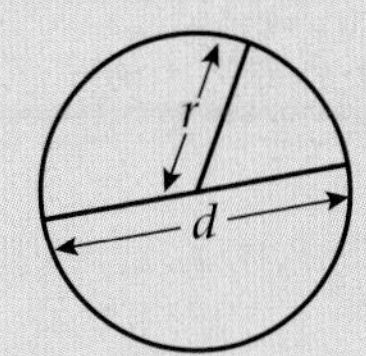

Circumference $= 2\pi r$ ***r*** **is the radius**

$= \pi d$ ***d*** **is the diameter**

Area $= \pi r^2$

Example 2

Work out **a** the circumference and **b** the area of a circle with diameter 18.2 cm.

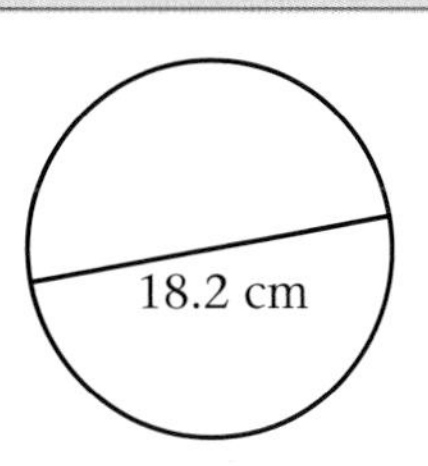

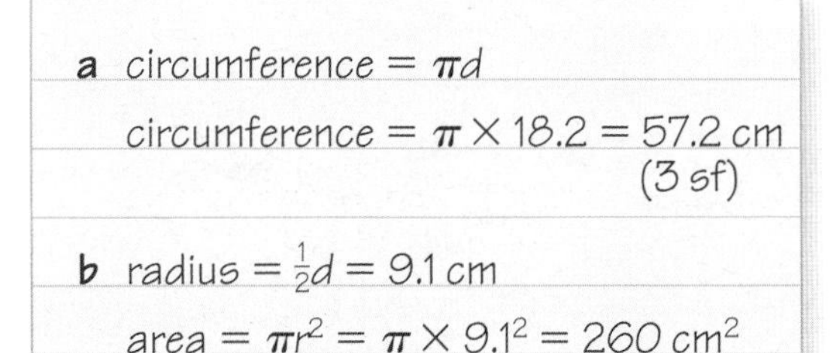

a circumference $= \pi d$

circumference $= \pi \times 18.2 = 57.2$ cm (3 sf)

Substitute 18.2 for *d*.

b radius $= \frac{1}{2}d = 9.1$ cm

area $= \pi r^2 = \pi \times 9.1^2 = 260$ cm^2 (3 sf)

The formula uses the radius.
r^2 means that cm^2 is the unit involved. As cm^2 is the unit of area, πr^2 is likely to be the area formula.

Example 3

A circular volcanic crater has a circumference of 285 metres.
What is the area of the top of the crater?

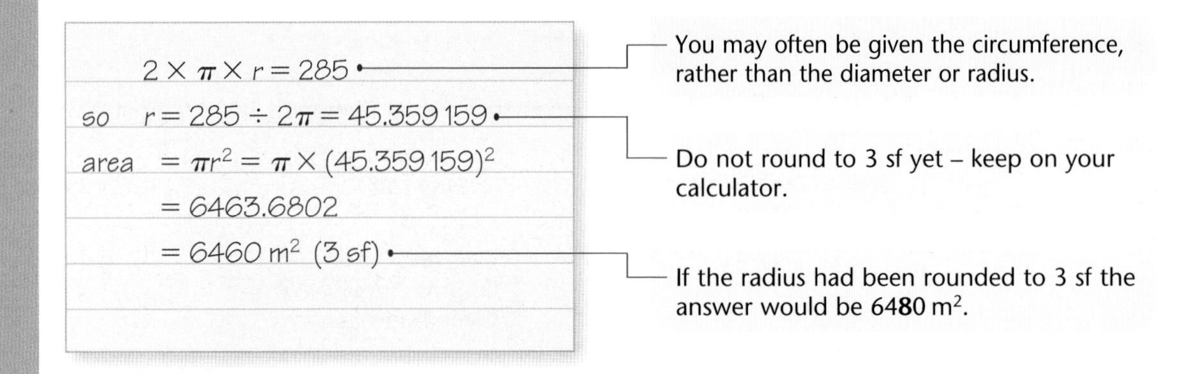

Example 4

The area of a circle is 53.3 cm². Work out **a** the radius and **b** the circumference.

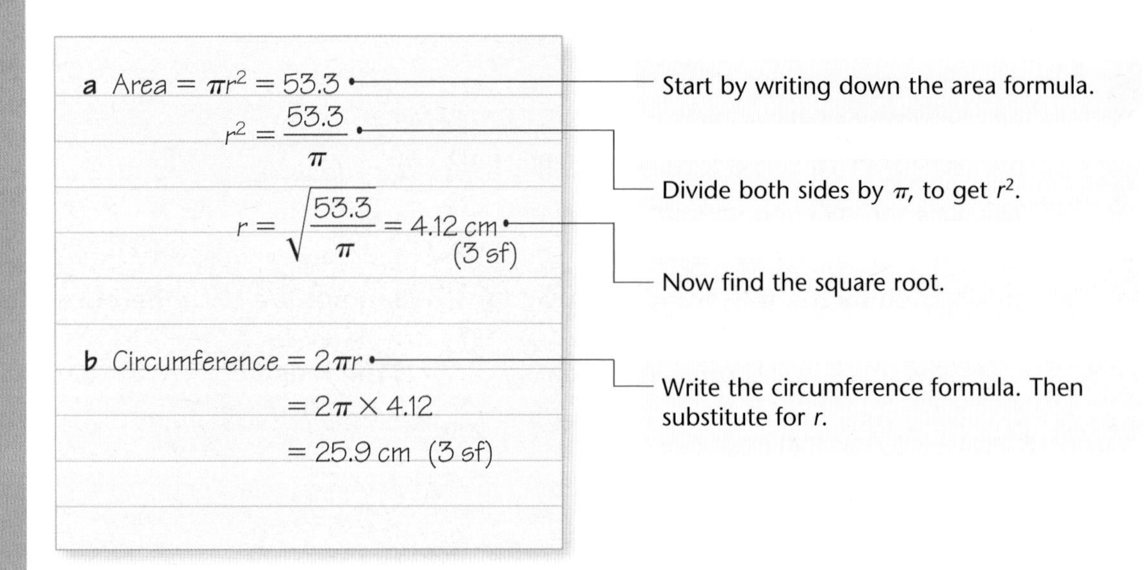

Example 5

Find the area and perimeter of the bath mat in the diagram.

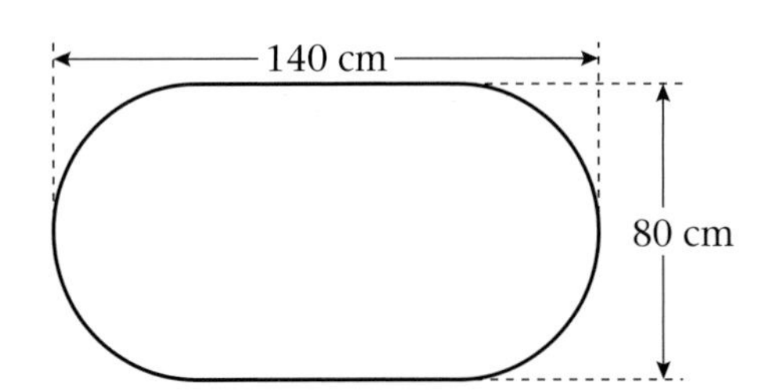

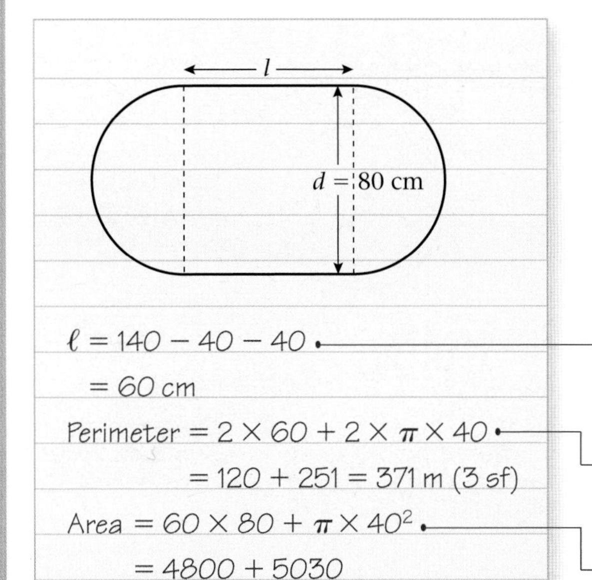

The shape is made from a rectangle and two semi-circular ends.

The total length loses a radius (40 cm) at each end.

2 straight sides and a circle.

Area = rectangle + circle.
$= \ell \times d + \pi r^2$

Example 6

Find the area and perimeter of this shape.

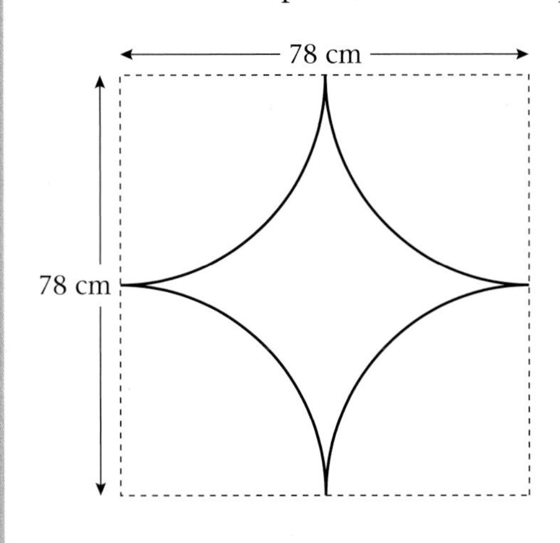

The shape is a square with four quarters of a circle removed.

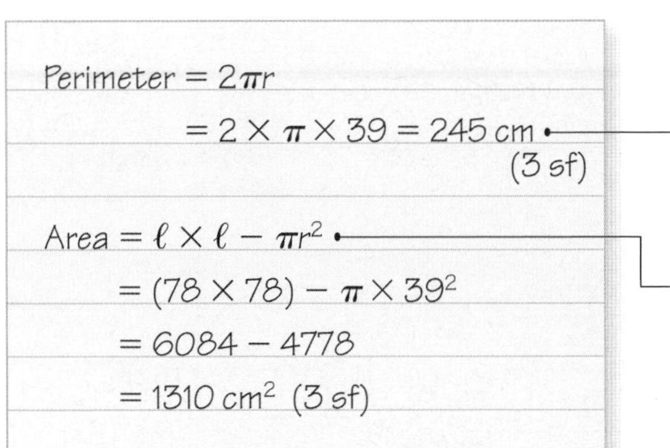

The perimeter is a circle with radius 39 cm

Area = area of square − area of circle

Practice 16B

1 Work out the circumference and areas of these circles.

a

3 cm

b

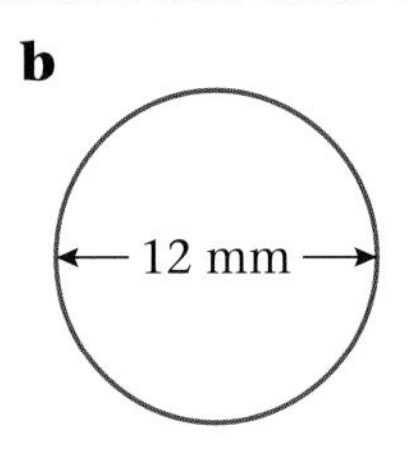

c

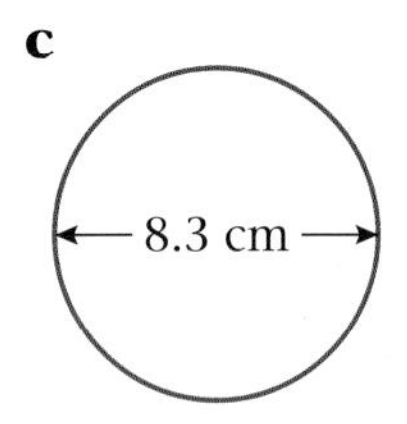

Hint: Look at Example 2, page 135.

2 The circumference of a circle is 83.7 metres.
Work out **a** the diameter and **b** the area of the circle.

Hint: Look at Example 3, page 136.

3 The area of a circle is 26.2 cm^2.
Work out **a** the circumference and **b** the diameter of the circle.

Hint: Look at Example 4, page 136.

4 The diagram shows a door which is in the shape of a rectangle with a semicircular top. Its width is 80 cm and its height is 200 cm. Work out the area of the door.

Hint: Look at Examples 5, page 136–137.

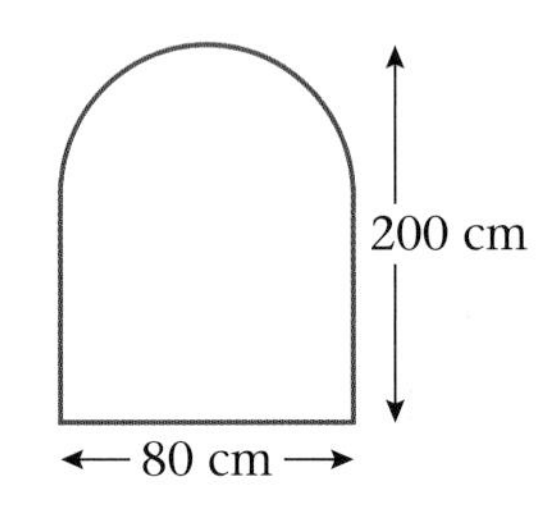

5 The shape is a square with side 12 cm with a semicircle removed.
Work out its area and perimeter.

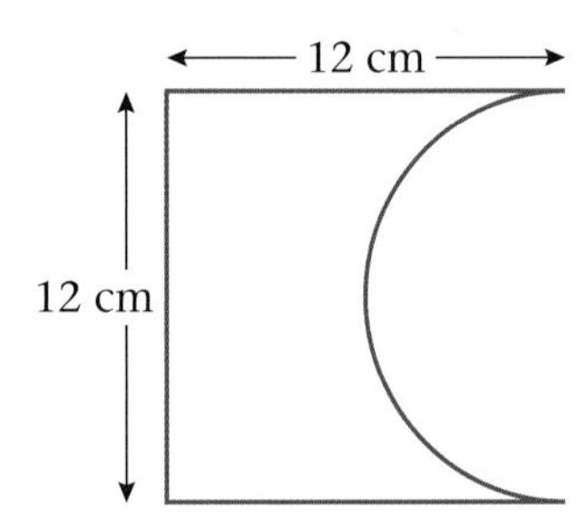

6 A circular washer has internal diameter 6.4 mm.
The external radius is 5.4 mm. Work out its area.

Hint: It may help to draw a diagram.

7 The shape is made of straight line segments and semicircular arcs.
Work out its perimeter and area.

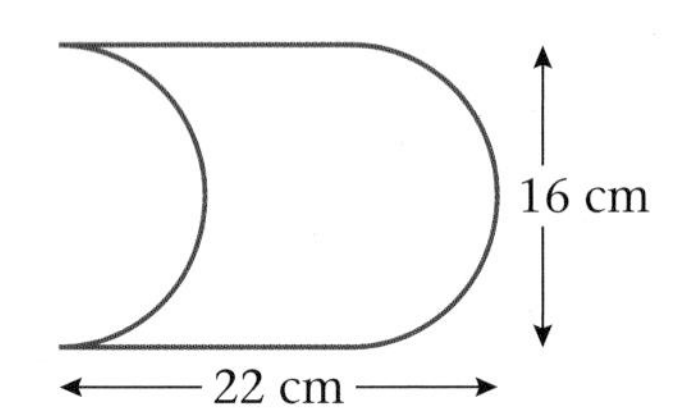

16.3 You can find the surface areas of cuboids and simple prisms by adding together the areas of all the faces.

Example 7

Find the surface area of the shape shown:

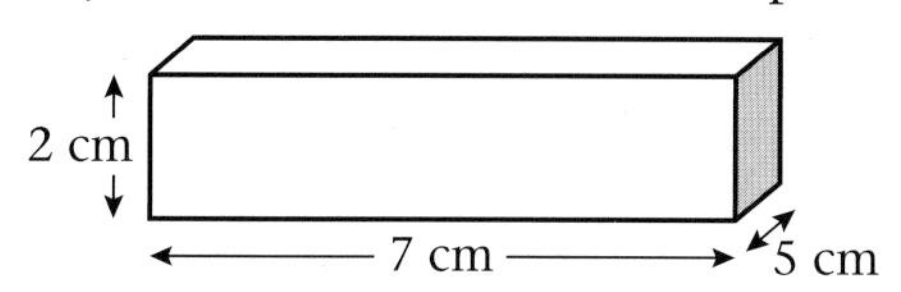

It may help to imagine the shape unfolded as in Example 4, page 108.

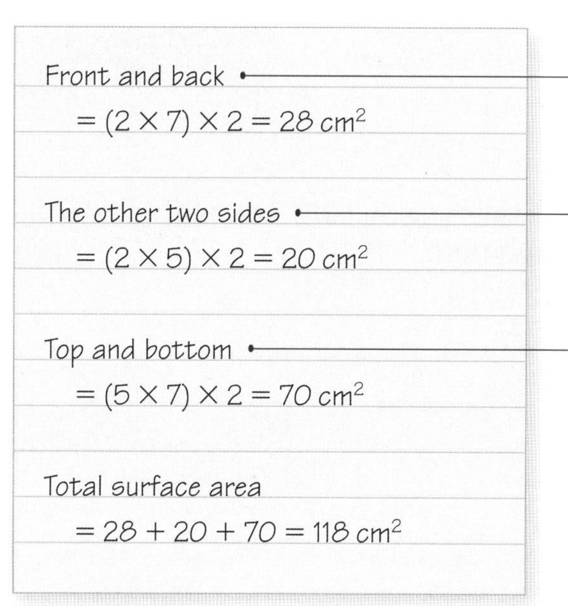

Front and back

$= (2 \times 7) \times 2 = 28\ \text{cm}^2$

The other two sides

$= (2 \times 5) \times 2 = 20\ \text{cm}^2$

Top and bottom

$= (5 \times 7) \times 2 = 70\ \text{cm}^2$

Total surface area

$= 28 + 20 + 70 = 118\ \text{cm}^2$

There are three pairs of opposite faces – work out the area of each.

Example 8

Find the surface area of this cylindrical tin.

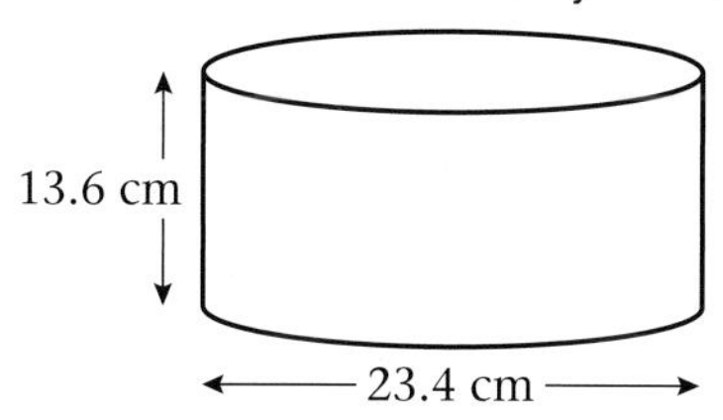

When unfolded, a cylinder consists of 2 circles and a rectangle.

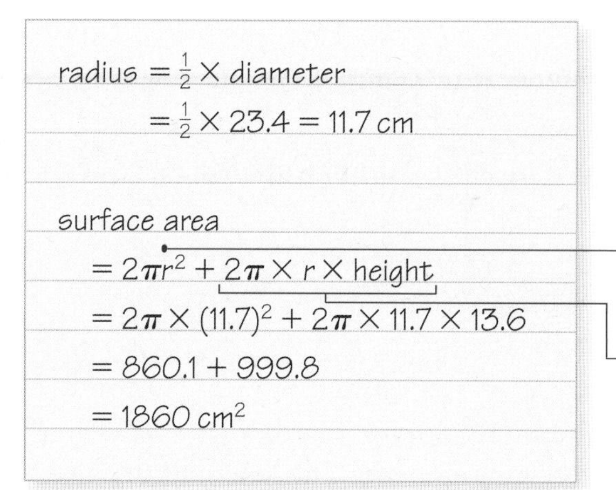

radius $= \frac{1}{2} \times$ diameter

$= \frac{1}{2} \times 23.4 = 11.7\ \text{cm}$

surface area

$= 2\pi r^2 + 2\pi \times r \times \text{height}$

$= 2\pi \times (11.7)^2 + 2\pi \times 11.7 \times 13.6$

$= 860.1 + 999.8$

$= 1860\ \text{cm}^2$

2 circular ends, each of area πr^2.

The side of the tin 'unfolds' to make a rectangle.

Practice 16C

1 A packet of salt is in the shape of a cuboid.
It measures 16 cm by 7 cm by 4 cm.
Work out its total surface area.

2 Copy and complete this table of data for cuboids.

	Length	Width	Height	Surface area
a	2 cm	0.5 m	0.5 m	
b	40 cm	25 cm	10 cm	
c	40 cm	40 cm	5 mm	
d	1.8 m	50 mm	40 mm	
e	3 m	15 cm	8 mm	

Hint: Make sure that length, width and height are expressed in the same unit, before working out the surface area.

3 A cylindrical tin, radius 11.2 cm and height 14.4 cm has a label which covers the curved surface. Work out the area of this label.

4 Work out the surface area of these prisms.

a

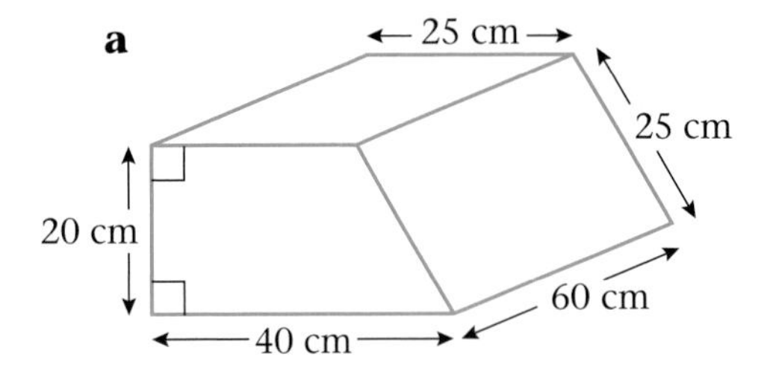

Hint: For Question 4a, look at Example 1, page 133.

b

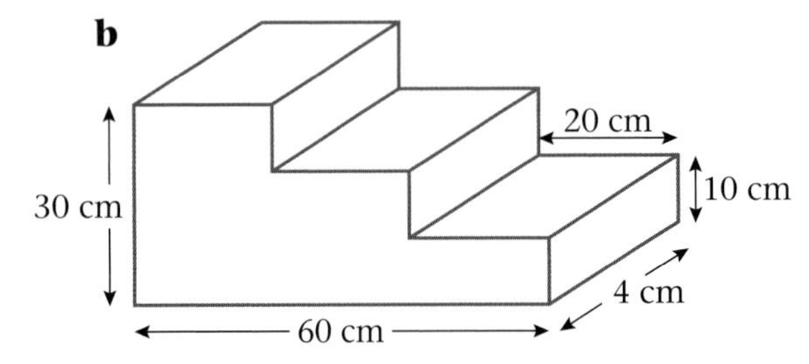

Hint: Assume all the 'steps' have the same dimensions.

5 Work out the total surface area of a cylinder with radius 18 cm and length 12 cm.

6 Work out the total surface area of a cylindrical steel drum with diameter 50 cm and height 1.6 m.

7 The curved surface area of a cylindrical tin is 500 cm^2.
Its height is 10 cm. Work out its radius.

8 The curved surface area of a length of pipe is 3200 cm^2.
The diameter of the pipe is 25 mm. Work out, in metres, the length of the pipe.

16.4

The volumes of cuboids and simple prisms can be found using the formula:

- **Volume = cross-section area × length**

This word formula works for cuboids, prisms and cylinders.

(There are algebraic formulas but, if you can work out cross-section areas, the word formula is all you need.)

Example 9

Find the volume of this cuboid.

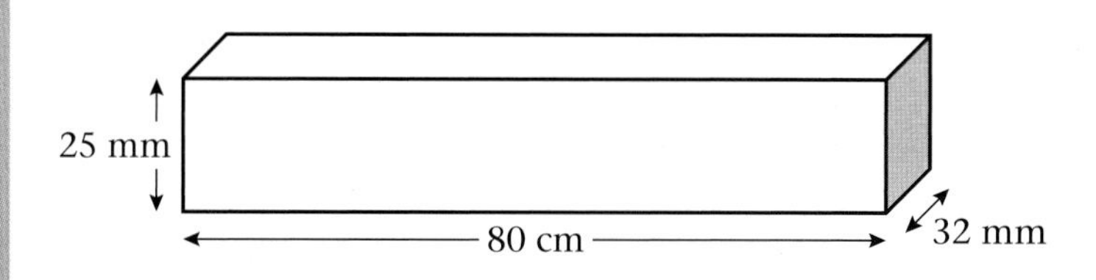

Height 25 mm = 2.5 cm

Width 32 mm = 3.2 cm

cross-section area = $3.2 \times 2.5 = 8\ \text{cm}^2$

Volume = $8 \times 80 = 640\ \text{cm}^3$

Make sure all the measurements are in the same units (in this case, cm).

Example 10

Find **a** the volume and **b** the capacity of this cylinder.

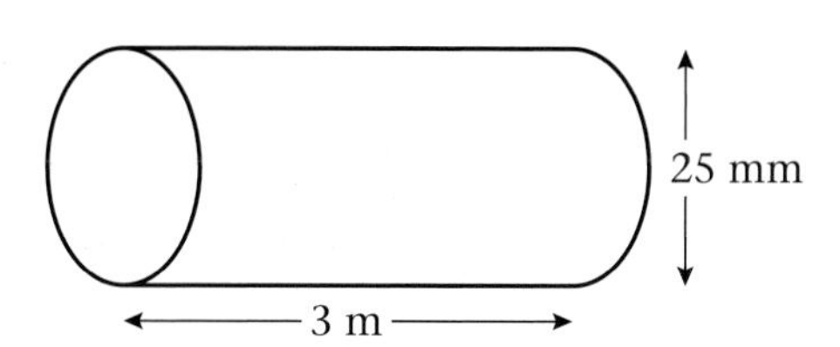

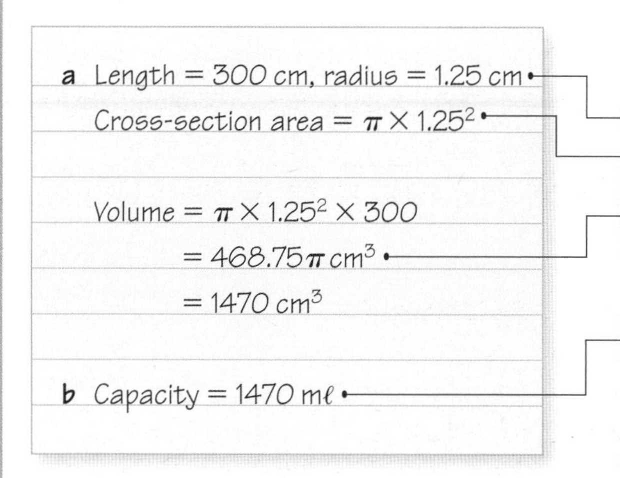

a Length = 300 cm, radius = 1.25 cm

Cross-section area = $\pi \times 1.25^2$

Volume = $\pi \times 1.25^2 \times 300$

$= 468.75\pi\ \text{cm}^3$

$= 1470\ \text{cm}^3$

b Capacity = 1470 mℓ

For this cylinder it makes sense to use centimetres.

Radius = half the diameter

Area of circle = πr^2

Sometimes an answer in terms of π is asked for.

Cylinders are often containers. Capacity measures how much you can get into them. Capacity is measured in litres (ℓ) and millilitres (mℓ).

1 mℓ = 1 cm^3

Example 11

How many boxes of candles measuring 6 cm by 4 cm by 14 cm can be fitted into a carton measuring 42 cm by 32 cm by 30 cm?

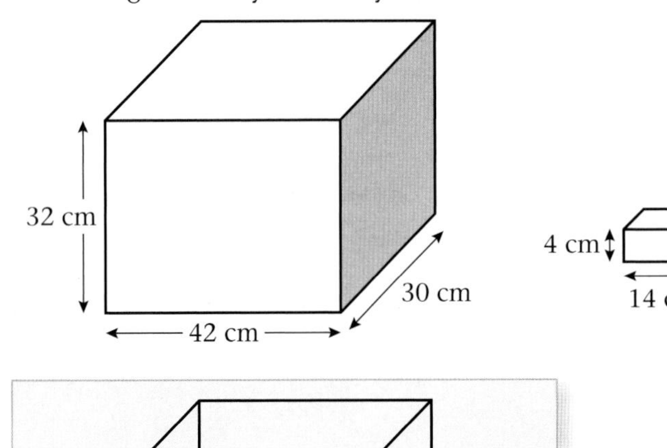

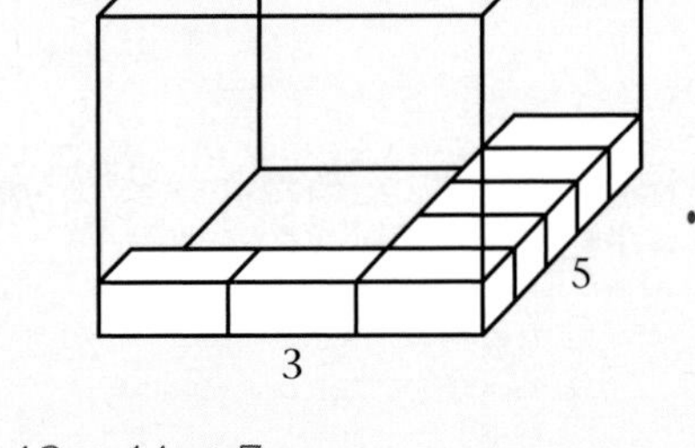

Draw a sketch showing how the boxes will fit.

$42 \div 14 = 3$

How many boxes can you get into a row 42 cm long?

The number of boxes in a layer is

$3 \times 5 = 15$

The rows go across 30 cm to make a layer.
6 cm goes into 30 cm 5 times without wasting any space.

There must be 8 layers.

The layers are 4 cm high and go up 32 cm.

Total number that can be packed in the carton $= 15 \times 8 = 120$

Example 12

Find the volume of the shed:

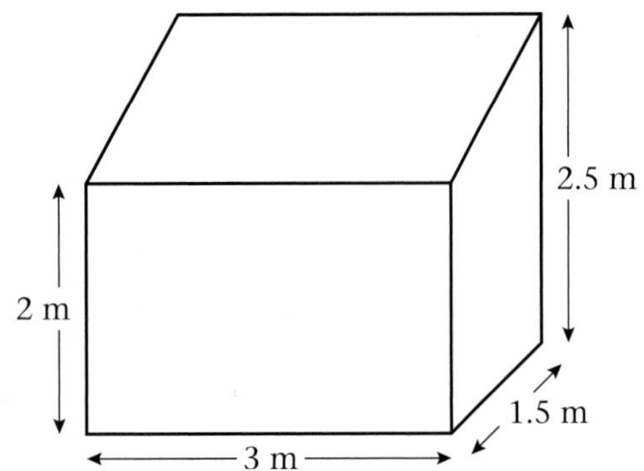

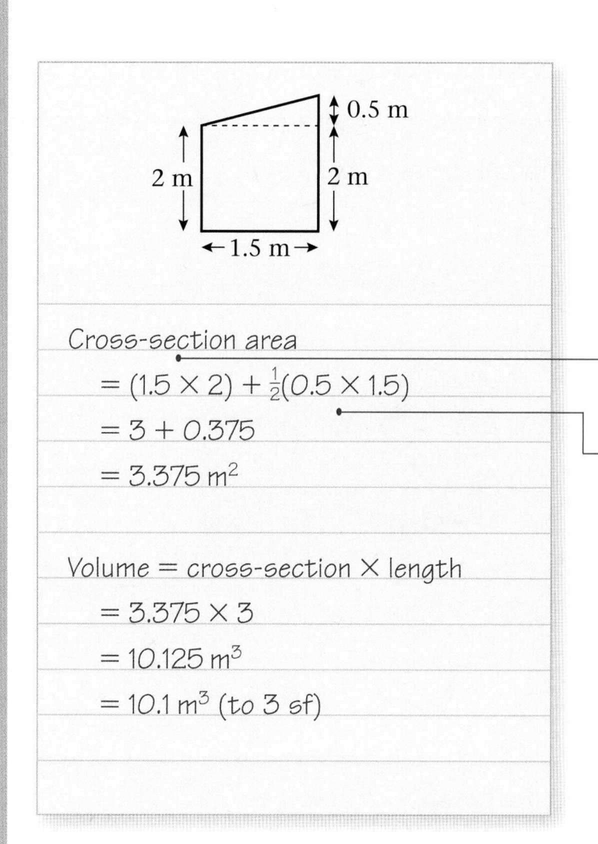

The cross-section is a trapezium similar to the shape in Example 1.

rectangle (width × height)

triangle ($\frac{1}{2}$ × base × perpendicular height)

Practice 16D

1 Copy and complete this table of values for cuboids.

	Length	Width	Height	Volume
a	5 cm	6 cm		90 cm³
b	14 cm		3 cm	231 cm³
c	1.2 m	15 cm		3600 cm³
e	2 m	10 cm	35 mm	
f		34 cm	25 mm	425 cm³
g	48 cm		40 mm	691.2 cm³

2 Packets of jelly measure 9 cm by 7 cm by 3 cm and are packed into a carton which measures 24 cm by 21 cm by 18 cm.
What is the highest number of packets that can be fitted into the carton?

3 A carton contains 72 matchboxes which measure 5 cm by 7 cm by 2 cm. The base of the carton measures 15 cm by 21 cm.
How high is the carton?

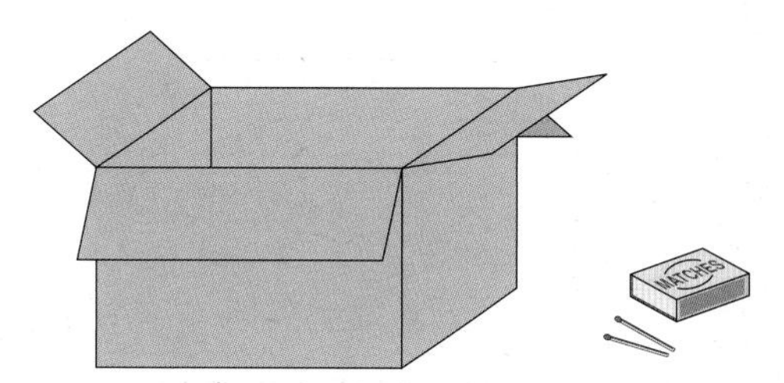

4 Work out the volume of a cylinder with radius 12.6 cm and height 18.2 cm.

5 A cylindrical pipe is 14.6 m long. Its internal diameter is 20.8 cm.
Work out the volume of the pipe.

6 Find the volume of the building shown.

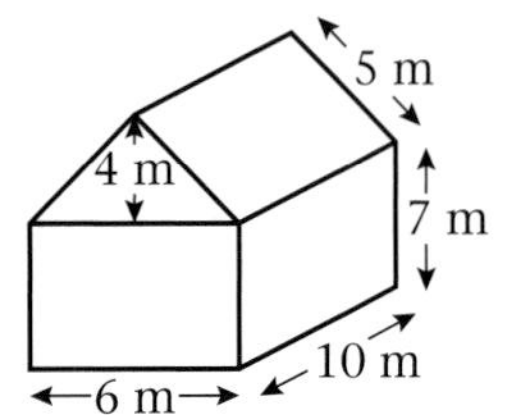

Hint: The formula *Volume of prism = area of cross-section × length* will be provided on the exam paper.

16.5

You need to remember these formulae, to be able to change the subject, and to substitute values.

- **Average speed = $\frac{\text{total distance travelled}}{\text{total time taken}}$**
- **Speed = $\frac{\text{distance}}{\text{time}}$**

Common units are m/s (metres per second) and km/h (kilometres per hour).

- **Density = $\frac{\text{mass}}{\text{volume}}$**

Common units are g/cm³ (grams per cm³), kg/m³ (kilograms per m³).

Example 13

An athlete runs 400 metres in 41.2 seconds.
What is his average speed in metres per second?

$$\text{Speed} = \frac{\text{distance}}{\text{time}}$$

$$= \frac{400}{41.2} = 9.71 \text{ m/s (3 sf)}$$

Always start by writing down the formula you will use.

Example 14

The density of glass is 2.6 g/cm^3
What is the mass of a sheet of glass measuring 1.4 m by 80 cm by 6 mm?

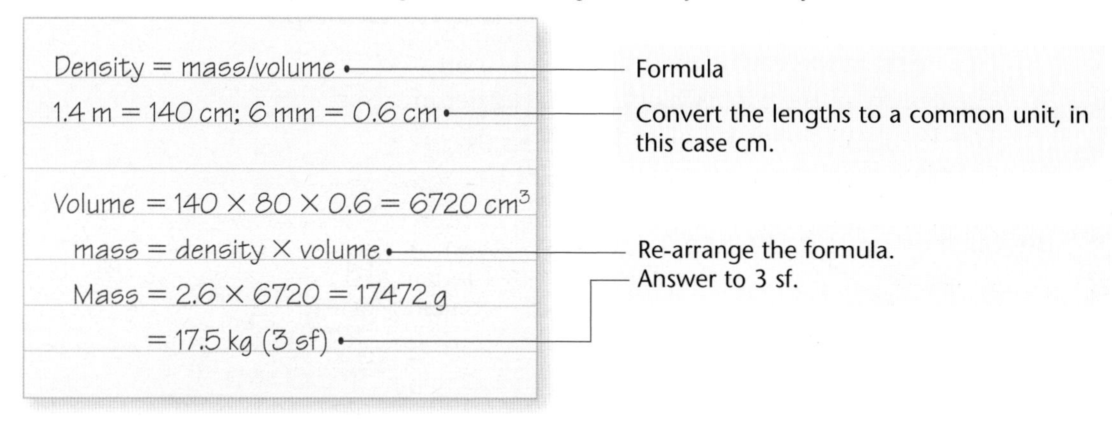

Hint: To re-arrange a formula, set out as a triangle, then cover up the value that you want as the subject:

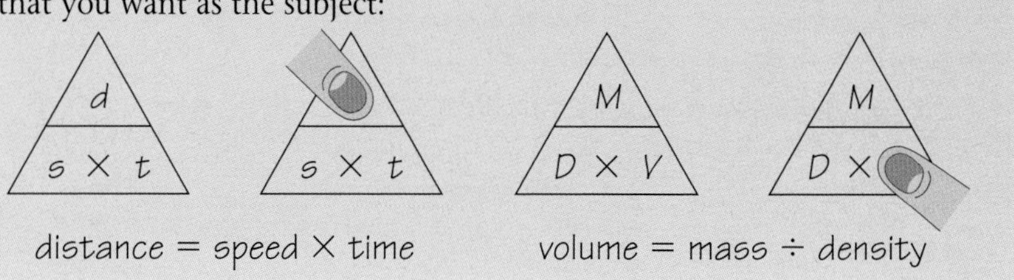

Practice 16E

1 In 1981 the record for the 1500 metres was 3 minutes 32.1 seconds.
Calculate the average speed in m/s.

2 Paula Radcliffe was the fastest woman in the 2002 London Marathon, with a time of 2 hours 18 minutes 56 seconds.
If the distance is 42 km, work out her average speed in m/s.

3 A motorcyclist averages a speed of 74.2 kph for 2 hours 15 minutes.
Calculate how far he travelled during this time.

4 A cyclist completes a 10 km time trial at an average speed of 17.6 m/s.
Calculate how long he took.

5 Peter sets off from Bristol and maintains an average speed of 45 kph for 3 hours. He stops for $1\frac{1}{2}$ hours. He then completes his journey to London by travelling a further 70 km which takes him 2 hours. Work out his average speed for the entire journey, including his stop.

6 A wooden ornament has a mass of 2.3 kg. Its volume is 3200 cm^3.
Work out the density of the wood.

7 The density of concrete is 2.3 g/cm^3.
What mass of concrete is required to complete a path which measures 10 m long by 1 m wide by 10 cm deep?

8 A quantity of lead pipe weights 3500 kg. Its density is 11400 kg/m^3.
It is melted down without losing any in the process. Calculate its volume.

9

NICKEL

Two metal blocks each have a volume of 0.5 m^3.
The density of the copper block is 8900 kg per m^3.
The density of the nickel block is 8800 kg per m^3.
Calculate the difference in the masses of the blocks.

E

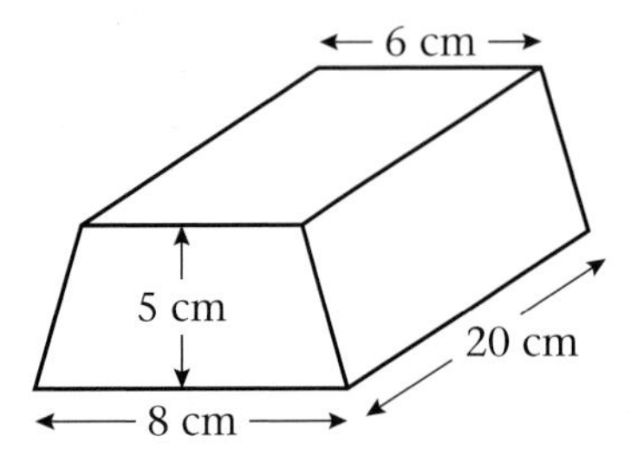

Diagram NOT accurately drawn

The diagram shows a prism.
The cross-section of the prism is a trapezium.
The lengths of the parallel sides are 6 cm and 8 cm.
The distance between the parallel sides of the trapezium is 5 cm.
The length of the prism is 20 cm.

a Work out the volume of the prism.

b The prism is made of gold. Gold has a density of 19.3 grams per cm^3.
Work out the mass of the prism.
Give your answer in kilograms.

16.6 You need to know how to convert between Imperial and metric units.

Metric	***Imperial***
8 km	**5 miles**
1 kg	**2.2 pounds (lb)**
25 g	**1 ounce (oz)**
1 litre (ℓ)	**1.75 pints**
4.5 ℓ	**1 gallon**
1 m	**39 inches**
30 cm	**1 foot**
2.5 cm	**1 inch**

You must learn these conversions.

Example 15

Change 273 miles into kilometres.

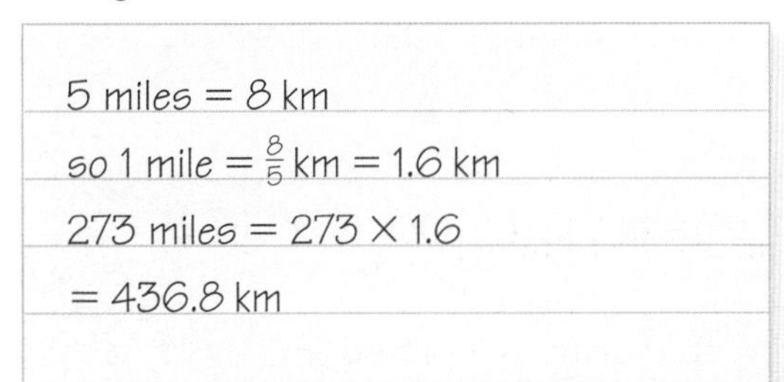

You know that 5 miles is 8 km, so divide that by 5 to get the equivalent of 1 mile.

Each mile is 1.6 km.

Kilometres are shorter so you need more of them.

Example 16

There are 14 lbs to a stone.
William weighs 12 stone 3 lbs.
What is his weight in kilograms?

William's weight is $(12 \times 14) + 3 = 171$ lb

2.2 lbs = 1 kilogram

$1 \text{ lb} = \frac{1}{2.2}$ kg

William weighs $171 \times \frac{1}{2.2} = 77.7$ kg

Convert William's weight to lbs.
12 stone = 12×14 lbs

Kilograms are heavier than pounds. Your weight in kg is about half your weight in pounds. Make a mental estimate, and use this to check that your answer is sensible.

Practice 16F

1 Change the following Imperial measurements into their metric equivalent:

a 23 lbs **b** 15 miles **c** 9 gallons **d** 6 inches **e** 5 ounces
f 6 pints **g** 4 feet **h** 103 miles **i** 46.2 lbs **j** 2.5 inches

2 Change the following metric measurements into their Imperial equivalent:

a 17 kg **b** 24 km **c** 15 cm **d** 3 litres **e** 105 cm
f 2.4 m **g** 200 g **h** 600 m **i** 750 mℓ **j** 850 g.

Hint: in questions **g** to **j** first write the units as kg, km, ℓ, kg

3 Mary weighs 7 stone 11 lbs.
1 stone is 14 lbs.
What is her weight in kilograms?

4 Alec is 5 feet 8 inches tall.
What is his height in centimetres?

5 The distance between Blackpool and Manchester by train is 48.25 miles.
How far is this in kilometres?

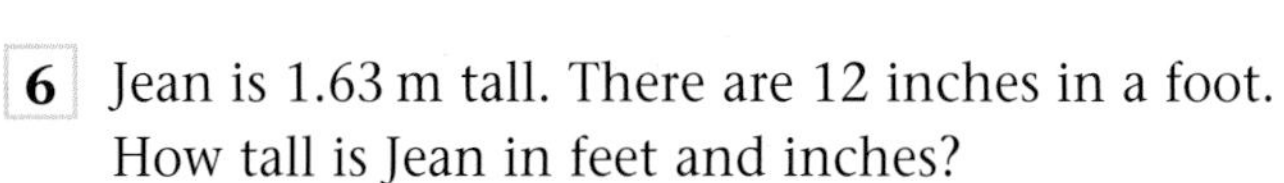

6 Jean is 1.63 m tall. There are 12 inches in a foot.
How tall is Jean in feet and inches?

7 Petrol costs 83.6p per litre.
How much is this per gallon?

8 The distance between Middlesbrough and Sunderland is 56 km.
How far is this in miles?

9 A drink dispenser holds 32 pints when full.
It dispenses drinks of 200 mℓ.
How many drinks can be obtained from a full container?

10 The speed limit on a road is 30 mph.
What is this speed in km/h?

(E14)

17 Pythagoras and trigonometry

In this chapter, you will use Pythagoras' theorem to work out the sides of a right angled triangle, and use sines, cosines and tangents to solve problems involving triangles. **(S11)**

17.1

Pythagoras' theorem states that, for a right-angled triangle, the sum of the squares of the shorter sides = the square of the longest side (hypotenuse).

$a^2 + b^2 = c^2$

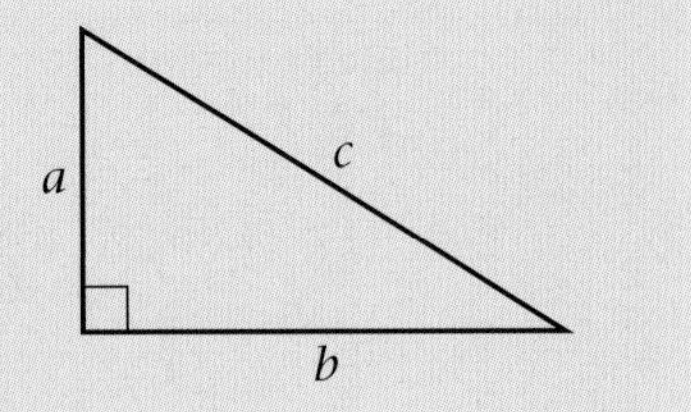

Example 1

Work out the length of AC.
Give your answer in surd form.

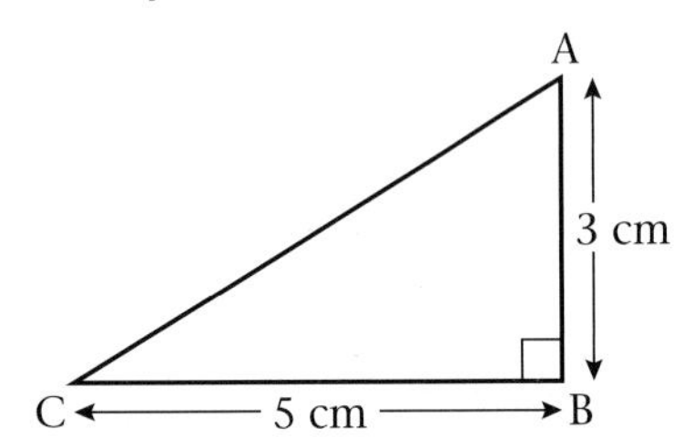

$AC^2 = BC^2 + AB^2$

$= 5^2 + 3^2$

$= 25 + 9 = 34$

$AC = \sqrt{34}$

From Pythagoras

On non-calculator papers you may be asked for answers in surd form. A number written exactly using a square root sign is a surd. For example $\sqrt{34}$ is in surd form.

Example 2

A 4.5 m ladder rests against a vertical wall.
The foot of the ladder is 1.5 m from the wall.
How far does the ladder reach up the wall?
Give your answer to 2 dp.

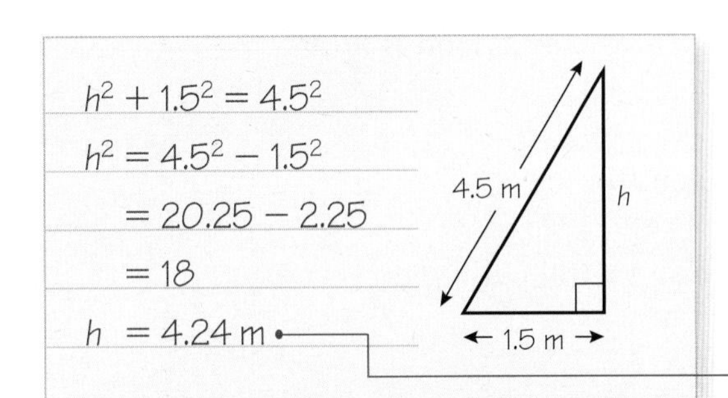

Use a calculator to work out $\sqrt{18}$, or on a non-calculator paper leave in surd form.

It is easy to make a mistake when doing the substitution. Check, by estimating, that your answer is sensible.

In this case, h must be less than 4.5 m.

Practice 17A

In Questions 1–4, work out the length of the unmarked side.

Give your answer in surd form.

1

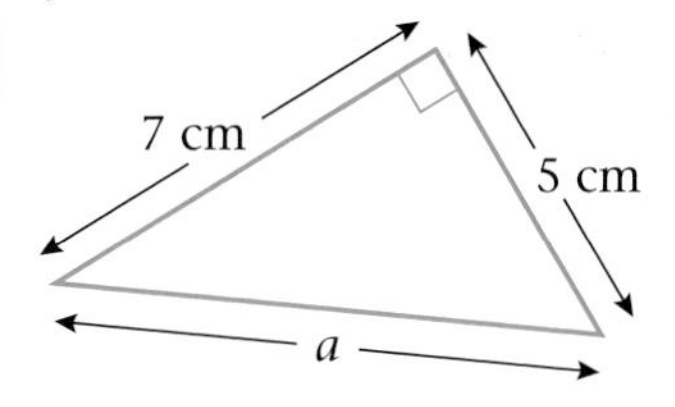

2

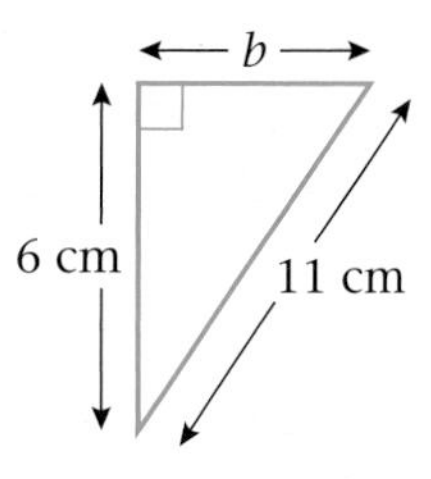

Hint: Check your answer. Does your answer leave the hypotenuse as the longest side?

3

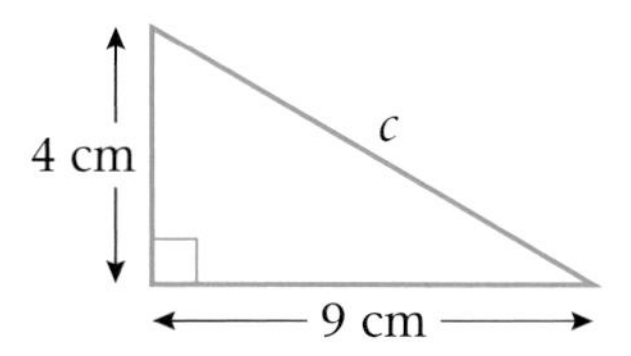

4

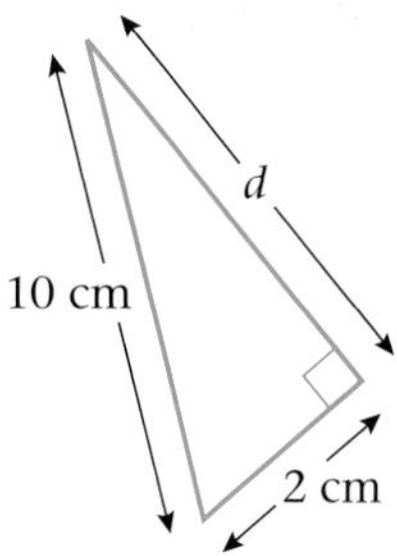

5 In triangle ABC, AB = 4 m and BC = 6 m. Work out the length of AC to the nearest centimetre.

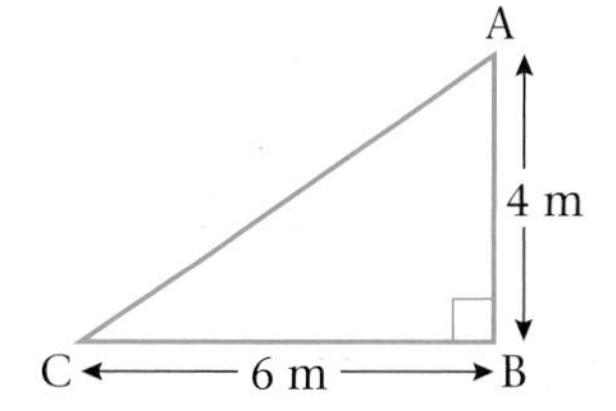

6 PQ is a ladder 3.8 m long resting against a vertical wall. The foot of the ladder is 1.6m from the wall. Work out how far up the wall the ladder reaches.

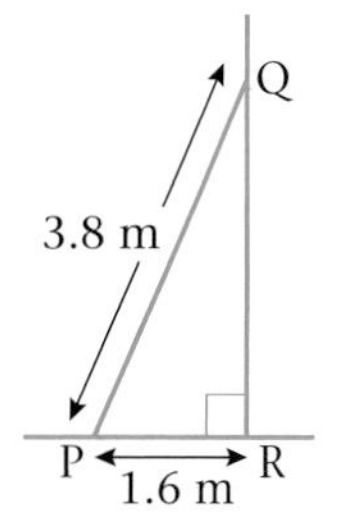

7 Southend (S) is 65 km East of London. Boston (B) is 175 km North of London. Work out the straight line distance between Southend and Boston.

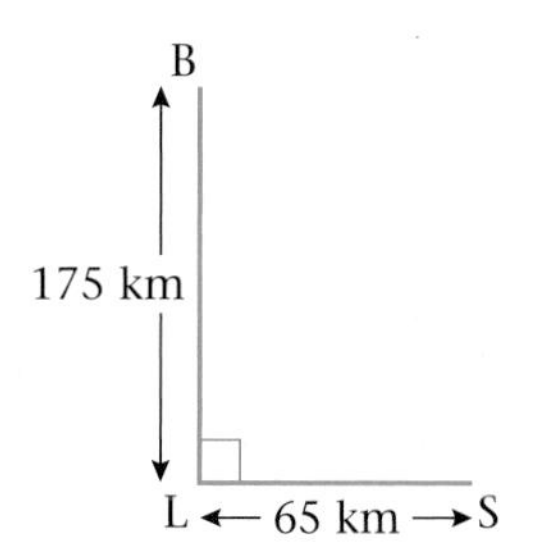

8 Calculate BC and BD.

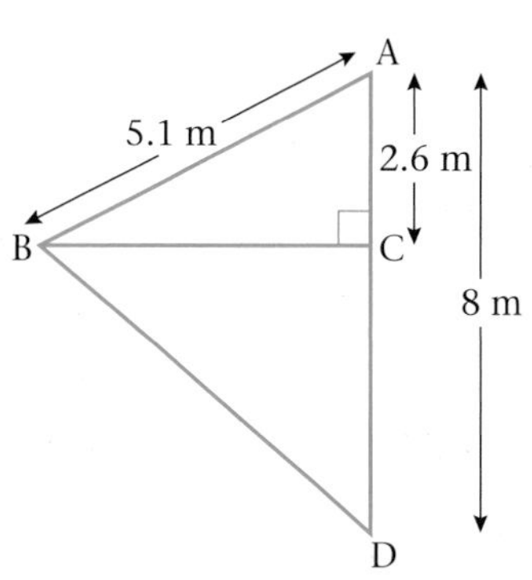

17.2 You need to know where to find angles of 90° (called right angles).

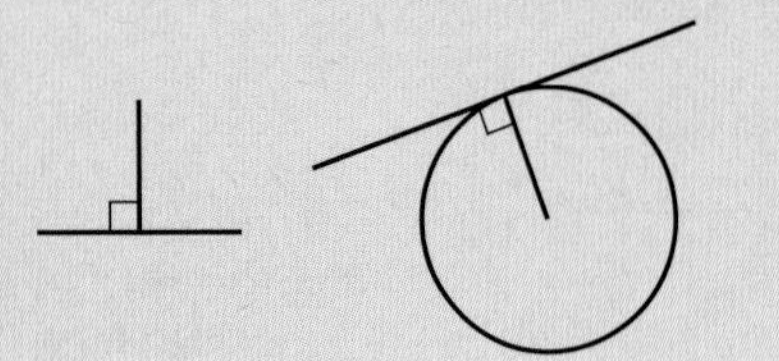
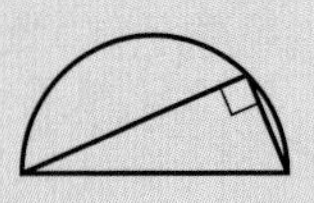
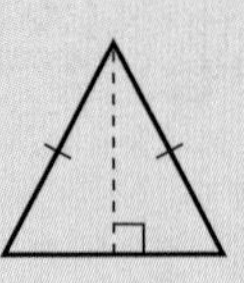

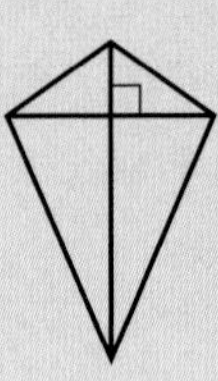

Example 3

Find the length of the sloping roof in the diagram.

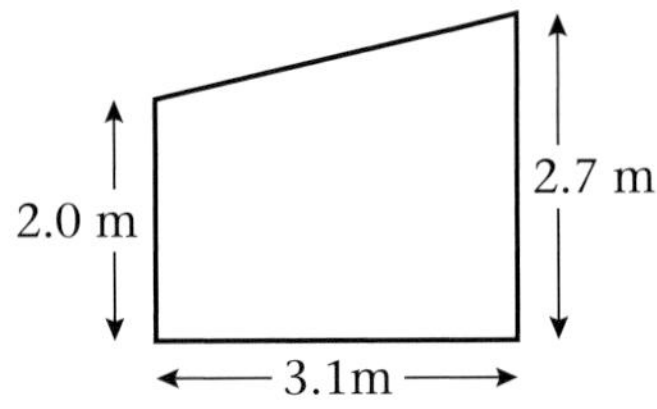

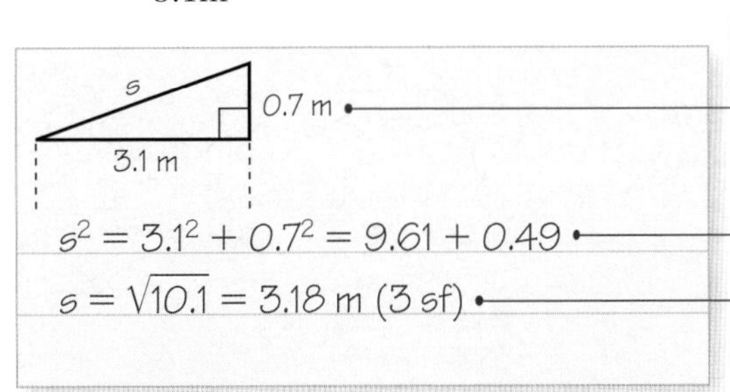

Draw in a horizontal line to make a right-angled triangle.
$2.7 - 2 = 0.7$
This is the top of the roof.

From Pythagoras

Answers to 3 sig figs are normal in this situation.

Practice 17B

1 BC is a diameter of a circle.
A is a point on the circumference.
AB = 5.6 cm; AC = 3.2 cm
Work out the radius of the circle.

Hint: It may help to draw a diagram. Diagrams are only helpful if the right-angle looks like 90°.

2 XYZ is an isosceles triangle with
XY = XZ = 9.2 cm and YZ = 12 cm.

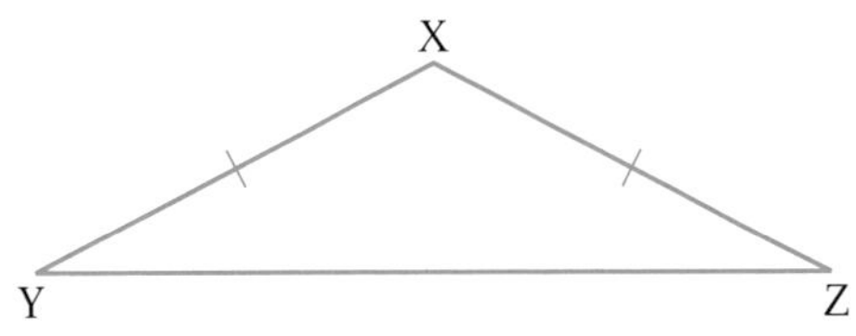

a Work out the height of the triangle.
b Work out the area of the triangle.

3 A rhombus has diagonals of length 8 cm and 12 cm. Work out the length of a side of the rhombus.

Hint: Diagonals of a rhombus bisect at 90°.

4 The length of a tangent from a point P to a circle, centre O, is 15 cm. The radius of the circle is 6 cm.
Work out the length of OP.

Hint: Radius and tangent meet at 90°.

5

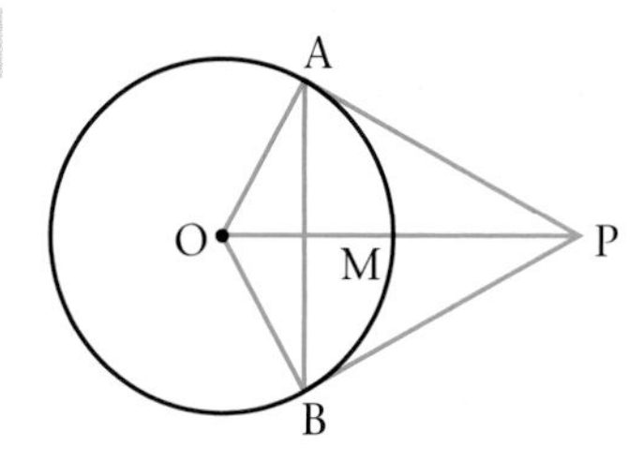

PA and PB are tangents to the circle, centre O, radius 7.5 m. PO = 12.5 m.

a Work out the length of the tangent PA.

AB = 12 m

b Work out the length PM.

6 The diagram represents the symmetrical cross-section of a barn. Work out the total height of the barn.

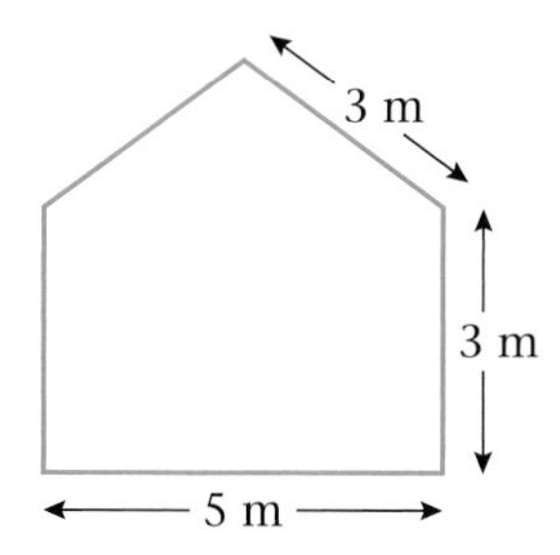

7 The diagram shows the cross-section of a ramp. Work out the height of the ramp.

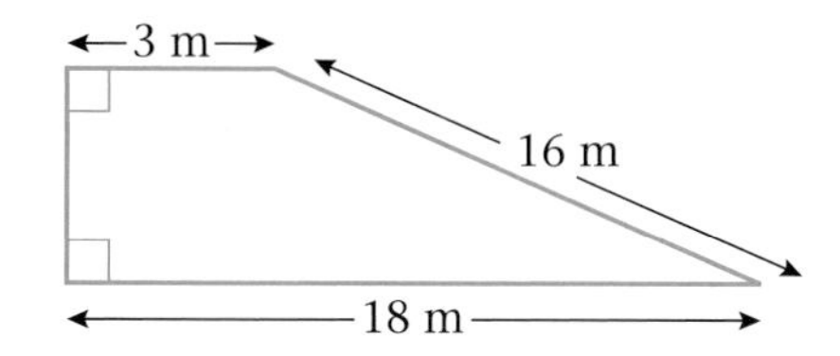

8 A plane flies 200 km in a straight line in a direction which is roughly North-east. It is now 110 km East of its starting point. How far North of its starting point is it?

17.3 The lengths of line segments on coordinate grids can also be found by constructing a right-angled triangle.

Example 4

Find the length of the line segment which joins A (2, −5) to B (6, 8)

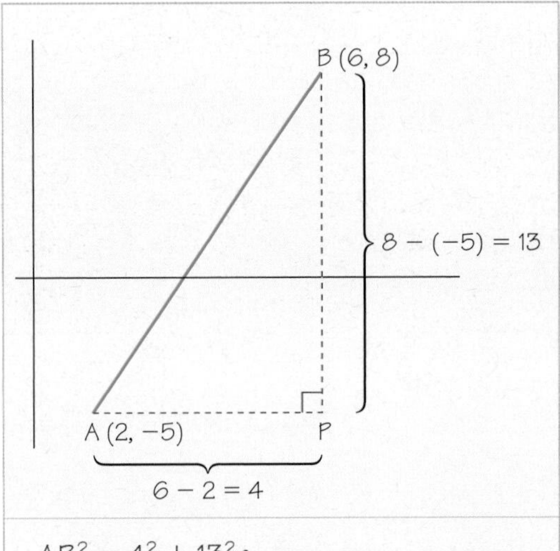

Make a right-angled triangle.

Work out the lengths of AP and BP.

$AB^2 = 4^2 + 13^2$ — Use Pythagoras' theorem: $AB^2 = AP^2 + BP^2$.

$= 16 + 169 = 185$

$AB = 13.6$ — 3 sf is usual.

Practice 17C

1 Work out the length of the line segment joining

a (2, 5) to (7, 17) **b** (3, 1) to (23, 20)

c (−1, 2) to (11, −3) **d** (−8, −13) to (0, 2)

2 Work out the length of the line segment joining

a $(3\frac{1}{2}, 2\frac{1}{2})$ to $(1\frac{1}{2}, 1)$ **b** (7, −3) to (−8, −9)

c (−4, −11) to $(-2\frac{1}{2}, -16\frac{1}{2})$ **d** (7, 6) to $(-3, -1\frac{1}{2})$

17.4 You need to remember how the sides of a right-angled triangle are named, and the formulae for working out sine, cosine and tangent.

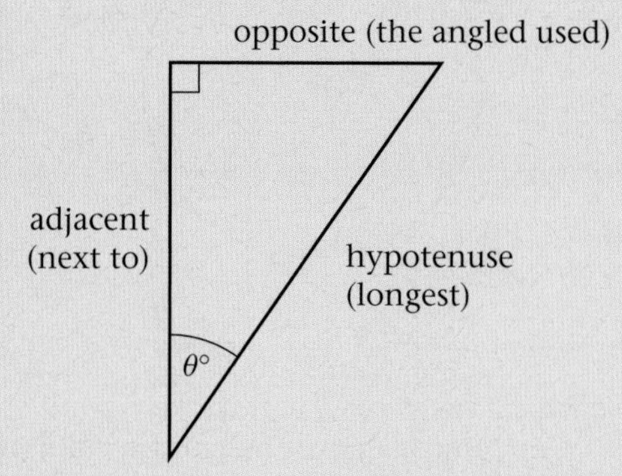

SOH

Sine $\theta = \dfrac{\textbf{O}\text{pposite}}{\textbf{H}\text{ypotenuse}}$

CAH

Cosine $\theta = \dfrac{\textbf{A}\text{djacent}}{\textbf{H}\text{ypotenuse}}$

TOA

Tangent $\theta = \dfrac{\textbf{O}\text{pposite}}{\textbf{A}\text{djacent}}$

Example 5

Find BC.

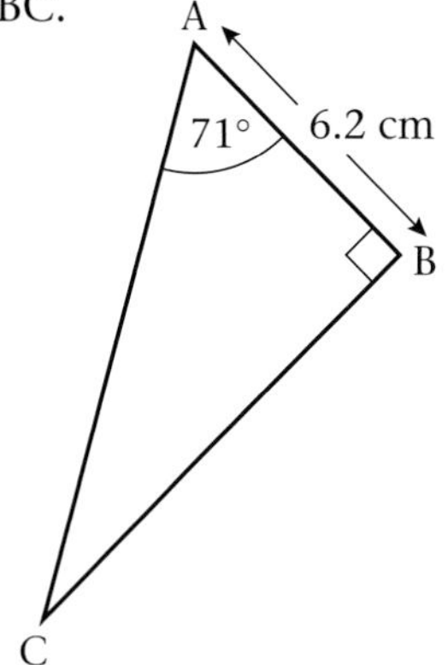

Make sure you have a right-angled triangle.

The angle in use is 71° so BC is the opposite side.

AB is given and is the adjacent side.

BC = opp.

6.2 cm = adj.

$\tan 71° = \dfrac{\text{opp}}{\text{adj}} = \dfrac{BC}{6.2}$ — So choose tangent = $\dfrac{\text{opposite}}{\text{adjacent}}$

$BC = 6.2 \times \tan 71°$ — Substitute 71° and 6.2

$= 18.0$

$= 18.0$ cm — It is usual to give answer to 3 sf, and don't forget the units!

Example 6

A circle has radius 5 cm.
Point P is 8 cm from the circle's centre. Find the angle between the two tangents from P to the circle.

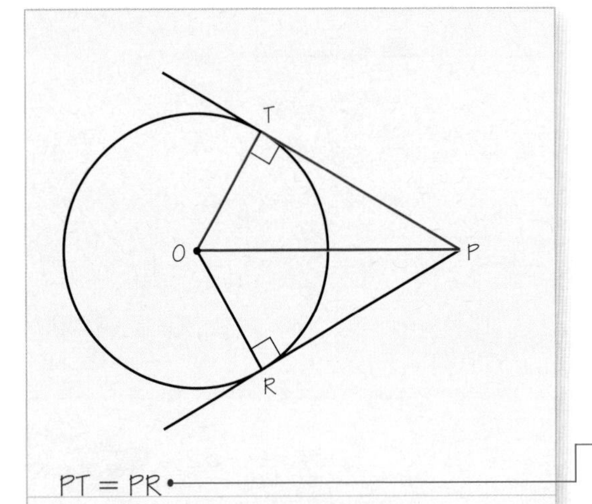

Copy the diagram, and construct a right-angled triangle.

$PT = PR$ — Tangents to a circle.

$\angle PTO = \angle PRO = 90°$ — A tangent is perpendicular to a radius. See Section 13.10.

$OP = 8$ cm

Re-draw the right-angled triangle. The angle in use is at P.

$\sin TPO = \frac{5}{8}$ — OT is opposite and OP is the hypotenuse, so choose the sine formula.

$= 0.625$

$\angle TPO = 38.682°$

Similarly, $\angle RPO = 38.682$

So $\angle TPR = 2 \times 38.682°$

$= 77.4°$ — Answers to 1 dp are usual.

Practice 17D

1 Calculate the lengths marked with letters in the diagrams.

a

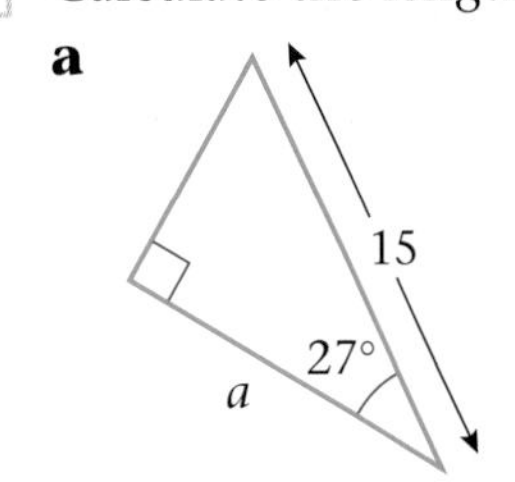

b

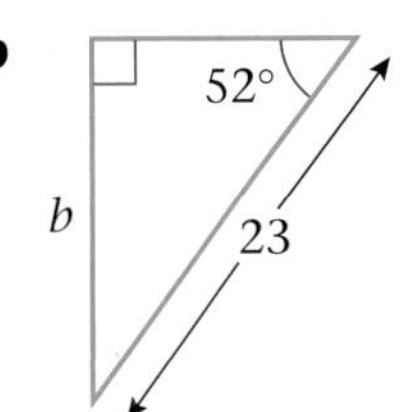

c

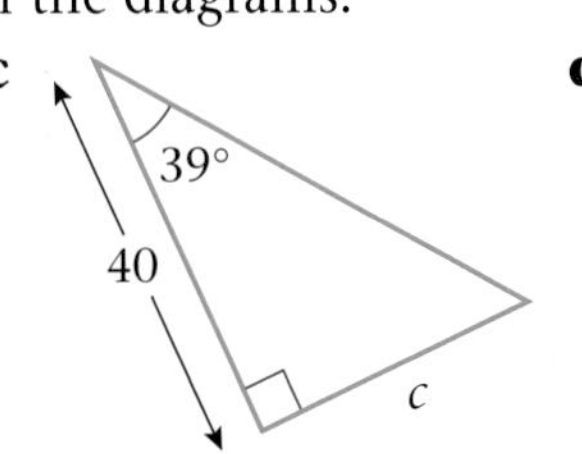

d

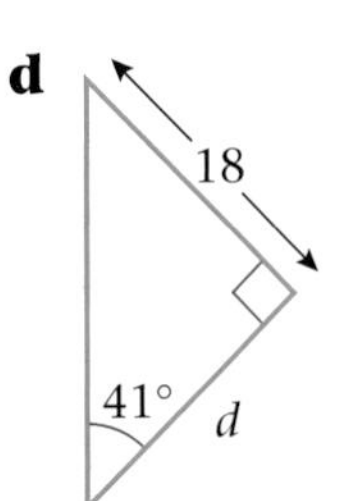

2 Calculate the named angles in the triangles.

a

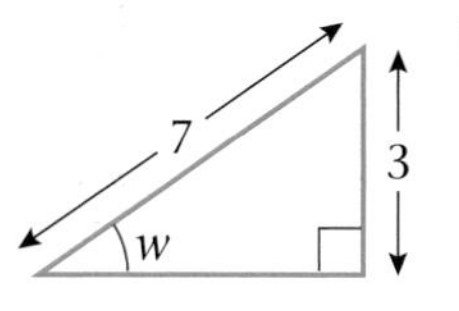

b

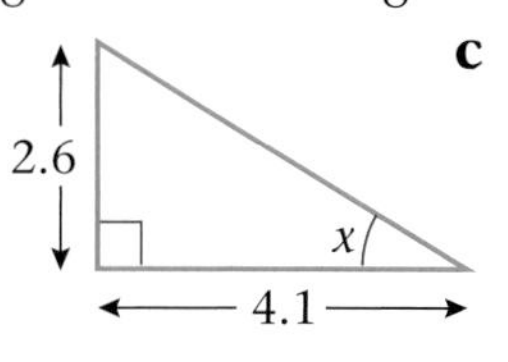

c

d

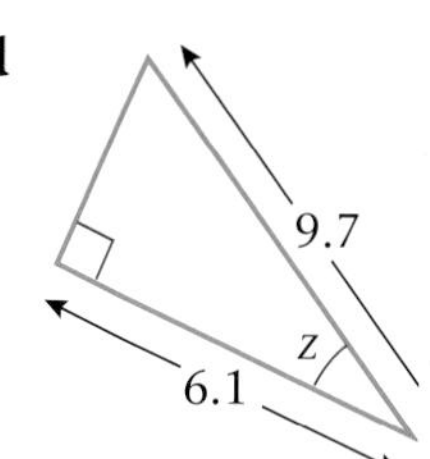

3 Calculate the lengths marked with letters in the diagram.

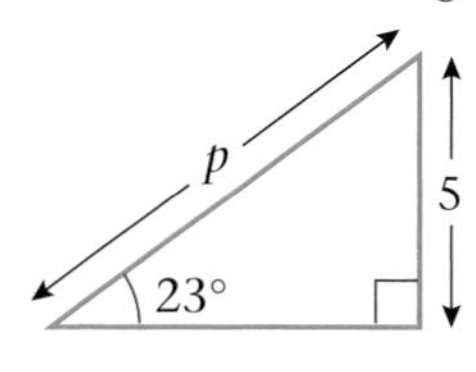

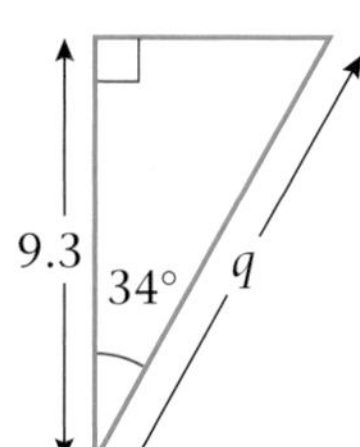

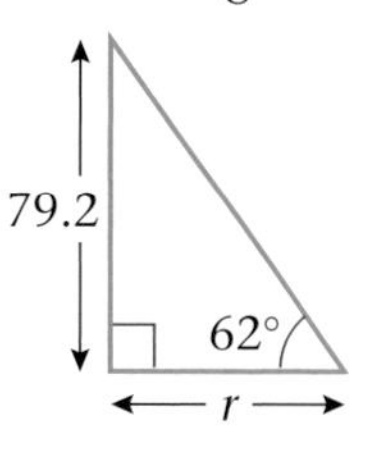

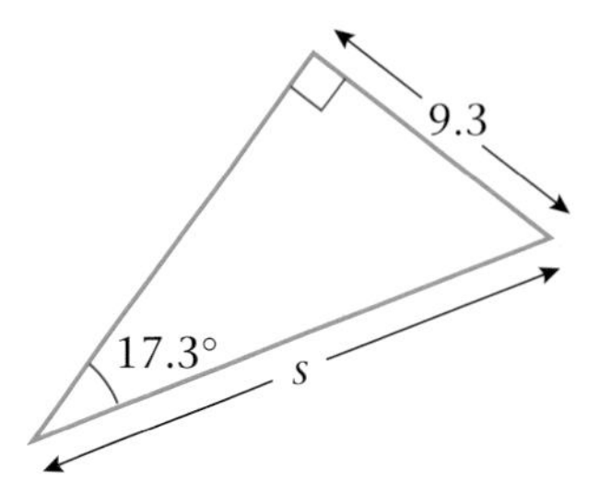

4 Work out the height AX of the triangle and angle ACB.

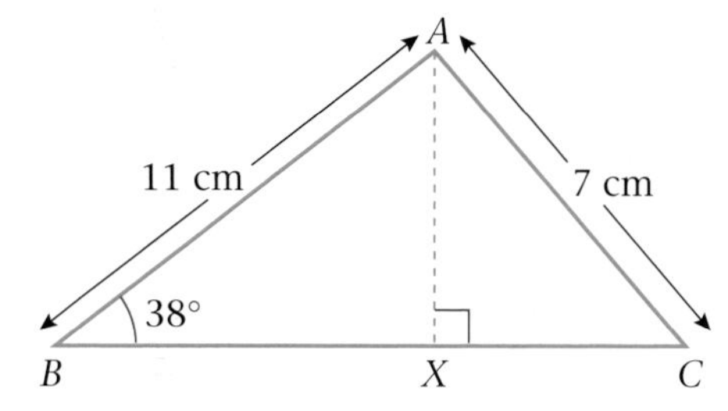

5 Two tangents, length 10 cm, are drawn from a point P to a circle, radius 6 cm. Work out the angle between the tangents.

17.5 You can use your trigonometry skills to solve real-life problems, such as those involving bearings and heights.

Example 7

A plane flies 260 km on a bearing of 297°
How far North has it flown?

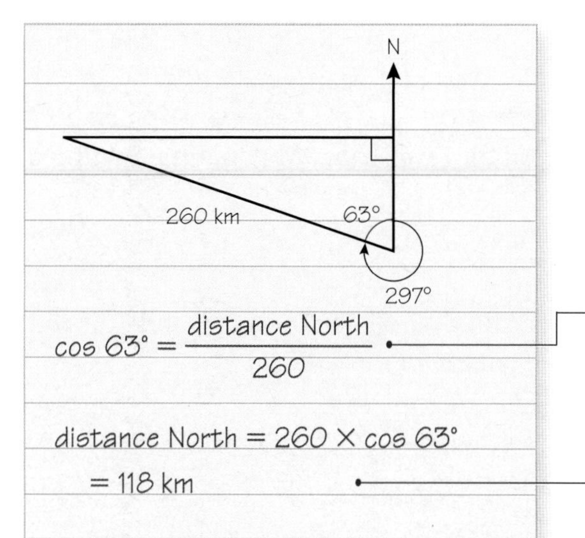

You are going to use 63°.
(360° − 297° = 63°)

The distance North is the adjacent side and 260 km is the hypotenuse, so choose the cosine formula ($\cos = \frac{\text{adj}}{\text{hyp}}$).

Answer to 3 sf.

Example 8

From the viewing platform, Angel Falls is 520 metres away.
The angle of elevation of the top of the falls is 50°.
The angle of depression of the foot of the falls is 20°.
Work out the height of Angel Falls.

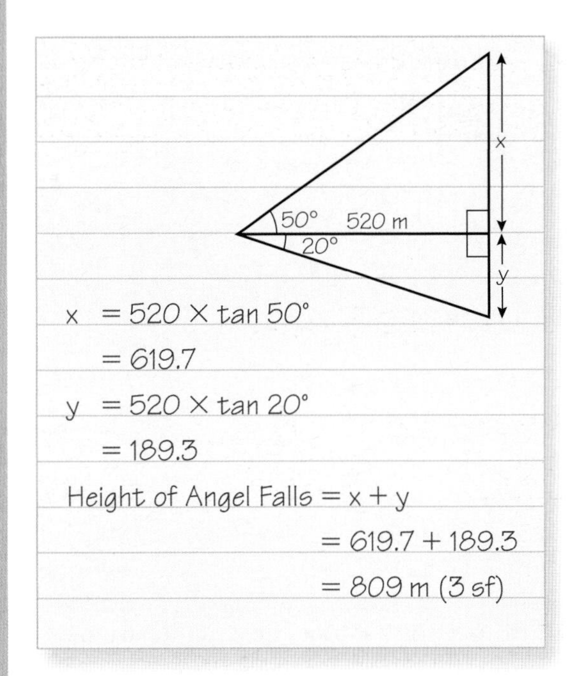

$x = 520 \times \tan 50°$
$= 619.7$
$y = 520 \times \tan 20°$
$= 189.3$
Height of Angel Falls $= x + y$
$= 619.7 + 189.3$
$= 809$ m (3 sf)

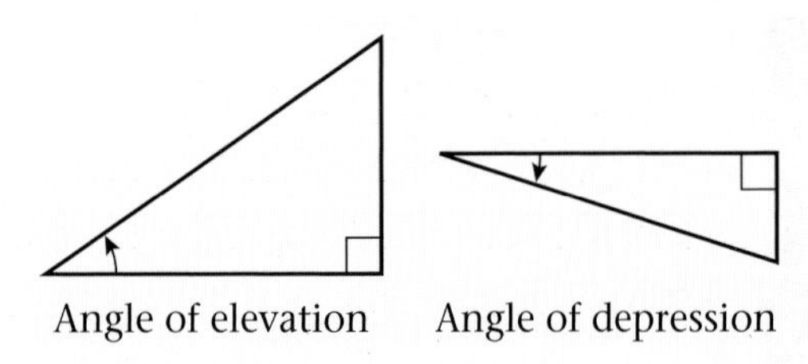

Angle of elevation   Angle of depression

Practice 17E

Hint: It often helps to draw a sketch first.

1 A ship sails on a bearing of 137° for 30 km.
Calculate how far East and South it has sailed.

2 Caerphilly is 21 km from Pontypool on a bearing of 230°.
Calculate how far West and South this is.

3 A plane is flying on a bearing of 312°.
How far has it flown when it is 35 miles West of its starting place?

4 Wickford is 26 km west and 32 km north of Faversham. Work out the bearing of
a Wickford from Faversham,
b Faversham from Wickford.

5 A man, eye height 1.65 m, stands on a cliff, 83 m high. The angle of depression of a boat out at sea is 21.5°. How far is the boat from the cliff?

6 The angle of elevation of a tower is 15.3° for an observer who is 220 metres away.
Calculate the height of the tower.
(E15)

18 Collecting and organising data

This chapter shows you how to collect and organise data so that the data is easy to work with and the resulting conclusions are as reliable as possible. **(S12)**

Organising data in a two-way table can give you more information than you started with.

Example 1

60 workers were asked about their lunch one day.

The table below provides some information about their responses.

Complete the two-way table.

	Canteen lunch	Sandwiches	Other	Total
Men	24	6		32
Women	16		4	
Total				60

The completed table is

	Canteen lunch	Sandwiches	Other	Total
Men	24	6	**2**	32
Women	16	**8**	4	**28**
Total	**40**	**14**	**2**	60

$24 + 6 + 2 = 32$
$32 + 28 = 60$
$16 + 8 + 4 = 28$
$24 + 16 = 40$

Practice 18A

1 This two-way table provides some information about the numbers of drinks sold at a snack bar.

	Small	Medium	Large	Total
Tea		40		150
Coffee	80	45	27	
Chocolate	30	45		
Total	135		135	400

Complete the table.

18.2

A data capture sheet is a blank frequency distribution upon which data can be recorded as it is collected.

Example 2

Dilara is going to carry out a survey to record information about the way students usually travel to college. Design a suitable data capture sheet she could use.

Method of travelling	Tally	Frequency
Walk		
Bus		
Car		
Cycle		
Motor cycle		
Train		
Other		

Using tally marks makes it easy to keep count. Make the fifth tally a bar through the previous 4, like this 𝍸.

The frequency is the total number of students using each method.

Practice 18B

1. Billy is going to carry out a survey to record information about the types of vehicles using a motorway car park. Draw a suitable data collection sheet that Billy could use.

18.3

It is not always possible to include a whole population in a survey, eg there may not be time to ask all the students at a college about their travel. We often need to take a sample.

- **A *population* is the complete set of items or people under consideration.**
- **A *sample* is part of the population selected for the survey.**
- **In a *random sample*, every member of the population has an equal chance of being selected each time a selection is made.**
- **A process to select a random sample must be unpredictable.**

Example 3

There are 800 GCSE students at Jordan Hill College.
As part of a survey into students' opinions of the restaurant facilities, the college principal wishes to interview a random sample of 10% of the GCSE students. Describe two methods by which the principal could select a 10% random sample.

a The principal could write the name of each student on a sheet of paper. Place the 800 pieces of paper in a hat, give the hat a good shake and then pick out 80 of the pieces of paper without looking.

b The principal could use a list of all 800 names in order and label them from 1 to 800. Use a random number generator on a scientific calculator. This will give a random number between 0 and 0.999. Multiply by 1000 and ignore numbers greater than 800.

Press SHIFT RAN#

If the calculator gives 0.647, then student numbered 647 would be chosen.

Practice 18C

1 Jenny is doing a survey of the heights and weights of the male and female students at Jordan Hill College. She takes a random sample of 60 males on the first year of the access course.

Suggest a suitable method Jenny could use to take this random sample.

18.4

When you are writing questions for a questionnaire:
- **Be clear about what you want to find out, and what data you need.**
- **Ask short, simple questions.**
- **Avoid questions which are too vague, too personal, leading or biased.**

Example 4

Lucy is undertaking a survey of the TV viewing habits of men and women, of varying ages, in her home town. Design part of the questionnaire Lucy could use.

Are you? Male ☐ Female ☐
What age are you? Under 21 ☐
21–40 ☐ 41–60 ☐ over 60 ☐
Which of these TV channels do you watch most often?
BBC1 ☐ BBC2 ☐ ITV ☐
Ch 4 ☐ Ch 5 ☐

Provide tick boxes to make answering easy.

Make sure you avoid bias by choosing people of different genders and a range of ages.

Look out for ways to improve your questions.

Practice 18D

1 As part of his statistics project, Declan is doing a survey into the types of films enjoyed most by men and women of different ages. He intends to design a questionnaire to help with the project. Design part of the questionnaire Declan could use.

2 Part of a questionnaire designed to collect students' opinions about a college restaurant facilities is shown below:

1 Do you agree that the college needs a new restaurant?

2 Are you a frequent user of the restaurant?

3 What do you think about the food served at the restaurant?

a Suggest reasons why each of these questions is not suitable.

b Write an improved question for each example. **(S13)**

19 Statistical representation

Statistical representations are diagrams and charts which communicate data in visual forms. This makes it easier to see patterns and trends, and to make comparisons.

In a pictogram a picture or symbol is used to represent a number of items.

Example 1

The incomplete pictogram drawn below provides some information about the numbers of doughnuts sold by Donna yesterday.

Plain	
Jam	
Lemon	
Chocolate	

= 8 doughnuts

a Work out the number of

i plain doughnuts sold

ii lemon doughnuts sold.

Donna sold 30 chocolate doughnuts yesterday.

b Complete the pictogram.

c Work out the total number of doughnuts sold by Donna yesterday.

a Donna sold 32 plain doughnuts and 10 lemon doughnuts.

$4 \times 8 = 32$

$1\frac{1}{4} \times 8 = 10$

b The completed pictogram is

Plain	
Jam	
Lemon	
Chocolate	

= 2 doughnuts

= 6 doughnuts

c The total number of doughnuts sold by Donna yesterday was 120.

$32 + 48 + 110 + 30 = 120$

Practice 19A

1 The pictogram shows the number of CDs sold by a shop last week.

a On which day were most CDs sold?

b Write down the numbers of CDs sold

i on Monday

ii on Tuesday.

Monday	
Tuesday	
Wednesday	
Thursday	
Friday	
Saturday	

Represents 10 CDs

19.2 Bar charts can be used to show up patterns in data. The bars may be horizontal or vertical.

Example 2

The frequency table for the numbers of doughnuts sold by Donna yesterday is given below.

Type of doughnut	Frequency
Plain	32
Jam	48
Lemon	10
Chocolate	30

Represent this information as a bar chart.

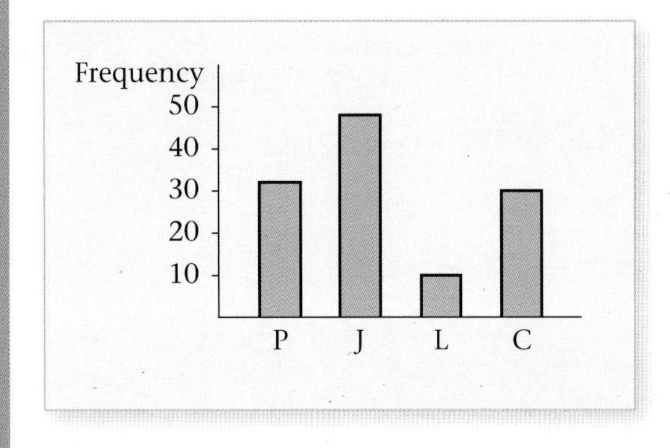

It would be equally correct to draw the bars horizontally, like this

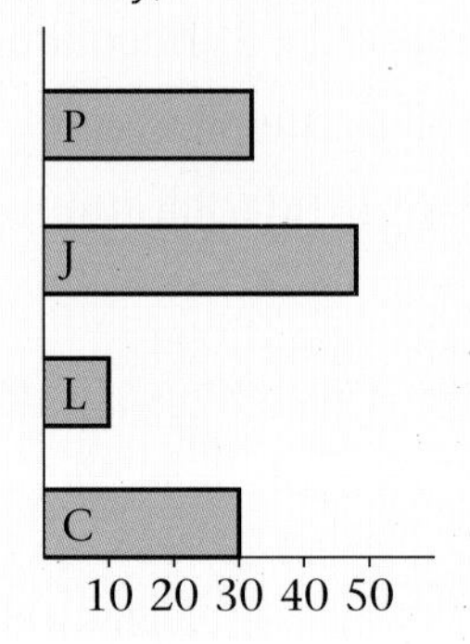

Practice 19B

1 The numbers of taxis used to bring students to Jordan Hill College for the 5 college days last week are shown in the table below.

Day	Number of taxis
Monday	8
Tuesday	5
Wednesday	4
Thursday	1
Friday	3

Draw a bar chart to show the information in the table.

19.3 A pie chart is a way of displaying data that shows how something is shared or divided.

Example 3

Represent the information given in Example 2 as a pie chart.

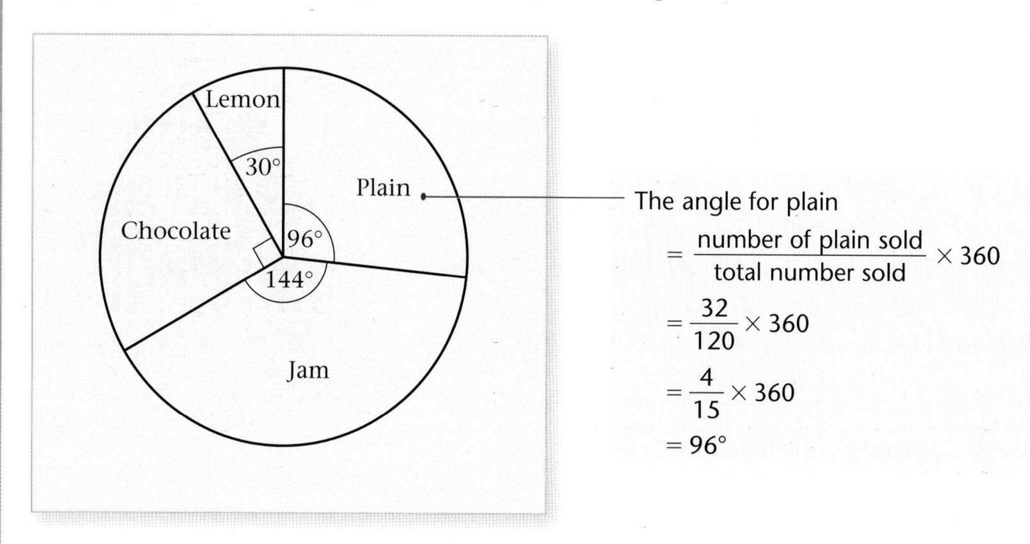

$$= \frac{\text{number of plain sold}}{\text{total number sold}} \times 360$$

$$= \frac{32}{120} \times 360$$

$$= \frac{4}{15} \times 360$$

$$= 96°$$

Practice 19C

1 The table gives information about the favourite subjects of 90 students.

Subject	Frequency	Angle
Art	18	
English	36	
Science		
Mathematics		

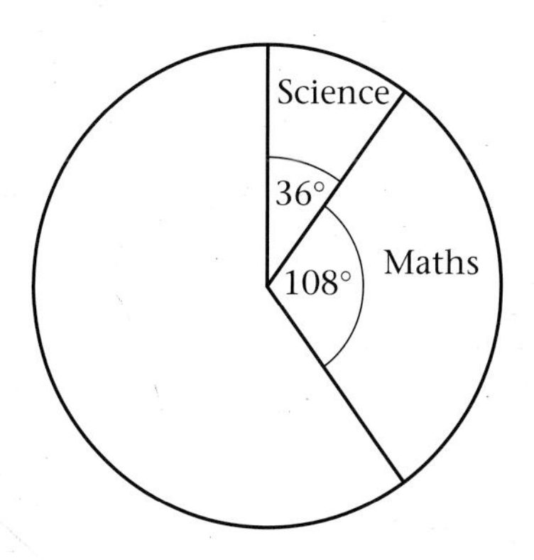

a Use the information in the table to complete the pie chart.

b Use the information in the pie chart to complete the table. **(S14)**

19.4 A dual bar chart is a method of displaying two sets of data so that a direct comparison can be made.

Example 4

The dual bar chart shows some information about the average daily maximum temperatures in London and the Costa del Sol during the summer months.

a Write down the average August temperature in
  i London
  ii the Costa del Sol.
  Give your answers in °C.

b During which month is the **difference** between the average daily maximum temperatures in London and the Costa del Sol at its lowest?

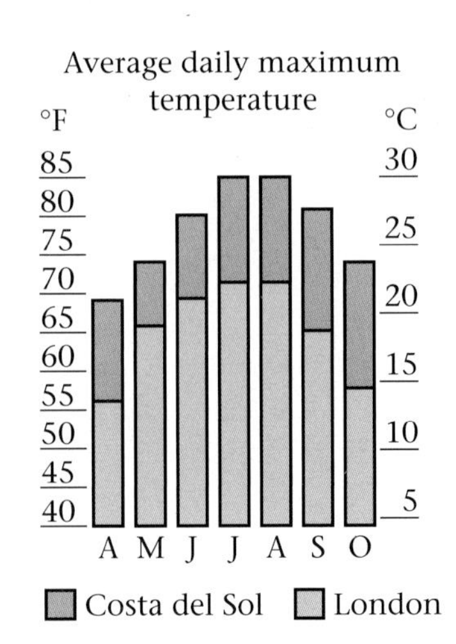

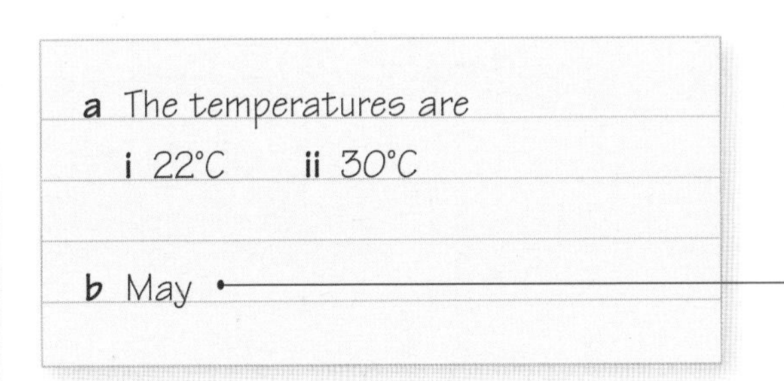

Look for the month that has the shortest distance between the top of the blue bar and the top of the red bar.

Practice 19D

1 The dual bar chart provides information about the average temperatures in London and Lanzarote for the months from October to April.

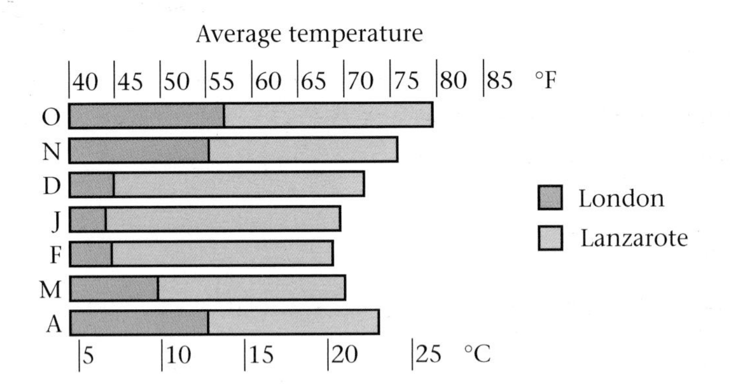

a Write down the average temperature, in °C, in Lanzarote during November.

b Write down the average temperature, in °F, in London during March.

c During which of the months from October to April is the difference between the average temperature in Lanzarote and London at its greatest?

19.5 Line graphs can be used to show continuous data.

Example 5

The table provides information about the average mid-day temperatures in Malta over a year.

Month	J	F	M	A	M	J	J	A	S	O	N	D
Temperature (°C)	15	17	18	20	21	24	28	29	27	23	20	17

Represent this data as a line graph.

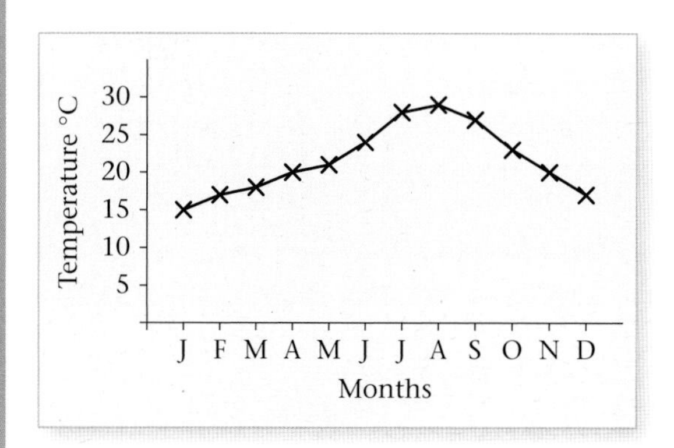

Line graphs can show trends over a period of time. The lines between the points help to show the trend, but don't always relate to actual observations.

Practice 19E

1 The table below provides information about the average hours of sunshine per day in Cyprus for each month of the year.

Month	J	F	M	A	M	J	J	A	S	O	N	D
Hours	6	7	8	9	11	13	13	12	11	9	7	5

Represent this data as a line graph.

19.6 A stem and leaf diagram retains the detail of the data and gives an idea of how the values are distributed.

Example 6

The ages of the forty members of an art class are given below.

22	34	40	19	9	54	56	63	66	13
31	54	33	41	48	20	10	27	44	51
36	28	29	30	26	25	24	58	29	42
62	27	8	17	37	29	36	41	51	46

a Represent this data as a stem and leaf diagram.

b State the range of the ages.

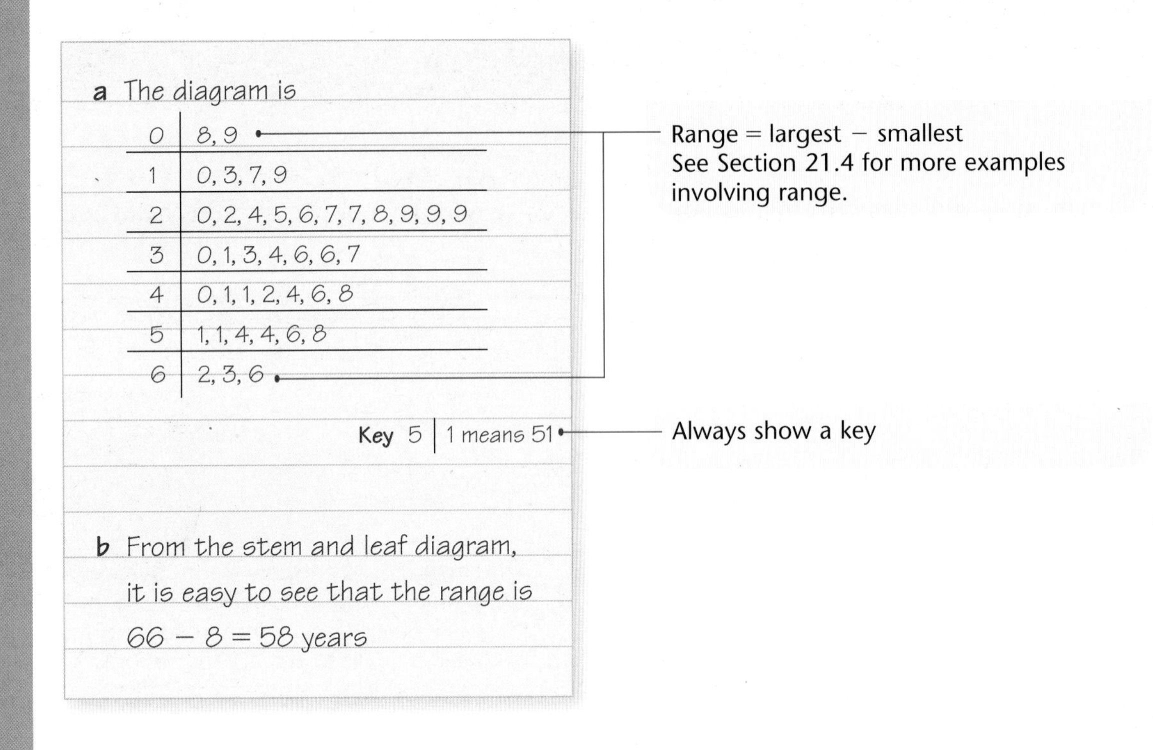

Practice 19F

1 There are 30 students in Mrs Bennett's mathematics class. Last week they took a mock examination. Their marks were

64	38	41	62	81	75	66	39	53	55
40	45	58	65	49	34	70	62	54	53
64	84	77	53	57	56	42	57	48	57

a Represent this information as a stem and leaf diagram.

b Work out the range of the marks.

2 Jack finds out the ages of 28 members of his family:

65	68	11	7	3	77	16
36	37	6	14	32	71	44
42	10	8	39	10	15	41
14	15	7	14	41	40	42

a Display this data in a stem and leaf diagram.

b Work out the range of the ages.

20 Scatter diagrams

This chapter covers the use of scatter diagrams to show whether two sets of data are related. You will also learn how to distinguish between positive, negative and zero correlation. **(S15)**

You can draw a scatter diagram to show whether two sets of data are related.

Example 1

The table below shows the average daily temperature and the maximum daily hours of sunshine in ten European holiday resorts.

Temperature (°C)	25	20	22	16	32	26	15	24	19	30
Sunshine (hours)	8	7	6	5	10	8.5	6	7.5	6.5	9.5

a Plot the data on a scatter diagram.
b Describe the relationship between the temperature and the hours of sunshine.

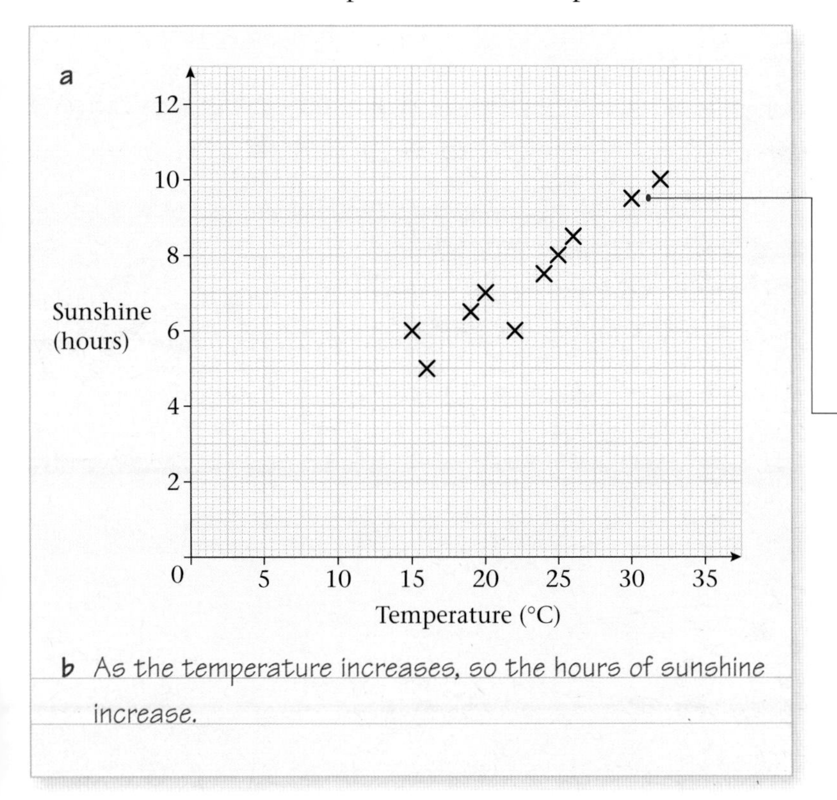

Plot temperature along the horizontal axis, and hours of sunshine along the vertical axis. Then plot the points. This point is (30, 9.5).

b As the temperature increases, so the hours of sunshine increase.

Practice 20A

1 The table shows the test results of a class of twelve pupils at a summer school.

French	24	20	28	17	37	18	30	40	9	24	27	15
German	33	36	34	20	38	13	23	45	25	26	38	30

Plot the data on a scatter diagram.

20.2

If the points on a scatter diagram lie almost along a straight line then there is correlation between the two sets of data. There are two different types of correlation: positive and negative.
If a straight line of best fit can't be drawn, there is no correlation.

Example 2

Comment on the relationship shown between x and y, s and t, and p and q in the scatter diagrams below.

a y against x **b** t against s **c** q against p

a Negative correlation — As x values increase, y values decrease.

b No correlation — The s and t values seem to be scattered about at random.

c Positive correlation — As p values increase, q values increase.

Example 3

Comment on the relationship between x and y.

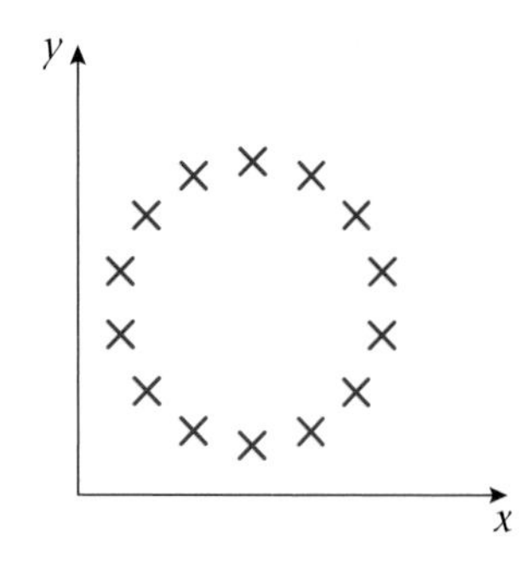

The points show no correlation but are clearly related because they all lie on the circumference of a circle.

No correlation between two sets of data does not necessarily mean they aren't related.

Practice 20B

1 Write down the type of correlation between the German and French test results in question 1 of Practice 20A.

2 Use one of the words **positive**, **negative** or **none** to describe the correlation on each of the scatter diagrams below:

a

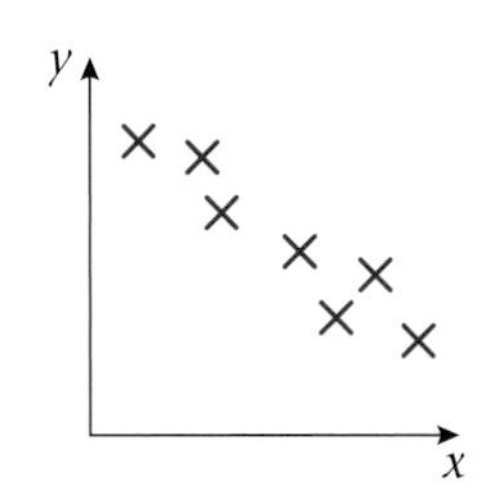

b

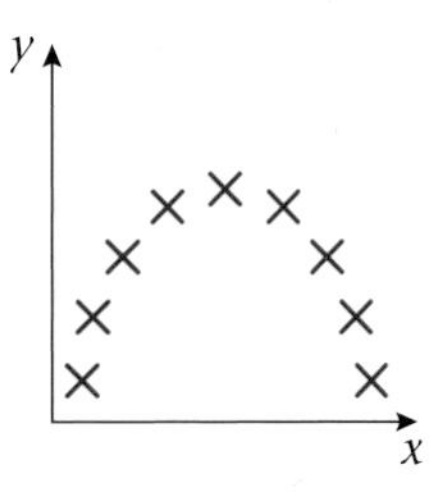

c

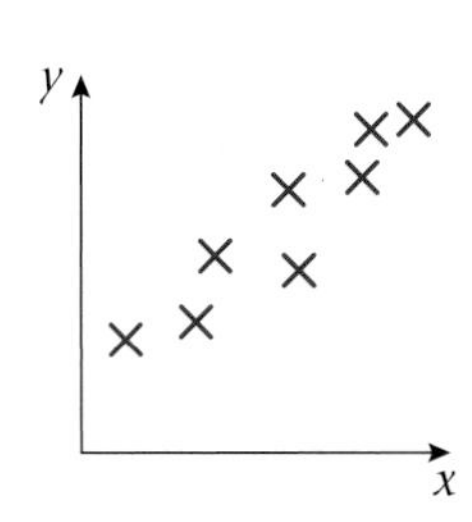

d

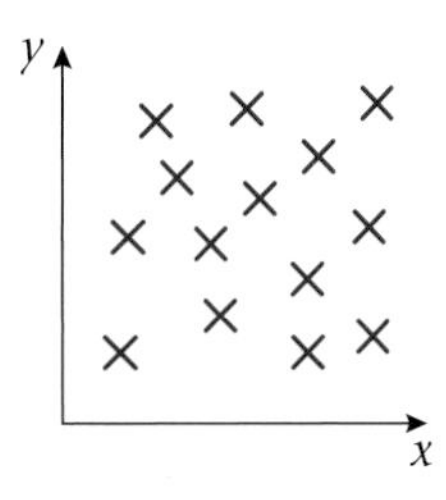

3 **a** Here is a scatter graph.
One axis is labelled 'weight'.

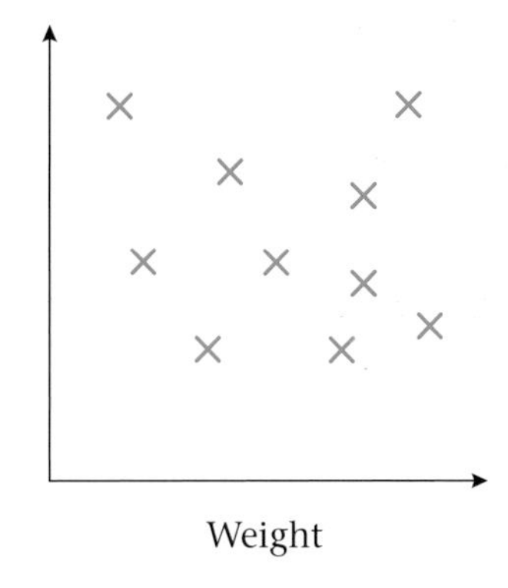

i For this graph state the type of correlation.
ii From this list choose an appropriate label for the other axis.
shoe size, length of hair, height, hat size, length of arm

b Here is another scatter graph with one axis labelled 'weight'.

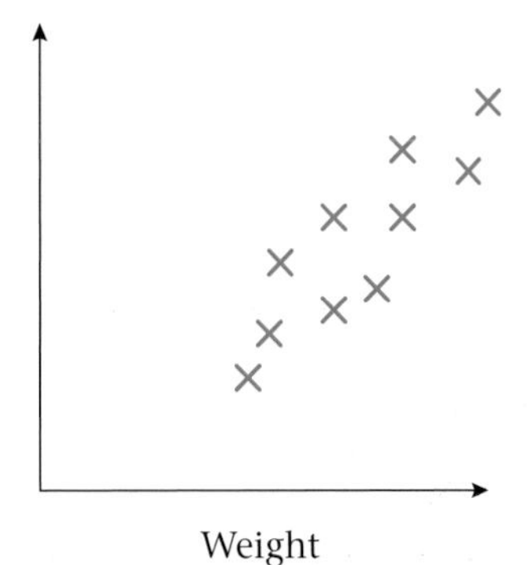

i For this graph state the type of correlation.
ii From this list choose an appropriate label for the other axis.
shoe size, distance around neck, waist measurement, GCSE Maths mark

E

20.3

A line which passes as close as possible to all the points on a scatter diagram is called a line of best fit.

Example 4

Using the data given in Example 1

a Draw the line of best fit on the scatter diagram.

b Use the line of best fit to estimate

i the hours of sunshine when the temperature is 19°C.

ii the temperature when there are 9 hours of sunshine.

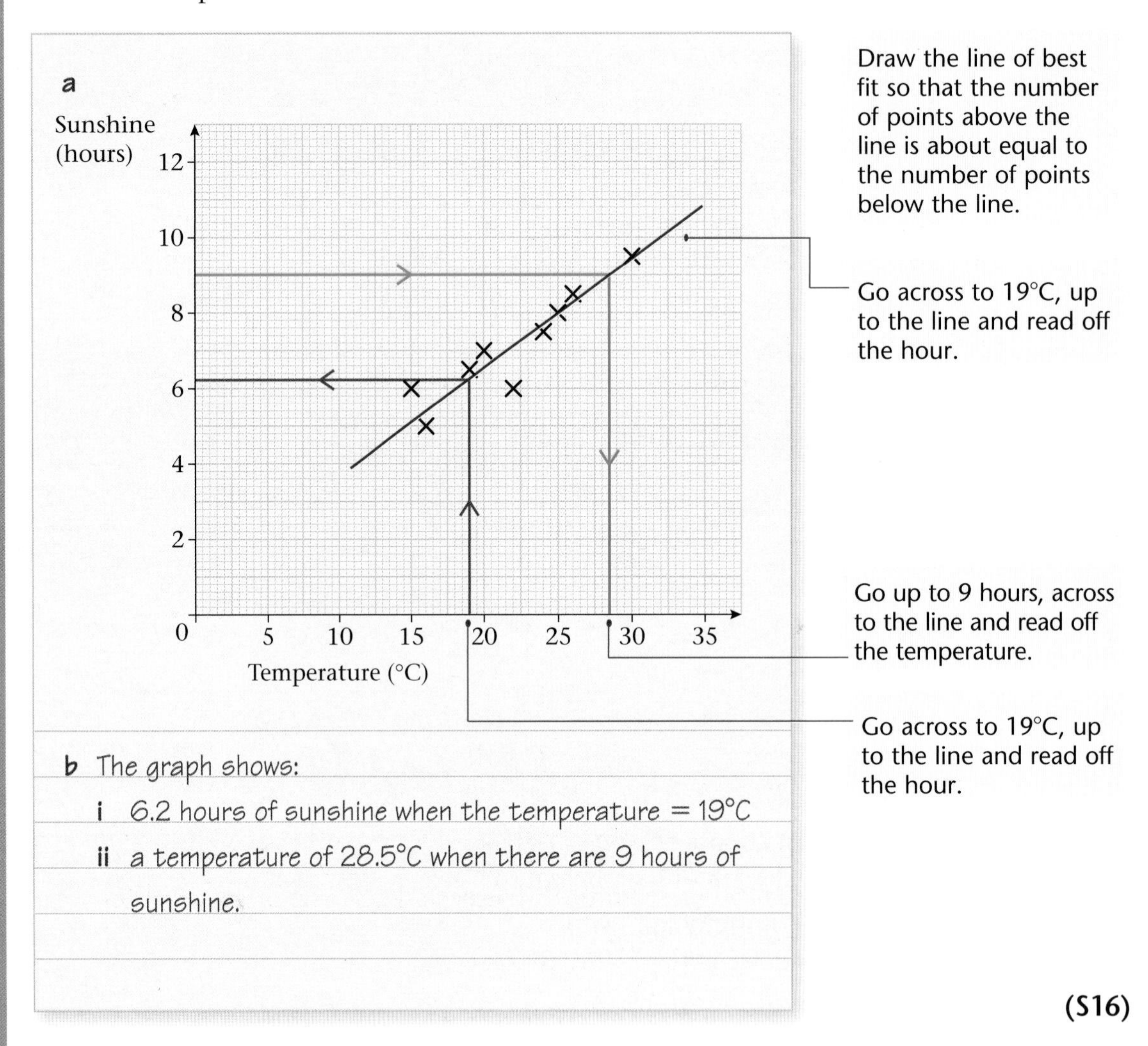

(S16)

Practice 20C

1 The table shows the hours of sunshine and rainfall, in mm, in ten towns last summer:

Sunshine (hours)	630	455	560	430	620	440	640	375	520	620
Rainfall (mm)	10	20	15	29	24	28	14	30	25	20

a Plot this information as points on a scatter diagram.

b Describe the relationship between the hours of sunshine and the rainfall.

c Draw a line of best fit on your scatter diagram.

d Use your line of best fit to estimate

i the rainfall when there are 450 hours of sunshine

ii the amount of sunshine when there are 18 mm of rainfall.

2 As part of her statistics coursework, Erica has some information about the ages and prices of second hand motorcycles.

Age (years)	3	1	4	5	8	2	6	3.5	1.5
Price (£1000s)	4.4	6.8	3.2	4.0	1.5	6.4	3.4	4.6	7.0

a Plot this information on a scatter diagram.

b Comment on the relationship between age and price of these motorcycles.

c Draw a line of best fit on the scatter diagram.

d Use your line of best fit to estimate

i the value of a motorcycle aged 7 years

ii the likely age of a motorcycle costing £5000.

3 The scatter graph shows information about eight countries.
For each country, it shows the birth rate and the life expectancy, in years.

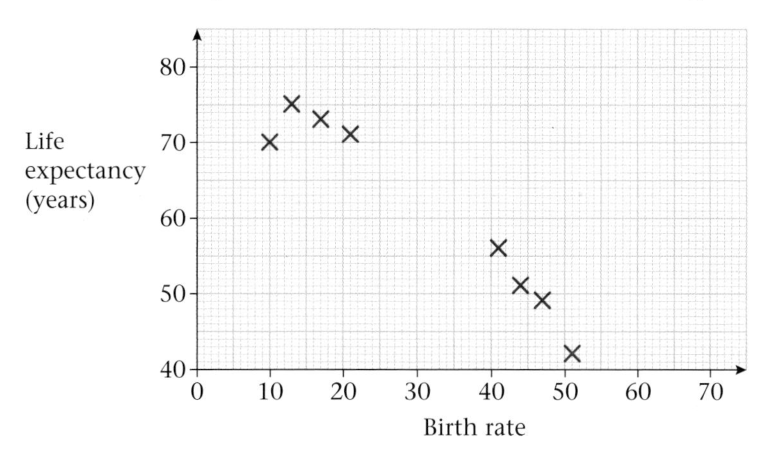

The table shows the birth rate and the life expectancy for six more countries.

Birth rate	25	28	30	31	34	38
Life expectancy (years)	68	65	62	61	65	61

a On the scatter graph, plot the information from the table.

b Describe the relationship between the birth rate and the life expectancy.

c Draw a line of best fit on the scatter graph.

The birth rate in a country is 42.

d Use your line of best fit to estimate the life expectancy in that country.

The life expectancy in a different country is 66 years.

e Use your line of best fit to estimate the birth rate in that country.

E

4 The table shows the number of units of electricity used in heating a house on ten different days and the average temperature for each day.

Average temperature (°C)	6	2	0	6	3	5	10	8	9	12
Units of electricity used	28	38	41	34	31	31	22	25	23	22

a Complete the scatter graph to show the information in the table.
The first 6 points have been plotted for you.

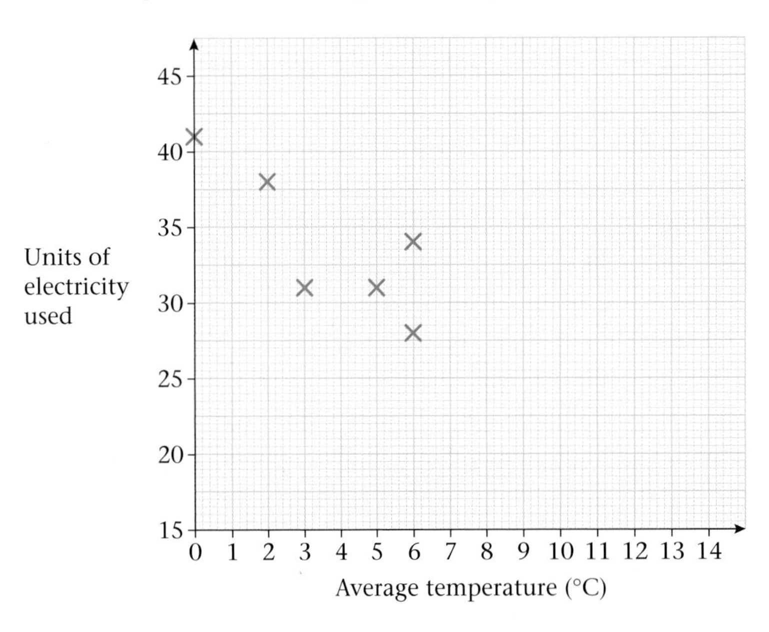

b Describe the **correlation** between the number of units of electricity used and the average temperature.

c Draw a line of best fit on your scatter graph.

d Use your line of best fit to estimate

i the average temperature if 35 units of electricity are used,

ii the units of electricity used if the average temperature is 7°C.

E

20.4 Sometimes a point on a scatter diagram clearly does not follow the general trend.

Example 5

The scatter diagram has been drawn for the prices and ages of the 12 cars on sale at a car showroom.

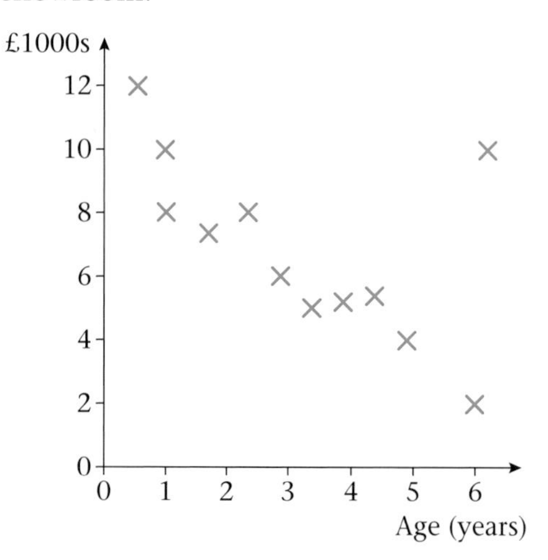

The exception is the oldest car which has a high price

Comment on the relationship between the prices and ages of these cars.

The scatter diagram shows negative correlation but there is one very clear exception to this general trend.

Practice 20D

1 Whilst working on a statistical project, Jon draws a scatter diagram showing information about the square metres of floor space and values of a sample of houses in his area.

Jon's scatter diagram is drawn below:

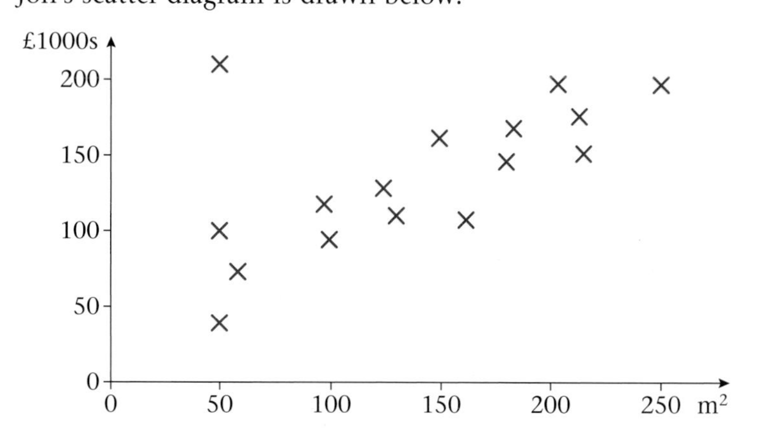

Comment on the relationship between floor space and house value.

2 The table below provides information about the heights (in cm) and the mass (in kg) of a random sample of 10 students at Jordan Hill High School.

Height (cm)	173	146	173	176	165	155	123	175	163	170
Mass (kg)	64	33	89	70	51	43	30	55	60	62

a Plot the information as points on a scatter diagram.

b Draw a line of best fit on your scatter diagram.

c Comment on the relationship between the heights and masses of these students.

d Use your line of best fit to estimate:

i the mass of a student who has a height of 168 cm

ii the height of a student who has a mass of 65 kg.

3 The table below provides information about the ages (in years) and value (in £'000s) of 12 cars for sale at Highfield Motors.

Age	3	$2\frac{1}{2}$	5	7	1	4	12	$1\frac{1}{2}$	2	4	3	$3\frac{1}{2}$
Value	7.0	8	4.1	2.6	12.0	3.8	23	10.5	7.3	4.4	4.7	5.2

a Plot the information as a scatter diagram.

b Write down the coordinates of the exceptional point.

c Draw a line of best fit on your scatter diagram.

d Comment on the correlation between age and value.

e Give an estimate for

i the value of a 6 year old car

ii the age of a car valued at £6000. **(E16)**

21 Averages and range

There are three averages used in mathematics. These are the mean, mode and median. The range is a measure of spread.

The mean of a set of data is the sum of the values divided by the number of values.

$$\text{Mean} = \frac{\text{sum of values}}{\text{number of values}} = \frac{\Sigma x}{\Sigma f}$$

The mean in symbols is written as $\bar{x}$. The symbol Σ means 'sum of all'.

Example 1

Sally did ten homeworks for science. Her marks were

6, 3, 8, 7, 5, 5, 6, 7, 8, 9

Work out the mean of these marks.

$$\text{Mean} = \frac{6+3+8+7+5+5+6+7+8+9}{10}$$

$$= \frac{64}{10} = 6.4$$

$\Sigma x = 64$
There are 10 values, so $\Sigma f = 10$

Practice 21A

1 There are 20 houses in Streetfield Close.
The number of letters delivered to each house last Monday were
3, 4, 1, 0, 3, 5, 1, 1, 3, 7, 2, 1, 0, 4, 2, 1, 1, 2, 3, 8

Work out the mean number of letters delivered on the Monday.

The mode of a set of data is the value which occurs most often.

Example 2

Find the mode for Sally's science homework marks.

Setting the marks in order gives
3, 5, 6, 6, 6, 7, 7, 8, 8, 9
So the mode is 6.

6 occurs most often.

Practice 21B

1 Use the data from Practice 21A to work out the mode.

21.3 The median is the middle value when the data is arranged in order of size.

Example 3

Find the median of Sally's science homework marks.

Putting the marks in order gives

3, 5, 6, 6, 6, 7, 7, 8, 8, 9

So the median is 6.5.

There are two middle numbers, 6 and 7.
So the meidan is $\frac{6+7}{2} = \frac{13}{2} = 6.5$

Practice 21C

1 Use the data from Practice 21A to work out the median.

21.4 The range of a set of data is the difference between the highest and lowest values.

Range = highest value − lowest value

Example 4

Work out the range of Sally's science homework marks.

Range = highest mark − lowest mark
= 9 − 3 = 6

The highest value is 9 and the lowest value is 3.

Practice 21D

1 Use the data from Practice 21A to work out the range.

2 The number of goals scored by Beckstown during the first 10 games of the season was 1, 2, 0, 3, 2, 1, 2, 4, 2, 5

a Find the mode of the number of goals scored.

b Work out the range of the number of goals scored.

c Find the median of the number of goals scored.

d Work out the mean number of goals scored by Beckstown over the 10 games.

(S17)

21.5

- **For frequency distributions, the mean is**
  $\bar{x} = \frac{\Sigma fx}{\Sigma f}$, **where Σfx is the sum of the frequencies x values, and Σf is the sum of the frequencies.**
- **The mode is the value with the highest frequency.**
- **The median is the value half way into the distribution.**

Example 5

The ages of the 50 members of Melissa's aerobics club are given in the frequency table below.

Age	20	21	22	23	24	25	26	27	28	29	30
Frequency	3	5	4	8	3	6	7	4	4	4	2

Work out the mean age of these members.

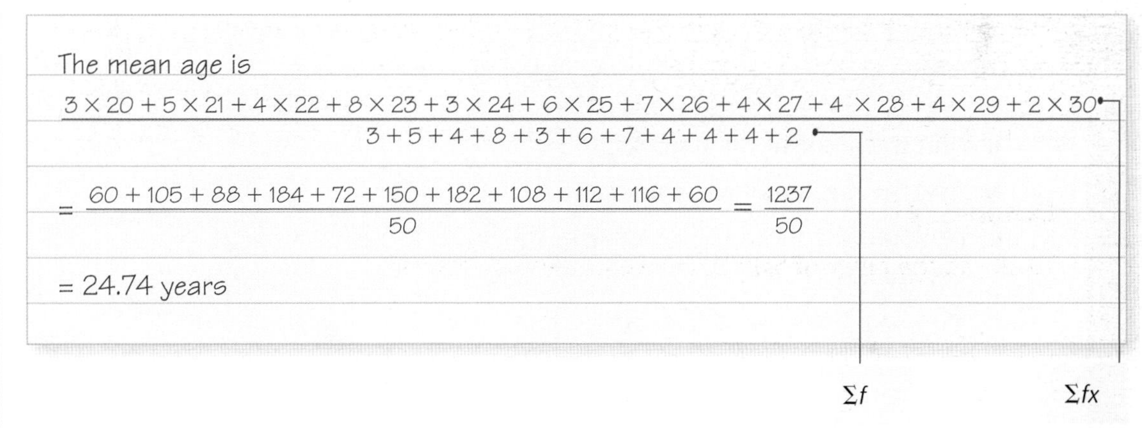

Σf Σfx

Example 6

Find the mode of the ages of the members of Melissa's aerobics class.

The mode is 23 years.

23 has the highest frequency, as there are 8 members aged 23.

Example 7

Find the median age of the members of Melissa's aerobics class.

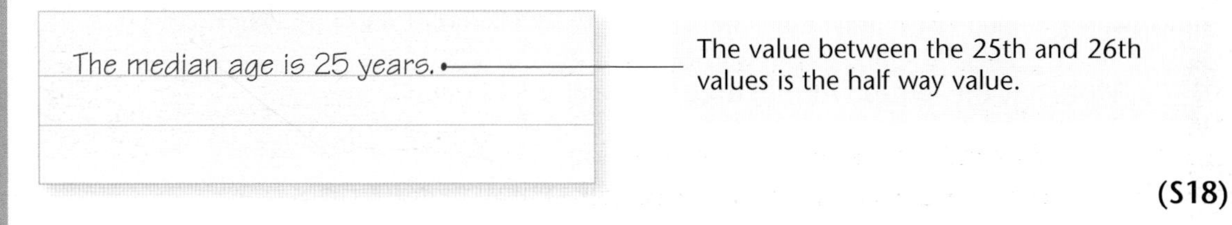

The value between the 25th and 26th values is the half way value.

(S18)

Practice 21E

1 During a science project, Jenny recorded information about the number of eggs in 30 bird's nests.
Her data is shown in the frequency table below:

Number of eggs	0	1	2	3	4	5	6
Number of nests	4	3	7	5	5	4	2

a Work out the mean number of eggs per nest.

b Find the modal number of eggs in a nest.

c Find the median number of eggs in a nest.

2 There are 26 students in a GCSE class.
Last week they took a test with a maximum mark of 10.
The distribution of marks is given in the table below:

Mark	3	4	5	6	7	8	9
Frequency	2	5	6	8	2	1	2

a Find the mode of these marks.

b Find the median of these marks.

c Work out the mean of these marks.

21.6

A moving average is calculated by taking a sequence of results and finding the average. You then move the sequence on by using the next unused value to replace the first value in the sequence.

Example 7

The table below shows the quarterly central heating costs, in £s, at Lucy's house for the period from 1999 to 2001.

Year	1999				2000				2001			
Quarter	1st	2nd	3rd	4th	1st	2nd	3rd	4th	1st	2nd	3rd	4th
Cost (£)	302	246	122	205	330	266	131	225	351	290	140	248

a Work out the four-point moving averages for this data.

b Plot the moving averages against time.

c Make three comments about trends in the central heating costs.

a The first moving average is

$$\frac{302 + 246 + 122 + 205}{4} = 218.75$$

Take the first four readings, as you are asked for a four-point average.

The second moving average is

$$\frac{246 + 122 + 205 + 330}{4} = 225.75$$

To find the second moving average, remove the first reading, but add on the fifth.

The third moving average is

$$\frac{122 + 205 + 330 + 266}{4} = 230.75$$

The other moving averages are

233, 238, 243.25, 249.25, 251.5, 257.25

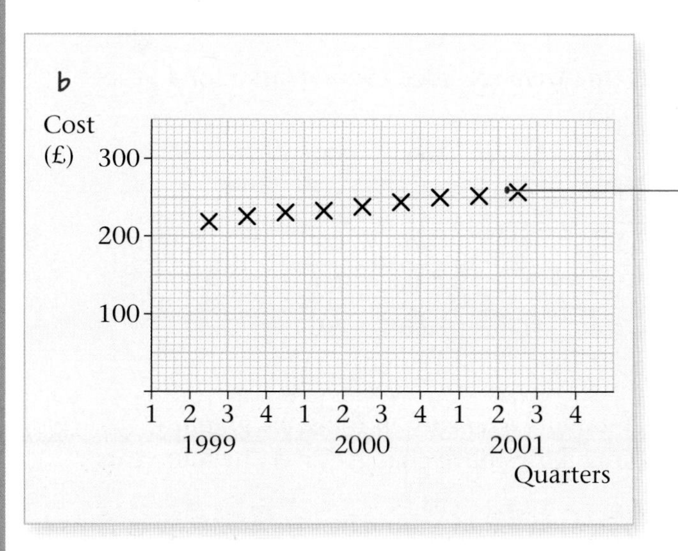

Start plotting points halfway between the 2nd and 3rd quarters for 1999.

c **i** The cost is at its highest in the 1st quarter.

ii The cost is at its lowest in the 3rd quarter.

These trends can be seen from the raw data.

iii There has been a steady increase in the central heating costs over the period of time.

This can be seen from the graph.

(S19)

Practice 21F

1 Mr Allis works as a sales representative.
His monthly petrol bill for last year is recorded below:

Month	Jan	Feb	Mar	Apr	May	Jun	Jul	Aug	Sept	Oct	Nov	Dec
Cost (£s)	280	276	302	186	283	304	294	38	288	188	302	230

a Work out the six-point moving averages for this set of data.
b The petrol bill for August was much lower than that for any other month. Suggest a reason why this might be so.

2 As part of a statistical project, Gemma recorded the amount spent on the telephone by her family. The quarterly telephone bills for the period from 1999 to 2001 are given below.

	1999				2000				2001			
Quarter	1st	2nd	3rd	4th	1st	2nd	3rd	4th	1st	2nd	3rd	4th
Cost	80	120	162	54	88	132	176	56	93	137	183	63

a Work out the four-point moving averages for the telephone costs.
b Plot the moving averages as a graph.
c Make three comments about the telephone costs.

Hint: Look at Example 7, page 179.

3 The takings, in £1000s, for a cinema were recorded every quarter for 4 years.
The results appear in the table below.

	1998	1999	2000	2001
1st quarter	178	201	208	222
2nd quarter	140	154	162	184
3rd quarter	125	132	138	150
4th quarter	166	176	188	200

a Work out the four-point moving averages for these takings.
b Plot the original data and the moving averages on the same graph.
c Comment on how the cinema's takings have changed over the four years.

4 Jon is paid a bonus twice a year, in March and September. His bonus payments (in £) from 1996 to 2002 were:

	March	September
1996	25 000	8 000
1997	27 200	9 100
1998	28 100	9 300
1999	29 200	10 400
2000	31 000	10 700
2001	33 500	11 000
2002	35 200	11 300

a Plot this information as a time series.
b Work out the two-point moving averages.
c Plot the moving averages on the same graph as the time series.
d Comment on how the bonus payments have changed over the years. **(E17)**

22 Modal class, frequency polygon and mean

In this chapter you will learn about dealing with data which has been presented in a grouped frequency table. In particular you will learn how to find the mean of the data, the modal class interval and how to represent the data as a frequency polygon. **(S20)**

Grouped data which is represented in a frequency table can also be represented as a histogram with equal lengths of class intervals. The class interval with the largest frequency is the modal class interval.

Example 1

A bag contains 100 potatoes.
Information about the distribution of the masses of these potatoes is given in the table opposite.

a Represent this data as a histogram.
b Write down the modal class interval.

Mass m (g)	Frequency
$0 < m \leqslant 100$	8
$100 < m \leqslant 200$	32
$200 < m \leqslant 300$	37
$300 < m \leqslant 400$	16
$400 < m \leqslant 500$	7

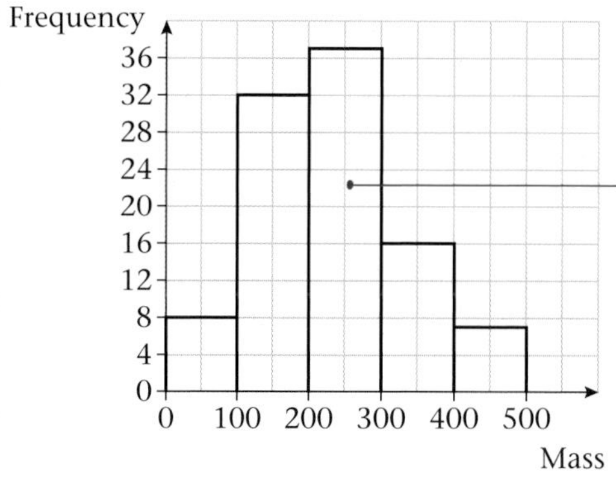

A histogram is different from a bar chart, because the data is continuous, so there can't be gaps between bars.

The modal class is the class with the largest frequency.

b The modal class is $200 < m \leqslant 300$

Practice 22A

1 At a supermarket, the manager recorded the lengths of time that 80 customers had to wait in the check-out queue.

The waiting times are grouped in the frequency table.

a Represent this data as a histogram.
b Write down the modal class interval.

Waiting time (t seconds)	Frequency
$0 \leqslant t < 50$	5
$50 \leqslant t < 100$	8
$100 \leqslant t < 150$	10
$150 \leqslant t < 200$	16
$200 \leqslant t < 250$	28
$250 \leqslant t < 300$	13

22.2

Frequency polygons can show the general pattern of data represented by histograms. A histogram can be converted into a frequency polygon by joining the mid-points of the tops of the bars with straight lines.

Example 2

Represent the data in Example 1 as a frequency polygon.

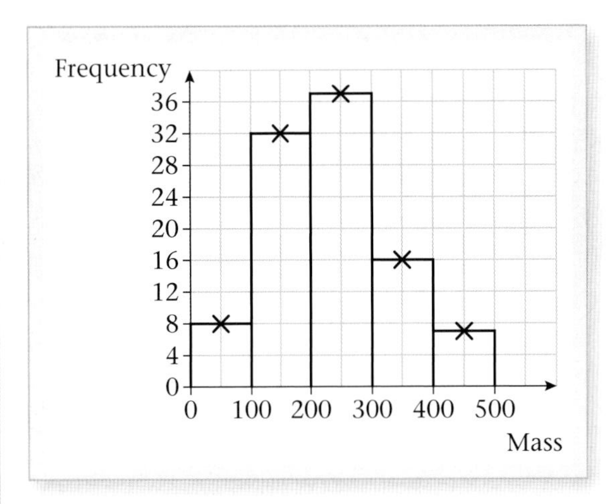

Mark the mid-points of the top of the bars.

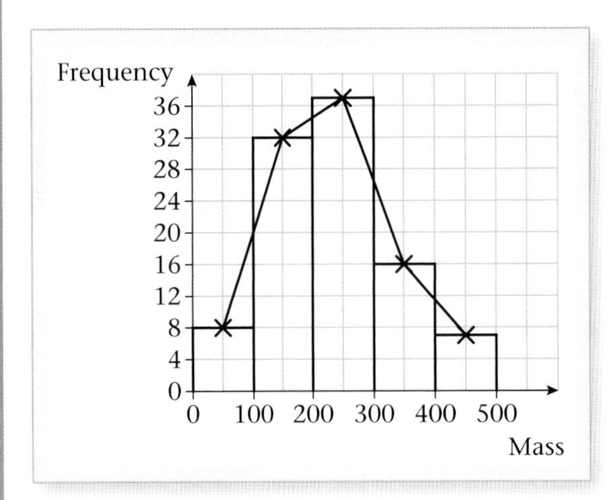

Join up the mid-points with straight lines.

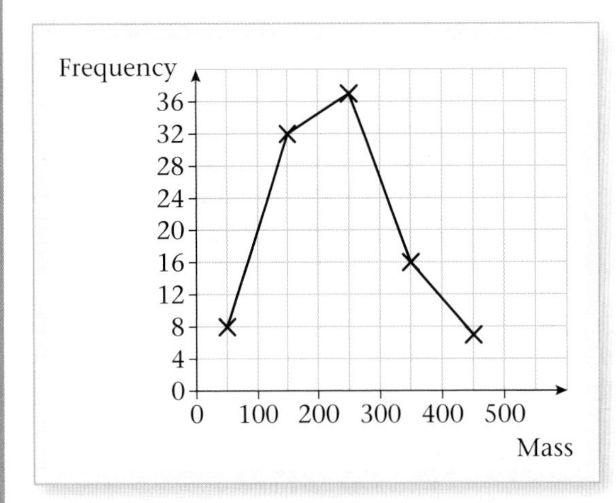

Finally remove the markings from the histogram.

Practice 22B

1 The masses, in grams, of 100 potatoes are shown in the table below

Mass (m) grams	Frequency
$0 < m \leqslant 20$	1
$20 < m \leqslant 40$	17
$40 < m \leqslant 60$	28
$60 < m \leqslant 80$	25
$80 < m \leqslant 100$	19
$100 < m \leqslant 120$	10

Represent this data as

a a histogram **b** a frequency polygon.

2 The diagram represents a histogram for the miles covered by a group of company cars.

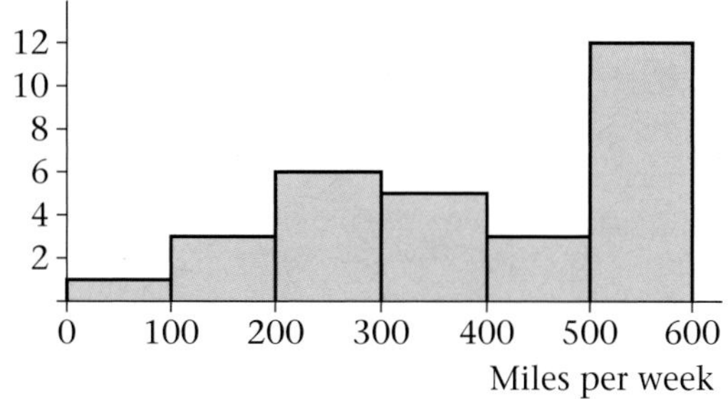

Draw the frequency polygon for this data.

22.3 Frequency polygons can be used to compare sets of data.

Example 8

The two frequency polygons provide information about house prices from a random sample of 160 houses in the South and North.

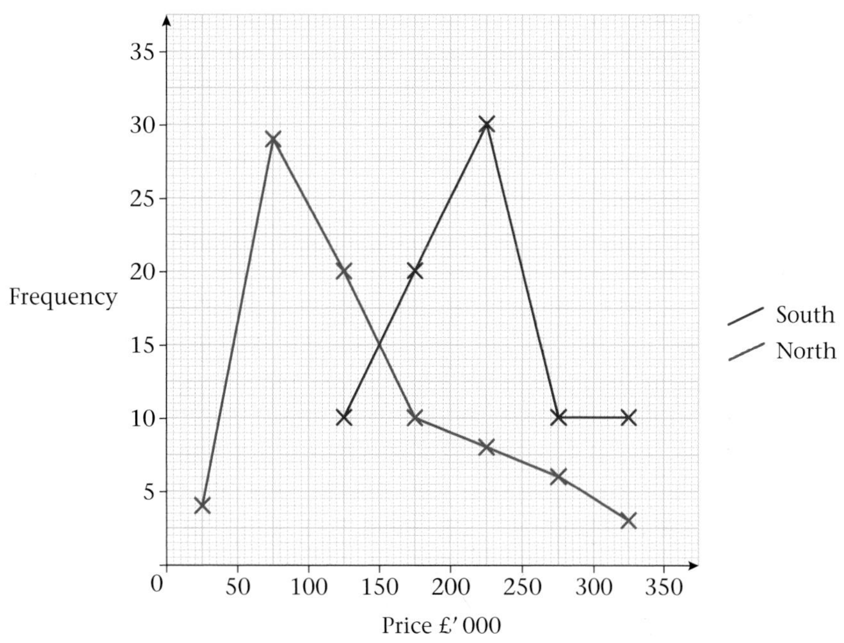

Based on this information, make 3 comments comparing the house prices in the South and North.

a The modal class interval in the South is £200 000 to £250 000 whilst in the North it is £50 000 to £100 000.

The modal class is the class interval with the highest frequency.

b In the South there are no houses with a price below £100 000, whilst in the North there are 4 houses for sale with a price below £50 000.

This can be read directly off the graph.

c The graphs clearly show that overall the prices in the South are much greater than those in the North.

Practice 22C

1 As part of a statistics project, Emma recorded the masses of 80 men and 80 women. She then represented the information on the frequency polygons below.

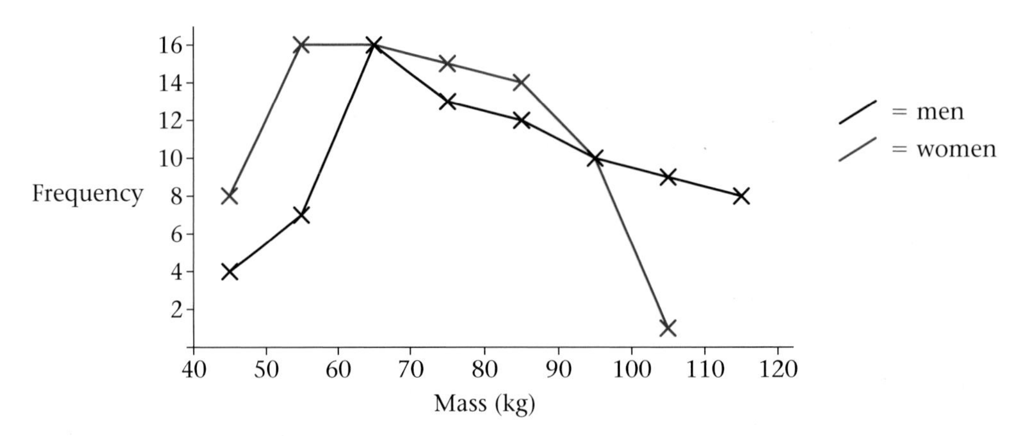

Make three comments, based on these frequency polygons, comparing the masses of the men and women.

22.4 An estimate for the mean of the data can be found by assuming that each value in a class interval takes on the value of the mid-point of the interval, and then using the formula

$$\text{Mean} = \frac{\Sigma(f \times \text{mid-point})}{\Sigma f}$$

Σ means 'sum of' and f represents the frequency.

Example 3

Work out an estimate for the mean mass of the potatoes in Example 1.

Mass m (g)	Frequency f	$f \times$ mid-point
$0 < m \leqslant 100$	8	$8 \times 50 = 400$
$100 < m \leqslant 200$	32	$32 \times 150 = 4800$
$200 < m \leqslant 300$	37	$37 \times 250 = 9250$
$300 < m \leqslant 400$	16	$16 \times 350 = 5600$
$400 < m \leqslant 500$	7	$7 \times 450 = 3150$
		Total = 23 200

$\text{Estimated mean} = \frac{23\,200}{100}$ ← $\Sigma(f \times \text{mid-point})$, Σf

$= 232\text{ g}$

Check that your answer is sensible. It is often (but not always) within the modal class.

Practice 22D

1 Sybil weighed some pieces of cheese.
The table gives information about her results.

Weight w (grams)	Frequency
$90 < w \leqslant 94$	1
$94 < w \leqslant 98$	2
$98 < w \leqslant 102$	6
$102 < w \leqslant 106$	1

Work out an estimate of the mean weight.

E

2 The table below shows the frequency distribution of the ages of the 40 members of Karen's aerobics class.

Age a (years)	Frequency
$10 < a \leqslant 20$	5
$20 < a \leqslant 30$	12
$30 < a \leqslant 40$	8
$40 < a \leqslant 50$	10
$50 < a \leqslant 60$	3
$60 < a \leqslant 70$	2

a Write down the modal class interval.
b Represent the data as a frequency polygon.
c Work out an estimate of the mean age of the members.

3 The grouped frequency table shows information about the number of hours worked by each of 200 headteachers in one week.

Number of hours worked (h)	Frequency
$30 < h \leqslant 40$	4
$40 < h \leqslant 50$	18
$50 < h \leqslant 60$	68
$60 < h \leqslant 70$	79
$70 < h \leqslant 80$	31

a Represent this information as a frequency polygon.
b Work out an estimate for the mean number of hours worked by the headteachers.
c Write down the modal class interval.

4 The table shows the frequency distribution of student absences for a year.

Absences (d) days	Frequency
$0 \leqslant d < 5$	4
$5 \leqslant d < 10$	6
$10 \leqslant d < 15$	8
$15 \leqslant d < 20$	5
$20 \leqslant d < 25$	4
$25 \leqslant d < 30$	3

Hint: The data in this question is discrete rather than continuous, but it can be dealt with in the same way.

a Work out an estimate for the mean number of student absences for the year.
b Represent the information as a frequency polygon.
c Write down the modal class interval for the number of student absences.

23 Cumulative frequency diagrams

Cumulative frequency diagrams can be used to estimate statistical data, such as medians, and to compare distributions.

The cumulative frequency is the running total of the frequency up to the end of each class interval.

Example 1

200 students at Jordan Hill College took the Mathematics GCSE Paper 1.
The paper has a maximum mark of 100.
The distribution of their marks on this paper is given below.

Mark range (m)	Frequency
$0 < m \leqslant 10$	8
$10 < m \leqslant 20$	12
$20 < m \leqslant 30$	15
$30 < m \leqslant 40$	20
$40 < m \leqslant 50$	32
$50 < m \leqslant 60$	57
$60 < m \leqslant 70$	34
$70 < m \leqslant 80$	12
$80 < m \leqslant 90$	7
$90 < m \leqslant 100$	3

a Complete a cumulative frequency table for this data.

b Draw the cumulative frequency curve.

a

Marks (m)	Cumulative frequency
$0 < m \leqslant 10$	8
$0 < m \leqslant 20$	20
$0 < m \leqslant 30$	35
$0 < m \leqslant 40$	55
$0 < m \leqslant 50$	87
$0 < m \leqslant 60$	144
$0 < m \leqslant 70$	178
$0 < m \leqslant 80$	190
$0 < m \leqslant 90$	197
$0 < m \leqslant 100$	200

$8 + 12 = 20$

$20 + 15 = 35$

$35 + 20 = 55$

$55 + 32 = 87$ and so on.

Check that this number is the same as the total in the sample.

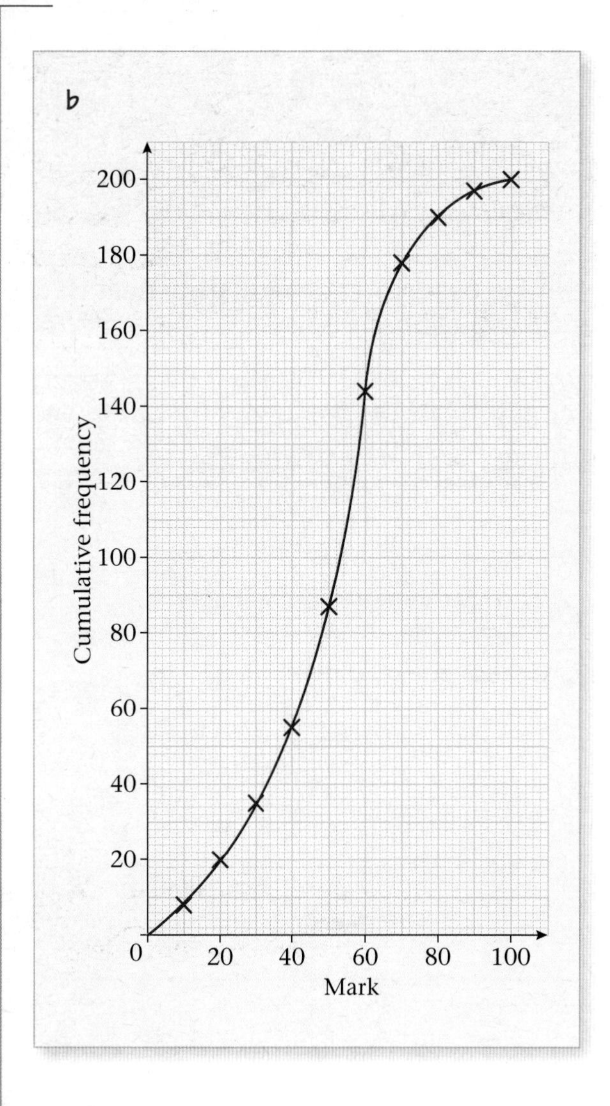

Always use the vertical axis for the cumulative frequency.
Plot the points: (10, 8) (20, 20) (30, 35) etc. from the table, then draw a smooth curve through the points.
The curve will usually be S-shaped.

Practice 23A

1 The data below gives the duration in seconds of the 60 most recent calls made on Fran's phone:

Duration d (seconds)	Number of calls (frequency)
$0 < d \leq 20$	5
$20 < d \leq 40$	10
$40 < d \leq 60$	18
$60 < d \leq 80$	12
$80 < d \leq 100$	9
$100 < d \leq 120$	3
$120 < d \leq 140$	3

a Produce a cumulative frequency table for this data.

b Plot a cumulative frequency curve.

c How many phone calls were less than or equal to 1 minute in duration?

23.2

In a cumulative frequency graph

- **the median is the middle value of the distribution**
- **the lower quartile is the value one quarter of the way into the distribution**
- **the upper quartile is the value three quarters of the way into the distribution.**

The interquartile range is the difference between the upper and lower quartiles;

Interquartile range = upper quartile − lower quartile.

Example 2

Use the cumulative frequency graph on page 188 to obtain estimates for

a the median

b the interquartile range.

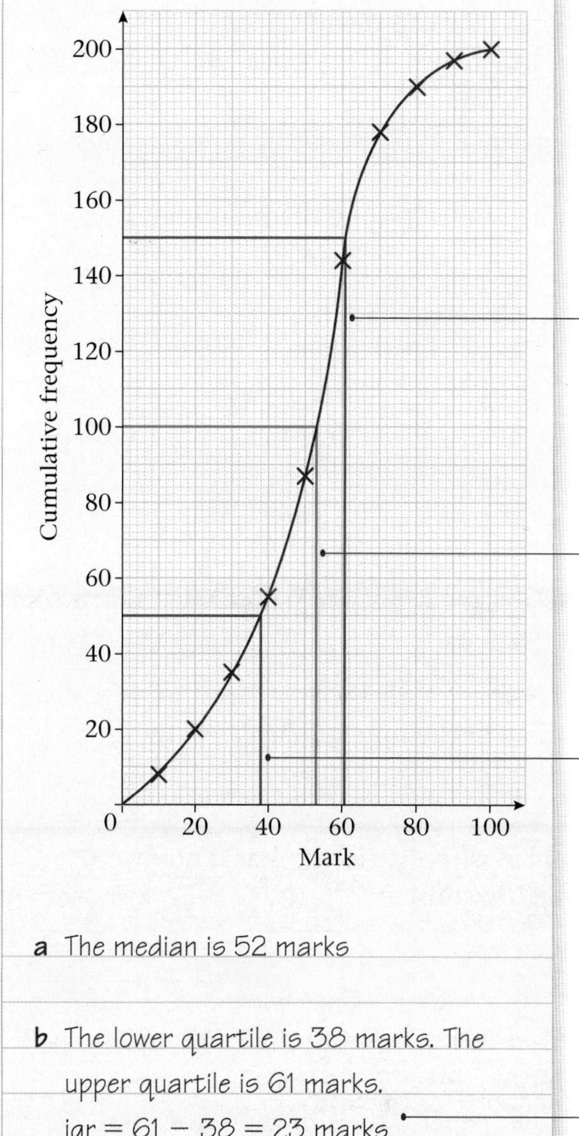

a The median is 52 marks

b The lower quartile is 38 marks. The upper quartile is 61 marks.
iqr = 61 − 38 = 23 marks

To find the **upper quartile**, go $\frac{3}{4}$ up to the maximum value ($\frac{3}{4} \times 200 = 150$), across to the curve, then down.

To find the **median**, go **halfway** up to the maximum value ($\frac{1}{2} \times 200 = 100$), then across to the curve, and down to meet the horizontal axis. The value on the horizontal axis is the median.

To find the **lower quartile**, go $\frac{1}{4}$ up ($\frac{1}{4} \times 200 = 50$), across to the curve, then down.

The interquartile range (iqr)
= upper quartile − lower quartile.

Practice 23B

1 The manager of a post office recorded the lengths of times that 120 customers had to wait in the queues at the counter. The distribution of waiting times is given in the table below.

Waiting time (t seconds)	Frequency
$0 < t \leqslant 50$	6
$50 < t \leqslant 100$	10
$100 < t \leqslant 150$	15
$150 < t \leqslant 200$	24
$200 < t \leqslant 250$	45
$250 < t \leqslant 300$	20

a Construct a cumulative frequency table.

b Draw a cumulative frequency graph for this distribution.

c Use the graph to find estimates for

i the median waiting time

ii the upper quartile, lower quartile and interquartile range for this distribution

iii the number of people who had to wait more than 2 minutes.

2 The table gives information about the ages, in years, of 100 aeroplanes.

Age (t years)	Frequency
$0 < t \leqslant 5$	41
$5 < t \leqslant 10$	26
$10 < t \leqslant 15$	20
$15 < t \leqslant 20$	10
$20 < t \leqslant 25$	3

a Work out an estimate of the mean age of the aeroplanes.

b Construct a cumulative frequency table.

c Draw a cumulative frequency graph for your table.

d Use your graph to find an estimate of the upper quartile of the ages. Show your method clearly.

E

23.3

A box plot represents the main features of a distribution as a diagram. Box plots can be used to compare distributions.

Example 3

a Draw the box plot for the distribution of marks shown in Example 2 on page 189. Assume the highest mark is 100 and the lowest mark is 0.

b The 200 students also took paper 2. On paper 2 their median mark was 57, the lower quartile was 44 marks and the upper quartile was 71.
Make three comments comparing the distributions of marks on the two papers.

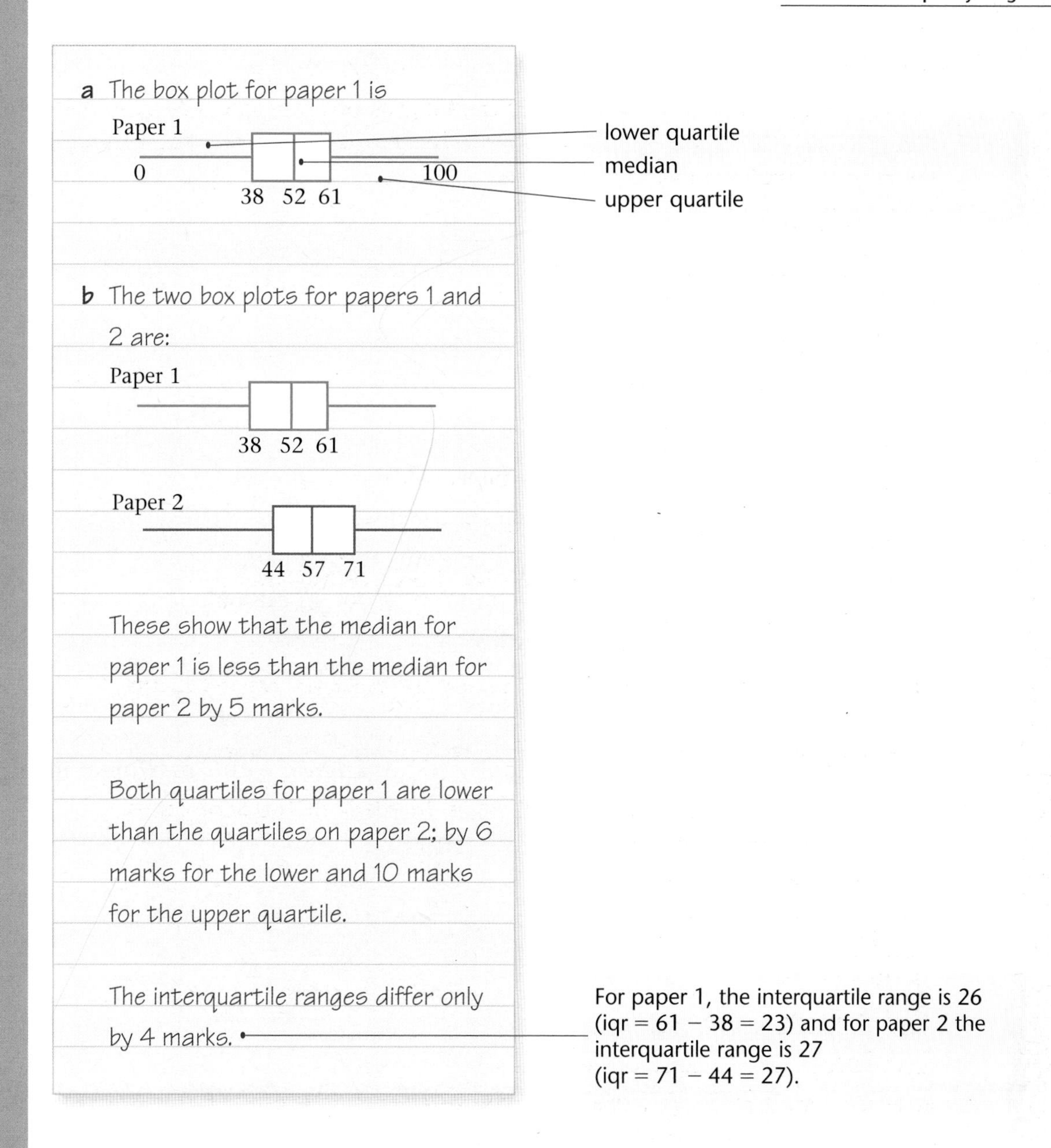

Practice 23C

1 For the data in Question 1, Practice 23B,

a draw the box plot for this distribution, assuming highest and lowest readings are unchanged.

The manager makes some changes to the organisation of the post office. After these changes, she discovers that

- the median waiting time is 190 seconds
- the lower quartile waiting time is 120 seconds, and
- the upper quartile waiting time is 210 seconds.

b Draw the box plot for the new distribution.

c Make three comments comparing the waiting times before and after the changes made by the manager.

2 The table gives information about the mass, in kilograms, of a sample of 100 people.

Mass (m kg)	Frequency
$60 < m \leq 65$	2
$65 < m \leq 70$	4
$70 < m \leq 75$	12
$75 < m \leq 80$	30
$80 < m \leq 85$	28
$85 < m \leq 90$	18
$90 < m \leq 95$	6

a Construct a cumulative frequency table.

b Draw a cumulative frequency graph for the distribution.

c Write down the class interval which contains the median.

d Use your cumulative frequency graph to work out an estimate for

i the interquartile range for the mass

ii the number of people in the sample with a mass greater than 88 kg.

e Draw the box plot for the distribution.

3 120 students at Jordan Hill College took a test in English with a maximum mark of 100.
The cumulative frequency graph for the distribution of their marks is drawn below.

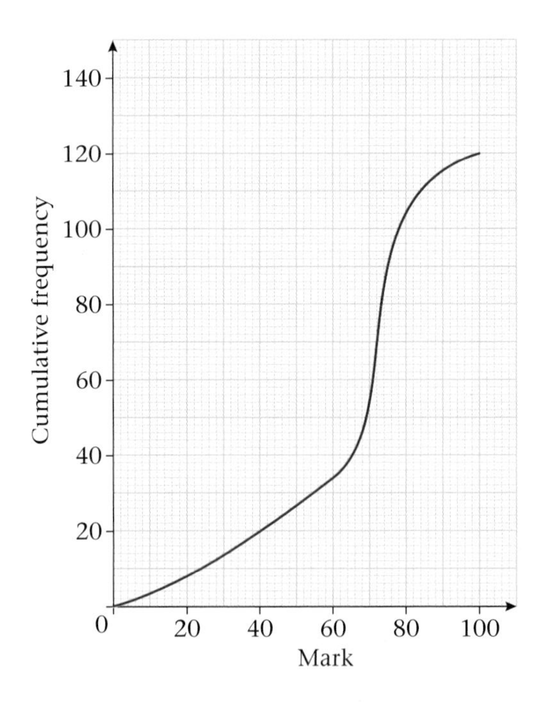

a Use the graph to find an estimate of the median mark.
The pass mark in the test was 55.

b Use the graph to find an estimate for the number of students who passed the test.

23.4 The shape of a cumulative frequency curve depends on the distribution of the data.

Example 4

The diagrams below represent three histograms.

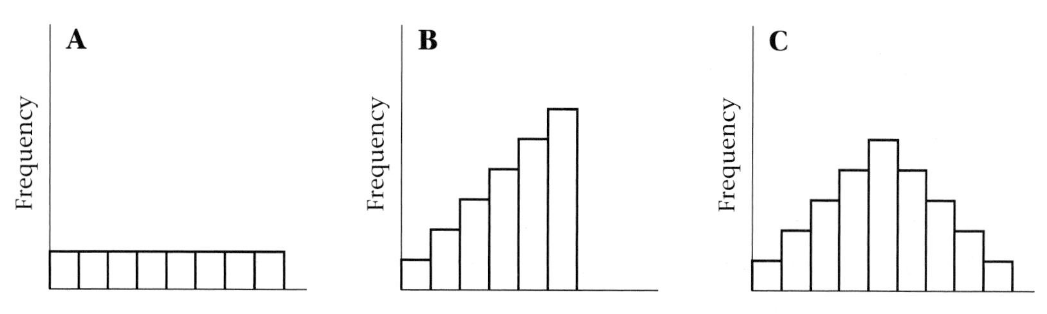

For each histogram, sketch the cumulative frequency graph.

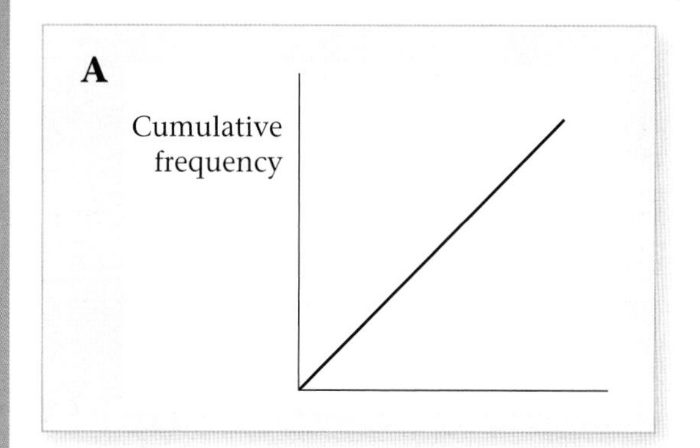

A goes up in 'equal steps'.

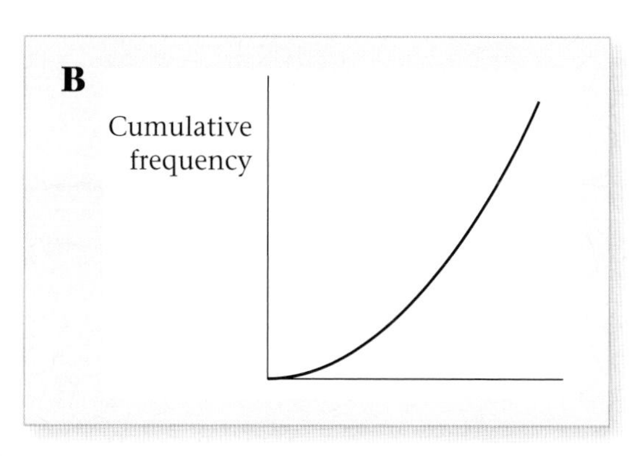

In **B** the 'steps' get bigger and bigger.

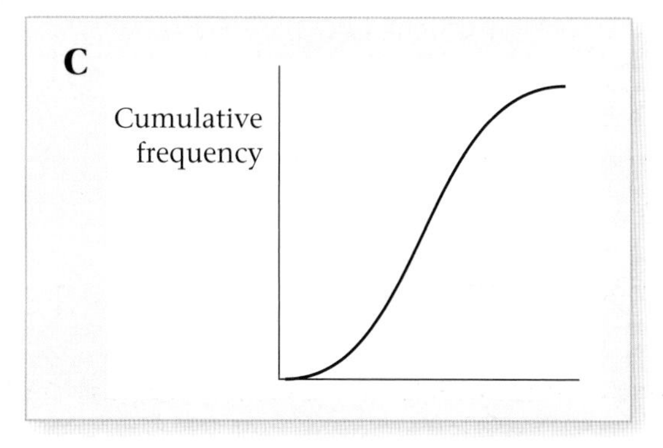

C is the more usual 'S' shape.

Practice 23D

1 The diagrams show frequency polygons and cumulative frequency graphs. Each cumulative frequency graph represents the same information as one of the frequency polygons.

Write down the letters of the pairs of diagrams which represent the same information.

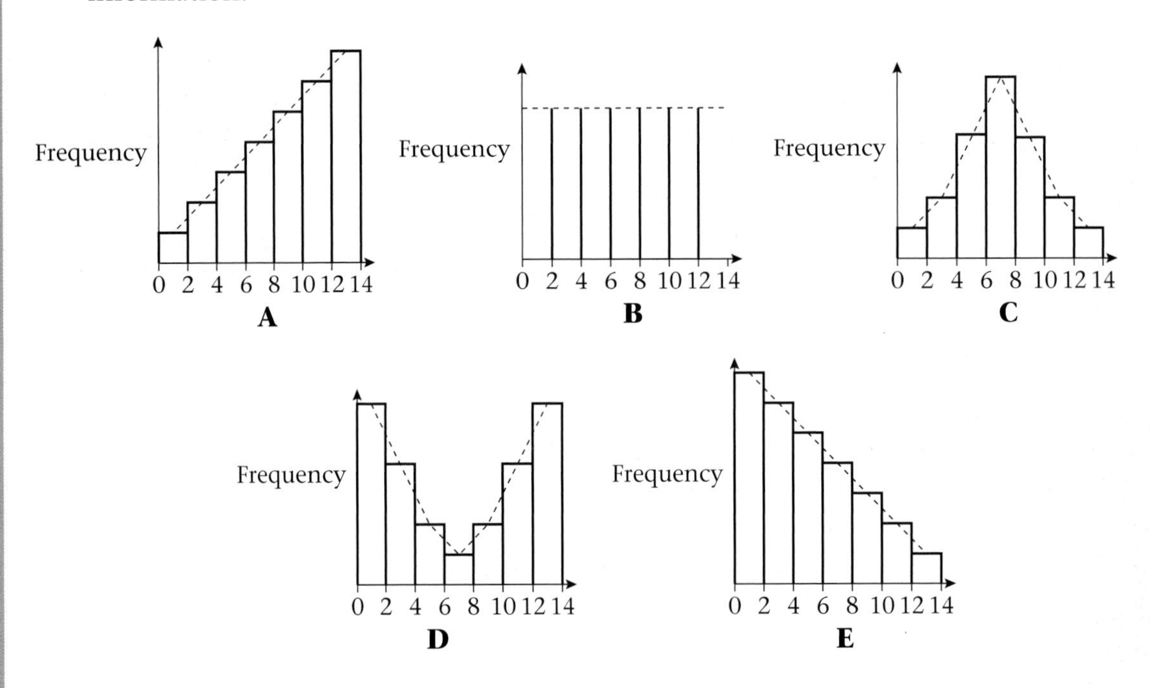

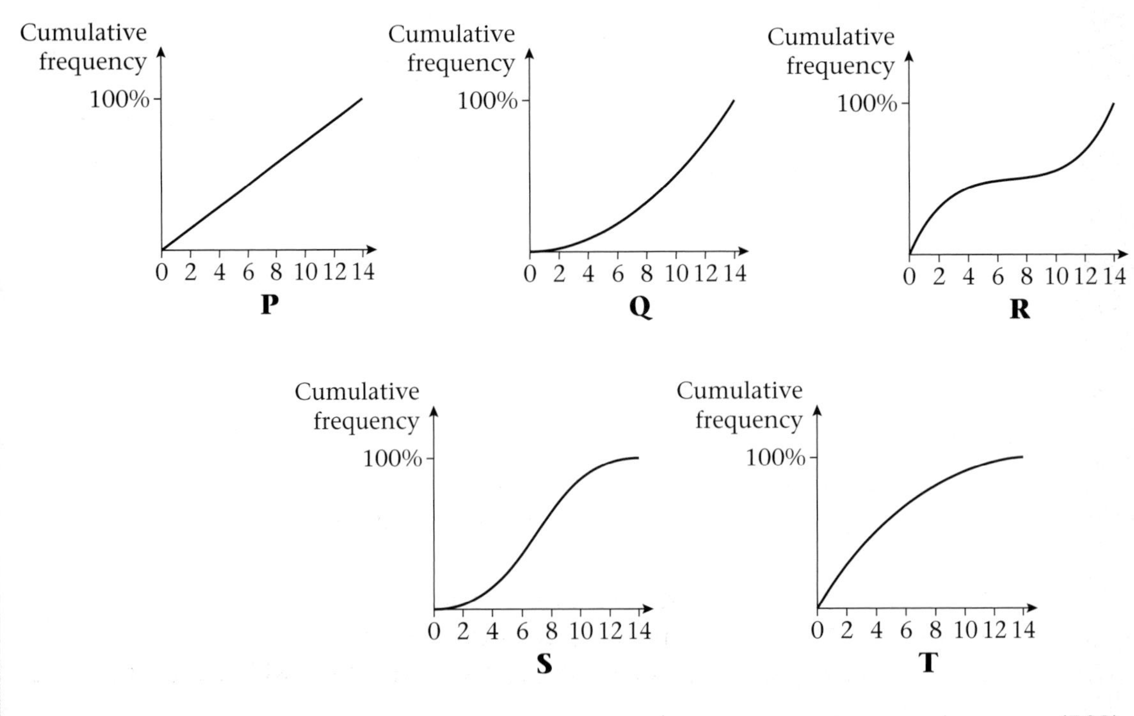

(E18)

24 Probability

This chapter tells you how to use probability scales; how to work out probabilities using tree diagrams; and how to identify mutually exclusive and independent outcomes.

Events, or outcomes of events, can have varying chances, or probability, of happening. These can be shown on a probability scale going from 0 to 1.

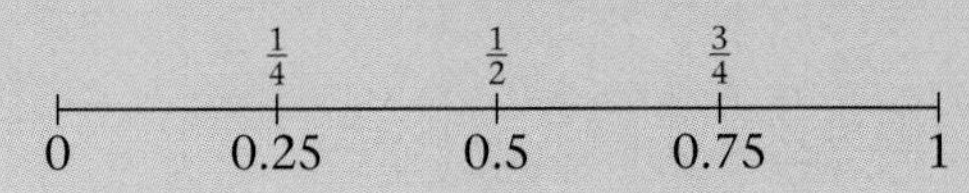

Example 1

Place the following events on a probability scale:

A The next baby to be born will be female.
B A pet rabbit will live forever.
C You will use a pen in college today.
D Snow will fall in London on 1 August next year.

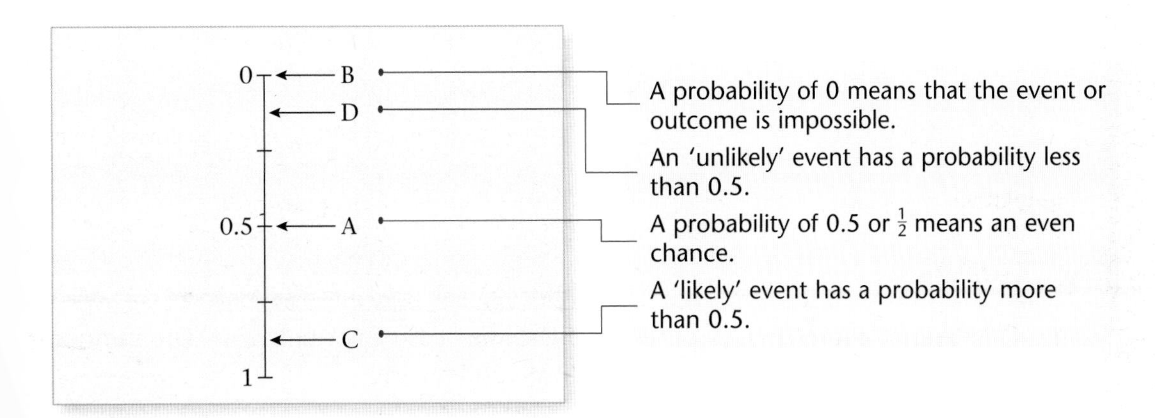

(S21)

Practice 24A

1 Record each of these on a probability scale:

A An ordinary 2p coin will land heads when it is tossed.
B A flower will live forever.
C It will be cold at the South Pole tomorrow.
D There will be at least one sunny day next August.
E The winning numbers in next week's National Lottery will be the same as the ones in this week's National Lottery.

24.2

The probability of an outcome of an event occurring is given by

$$\frac{\text{number of successful outcomes}}{\text{total number of possible outcomes}}$$

Example 2

The diagram represents a spinner in the shape of a regular pentagon. The spinner is to be spun once.
Write down the probability that it will stop on

a a red section
b a yellow section
c a blue section
d a red or yellow section.

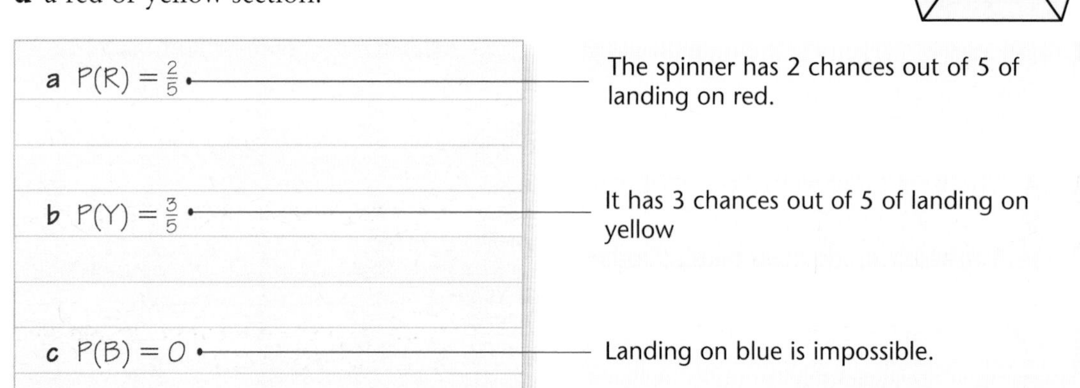

Practice 24B

1 The diagram represents a spinner in the shape of a regular hexagon. The spinner is to be spun once. Find the probability that it will land on

a a blue section
b a red section
c a green section
d a red or white section.

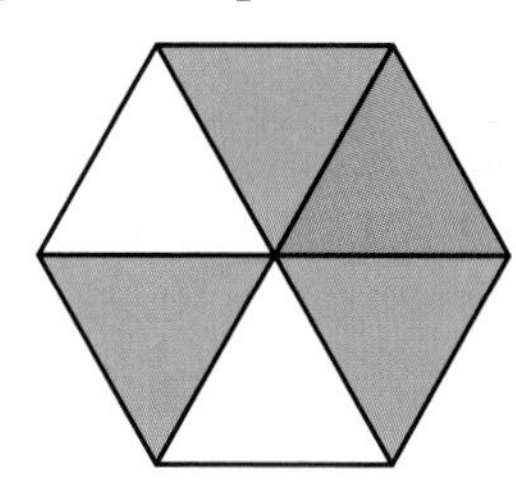

2 Joan has a bag of 12 chocolates.
6 are plain, 4 are milk and 2 are white.
Without looking into the bag, Joan selects a chocolate.
Find the probability of the chocolate she selects being

a milk
b plain
c not white
d milk or plain
e a liquorice allsort.

24.3

When one outcome excludes another outcome from happening then the two outcomes are mutually exclusive.

The probabilities of all the mutually exclusive outcomes add up to 1.

For two outcomes, A and B, which are mutually exclusive, P(A or B) = P(A) + P(B). This is called the OR rule.

Example 3

A bag contains 20 beads. 8 of the beads are white, 7 are red and 5 are blue. A bead is selected at random.

Find the probability of the selected bead being

a white

b white or blue

c not red.

a $P(W) = \frac{8}{20} = \frac{4}{10} = \frac{2}{5}$ — White has 8 chances out of 20 of being selected.

b $P(W \text{ or } B) = P(W) + P(B)$

$= \frac{8}{20} + \frac{5}{20} = \frac{13}{20}$ — White has 8 chances out of 20, and blue has 5 chances out of 20.

c $P(\text{not red}) = \frac{13}{20}$ — 7 out of 20 are red so 13 out of 20 are not red.

Practice 24C

1 The diagram shows a spinner in the shape of a regular hexagon. The spinner is spun once. Find the probability of it landing on

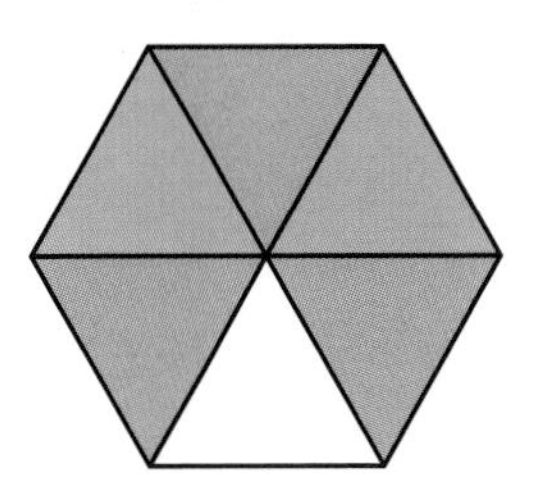

a a red section

b a red or white section

c a section that is not red.

24.4

The probability of something *not* happening is 1 − the probability of it happening.

If the probability of something happening is p then the probability of it not happening is $1 - p$.

Example 4

The probability of a newly laid egg being cracked is 0.02. Work out the probability of a newly laid egg not being cracked.

$P = 1 - 0.02$

$= 0.98$

P(not cracked) = 1 − P(cracked)

Example 5

The diagram represents a biased spinner.

When the spinner is spun once, the probabilities of it stopping on some of the sections are given in the table below.

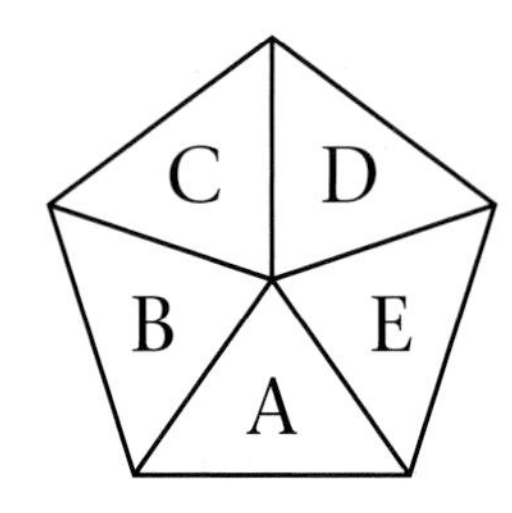

Section	A	B	C	D	E
Probability	0.23	0.22	0.17	0.26	

Work out the probability that the spinner will land on section E when it is spun once.

$P(A) + P(B) + P(C) + P(D) + P(E) = 1$

so

$0.23 + 0.22 + 0.17 + 0.26 + P(E) = 1$

so

$0.88 + P(E) = 1$

$P(E) = 1 - 0.88$

$P(E) = 0.12$

The outcomes are mutually exclusive, so the sum of their probabilities is 1, since these are all the possible outcomes.

If outcomes are mutually exclusive, only one outcome can happen at any one time.

Practice 24D

1 James has a bag of 50 mixed nuts.
20 are walnuts, 10 are brazil nuts, 15 are almonds and 5 are hazelnuts.
He selects a nut at random from the bag.

Find the probability that the selected nut is

a a brazil nut

b either a walnut or a hazelnut

c not a walnut.

2 The probability that a new battery being faulty is 0.003.
Work out the probability of a new battery not being faulty.

3 The train can be either late, on time or early.
The probability of it being late is 0.26.
The probability of it being early is 0.10.
Work out the probability of it being on time.

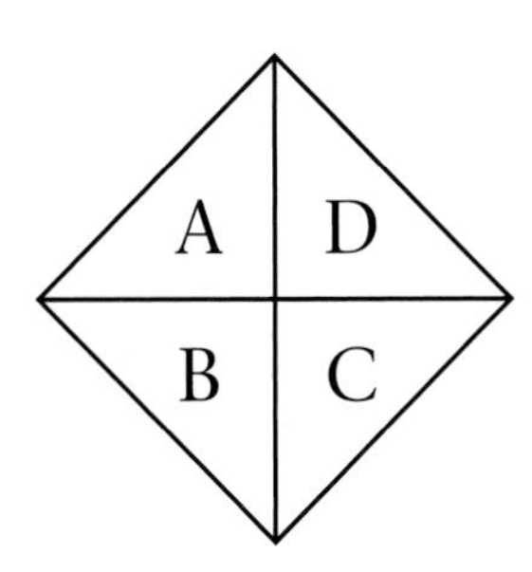

The diagram represents a biased spinner in the shape of a square.
When the spinner is spun once, the probabilities of it stopping on the sections A, B and C are given in the incomplete table below.

Section	A	B	C	D
Probability	0.2	0.31	0.17	

Work out the probability of the spinner stopping on section D when it is spun once.

5 Here is a biased four-sided spinner:

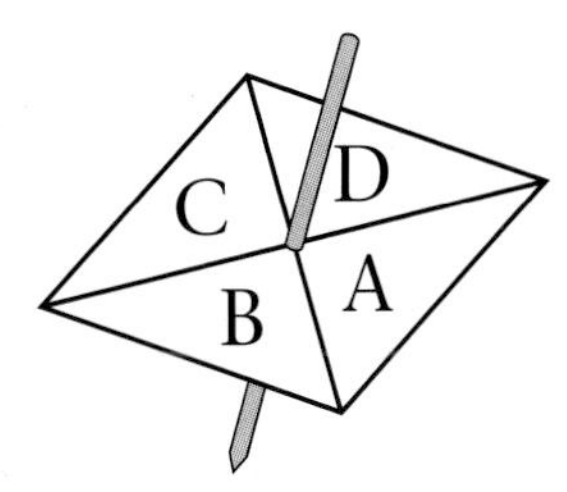

The sides are labelled A, B, C and D.
The probability that the spinner, when it is spun once, will land on each of the letters from A to C, is given in the table.

Letter	A	B	C	D
Probability	0.4	0.1	0.2	

The spinner is to be spun once.

a Work out the probability that the spinner will land on D.

b Write down which letter, from A, B, C or D, the spinner is least likely to land on.

c Work out the probability that the spinner will land on a consonant.

6 The probability of Joyce passing her driving test at the first attempt is 0.85.
Work out the probability of her failing her driving test at the first attempt.

24.5

The estimated probability of an event or outcome is

$$\frac{\text{number of successful trials}}{\text{total number of trials}}$$

and is often also called the relative frequency.

The greater the number of trials then the better the relative frequency as an estimate of the probability.

Example 6

There are four candidates (Anderson, Burnside, Catton and Darby) standing at an election. Shortly before the election, a market research company surveyed the voting intention of 1200 voters. The results of this survey are given in the table below:

Intending to vote for	Anderson	Burnside	Catton	Darby
Frequency	186	205	432	377

a Work out the best estimate for the probability of a randomly selected voter voting for Catton. On the day of the election 18 504 votes were cast.

b Estimate the most likely number of votes cast for Catton.

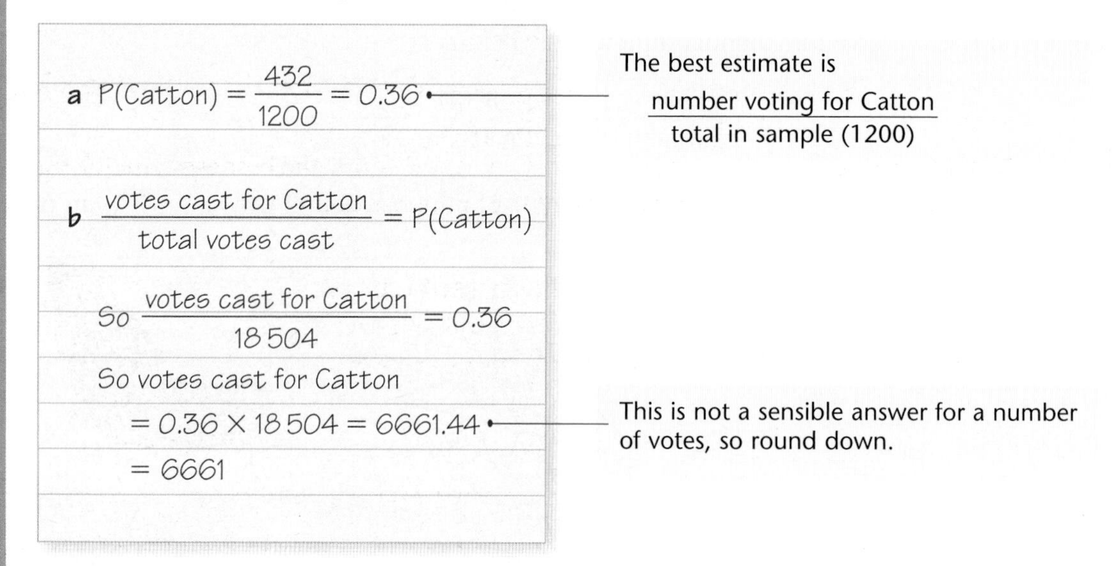

Practice 24E

1 The school bus was late on 12 days out of 195. Work out the best estimate of the probability of the school bus being late on any given day. Give your answer correct to 2 dp.

2 There are 3800 students at Jordan College who vote in the election for the president of the students' union.

Three students, named King, Malcolm and Nicholson, are standing for election.

On the day of the election, Erica asked 100 randomly selected students to tell her the name of the candidate for whom they will cast their vote. The results of her enquiry are given in the table below.

Candidate	King	Malcolm	Nicholson
Frequency	27	42	31

When all 3800 votes have been cast, work out the best estimate for the number of votes cast for Malcolm.

3 The diagram represents a biased spinner.

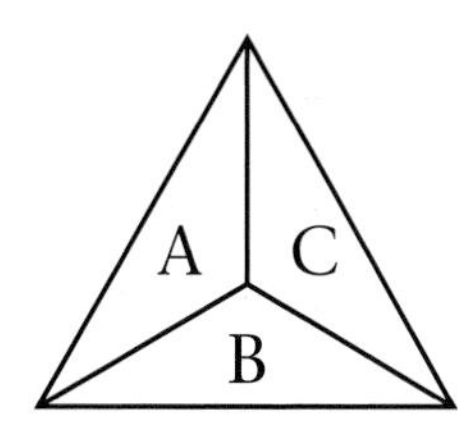

Jim spun the spinner 200 times and recorded the sections upon which it stopped.

His results were:

Section	A	B	C
Frequency	52	86	62

The same spinner was spun 300 times by Emma, who also recorded the sections upon which it stopped.

Her results were:

Section	A	B	C
Frequency	80	132	88

a Use the above information to work out the best estimate of the spinner stopping on section B when it is spun once.

b Kathy is due to spin the spinner 1200 times. Work out the best estimate for the number of times the spinner is likely to stop on section C.

4 The probability of a newly laid egg being cracked is 0.003. A supermarket receives a batch of 15 000 newly laid eggs. Work out the best estimate for the number of these eggs likely to be cracked.

24.6

When the outcome of one event does not affect the outcome of another event, the two events are called independent events.

Tree diagrams can be used to help list the outcomes of combined events.

Example 7

Lisa has a fair coin and an unbiased spinner.

The spinner is in the shape of an equilateral triangle.

She flips the coin once and spins the spinner once.

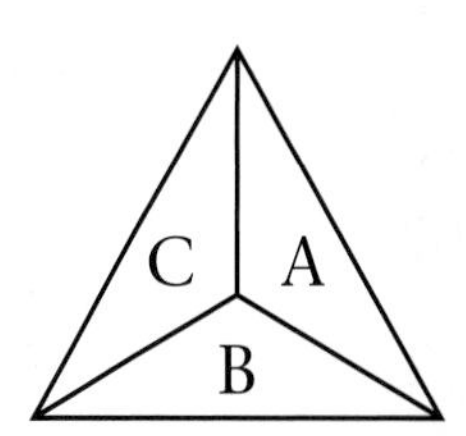

a Using a tree diagram, list all the possible joint outcomes.

b Find the probability of the joint outcome
'the coin lands heads and the spinner stops on B'.

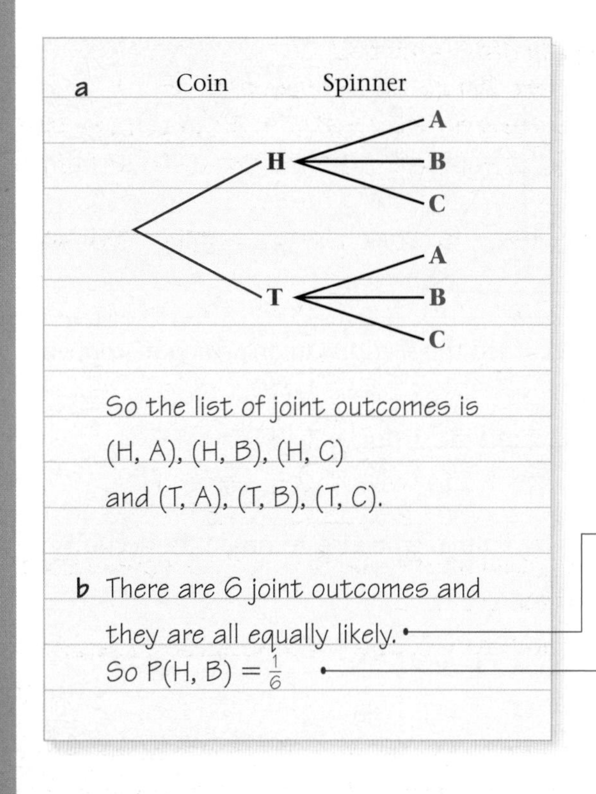

Going along a branch or

H — A

Gives the joint outcome (H, A).

Equally likely means each outcome has the same chance.

P(H, B) means the joint outcome of heads on the coin and B on the spinner.

(S22)

Practice 24F

1 Emma has a bag which contains three equal sized balls. Each ball is labelled with one of the numbers 1, 2 or 3.

She also has a triangular shaped spinner as shown below.

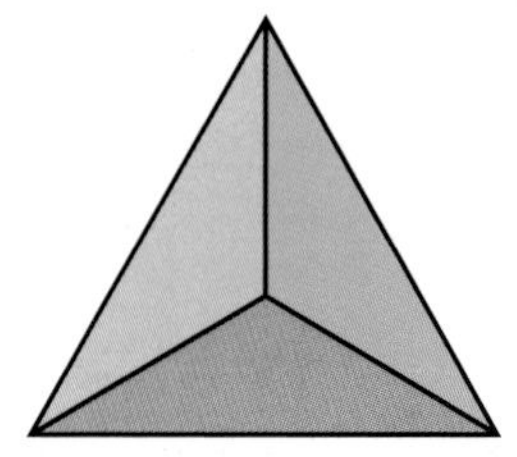

Emma selects one ball at random from the bag and spins the spinner once.

She records the joint outcome of the numbered ball selected and the colour of the section upon which the spinner stops.

a Using a tree diagram, or otherwise, list all the possible joint outcomes.

b Find the probability of Emma obtaining the joint outcome '2 and Green'.

2 Joe flips an ordinary coin and rolls an ordinary dice.

a Using a tree diagram, or otherwise, list all the joint outcomes.

b Find the probability of obtaining the joint outcomes:

  i the coin lands on tails and the dice has a 6 on its uppermost face

  ii the coin lands heads and the dice has an odd number on its uppermost face

  iii the coin lands heads and the dice has a number greater than 4 on its uppermost face

  iv the coin lands tails and the dice has a prime number on its uppermost face.

3 Jasmin has two spinners, as shown below.

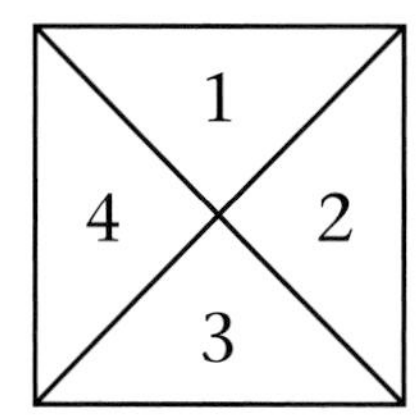

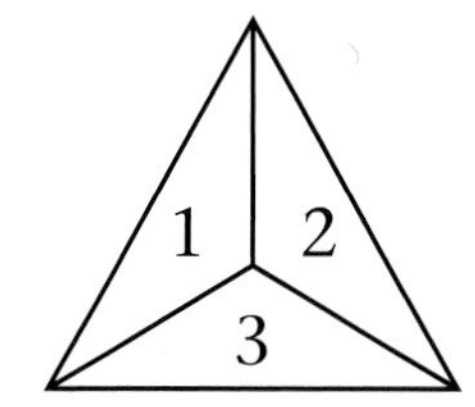

She spins each spinner once and adds together the numbers of the sections upon which each one stops to create the **total score**.

Using a tree diagram, or otherwise, find the probability of the total score being 5.

(E19)

24.7 You can use estimated probability in interpreting and explaining data.

Example 8

Yvonne has a spinner in the shape of a square.

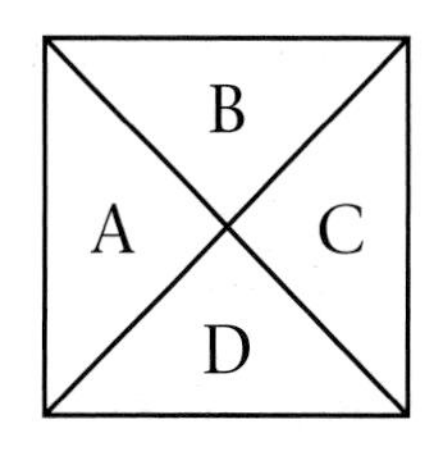

She has spun the spinner 200 times and recorded the section upon which it stopped each time. Her results are listed below.

Section	A	B	C	D
Frequency	42	78	37	43

Explain whether or not this evidence suggests that the spinner is biased.

If the spinner was unbiased then each frequency should be approximately $\frac{1}{4}$ of 200, or 50.

Since the frequency for B is much greater than 50, the evidence suggests that the spinner is biased.

It would be reasonable to expect 50 ± 10% for an unbiased dice.

Practice 24G

1 The diagram represents an ordinary cubic dice.
It is rolled 1200 times and the number on the uppermost face is recorded.
If the dice is unbiased, how many times, approximately, should the number 3 be recorded?

2 A football manager accused a referee of being biased towards the home teams. The league conducted an enquiry and found that the referee had awarded 64 yellow cards during the season.
31 of these cards had been shown to players in the home teams.
Explain whether or not this evidence supports the manager's accusation.

3 Shanie has a triangular spinner, as below.

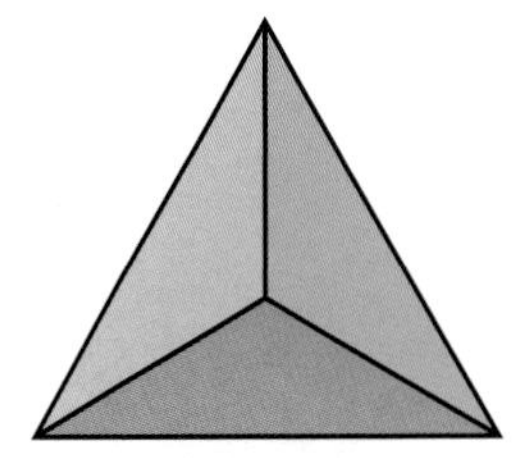

She spun it 300 times, recording the colour of the section upon which it stopped. Her results are shown below.

Colour	Red	Blue	Green
Frequency	152	77	71

Explain whether or not the evidence suggests that the spinner is biased.

4 The diagram represents a spinner.

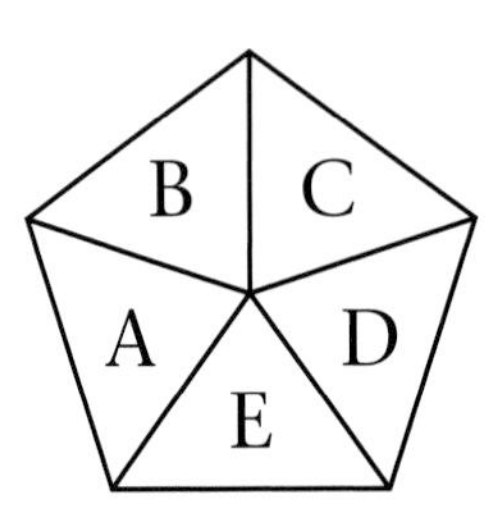

Jon spins the spinner 400 times and records the letter of the section upon which it stops each time. His results are:

Section	A	B	C	D	E
Frequency	72	42	85	112	89

a Explain why this evidence suggests that the spinner is biased.
b Lizzie spins the spinner 600 times and again records the letter upon which it stops each time. Her results are:

Section	A	B	C	D	E
Frequency	104	65	123	168	140

Work out the best estimate for the probability of the spinner stopping on section D when it is spun once.

Answers

Practice 1A

1 **a** 2 **b** −7 **c** 16 **d** 24
e −14 **f** −26 **g** 19 **h** 68
i −43 **j** −9

2 **a**

+	−2	3	−4
6	4	9	2
−2	−4	1	−6
−5	−7	−2	−9

b

−	−4	0	5
−3	1	−3	−8
8	12	8	3
3	7	3	−2

Practice 1B

1 **a** 12 **b** −24 **c** −9 **d** −3
e 3 **f** 36 **g** −24 **h** −8
i −6 **j** −48

2 **a**

×	−3	2	−6
8	−24	16	−48
−6	18	−12	36
4	−12	8	−24

b

÷	−2	−3	6
−12	6	4	−2
36	−18	−12	6
−18	9	6	−3

Practice 1C

1 **a** 590 **b** 8970 **c** 75 000 **d** 6000
e 240 **f** 60 000 **g** 63.3 **h** 6.8
i 0.04 **j** 0.0028

2 **a** 600 **b** 6000 **c** 90
d 5 **e** 0.4

3 **a** 79 000 **b** 6340 **c** 78.4
d 6.35 **e** 0.00285

4 **a** **i** 600×30 **ii** 18 000
b **i** $9000 \div 300$ **ii** 30
c **i** $\frac{40 \times 300}{60}$ **ii** 200
d **i** $\frac{7000 \times 400}{40}$ **ii** 70 000
e **i** $\frac{600 \times 60}{30 \times 50}$ **ii** 24

5 26 pints

6 10 pounds

Practice 1D

1 **a** 1 **b** 9 **c** 36 **d** 81
e 25 **f** 121 **g** 49 **h** 4
i 100 **j** 64 **k** 16 **l** 144
m 196 **n** 169 **o** 225

2 **a** 27 **b** 125 **c** 8
d 1000 **e** 64

3 **a** ±5 **b** ±6 **c** ±4 **d** ±14
e ±10 **f** ±15 **g** ±12 **h** ±7
i ±3 **j** ±11 **k** ±13 **l** ±2
m ±9 **n** ±1 **o** ±8

4 **a** 67.24 **b** 32.768 **c** 1.1 **d** 2.1
e −1.2 **f** −8 **g** 22.09 **h** 250.047
i −12.167 **j** 1.69 **k** −2.3 **l** −1.1

Practice 1E

1 **a** 34 **b** 7.5 **c** 28 **d** 3
e 27 **f** 81 **g** 51 **h** 5
i 19 or 13 **j** 4 **k** ±1 **l** 11

2 **a** $6 \times (8 - 4) = 24$ **b** $(3 + 2) \times (3 + 2) = 25$
c $7 \times (7 \div 7) \times 7 = 49$ **d** $(1 - 2) \times 3 - 4 = -7$

Practice 1F

1 11, 13, 17, 19, 23, 29

2 **a** 3 **b** 2 **c** 8 **d** 6
e 5 **f** 6

3 **a** 21 **b** 36 **c** 30 **d** 12
e 70 **f** 30

Practice 2A

1 **a** $\frac{3}{4}$ **b** $\frac{2}{3}$ **c** $\frac{3}{4}$ **d** $\frac{5}{6}$
e $\frac{4}{7}$ **f** $\frac{2}{3}$ **g** $\frac{3}{5}$ **h** $\frac{4}{7}$

2 **a** $\frac{2}{3} = \frac{4}{6}$ **b** $\frac{4}{5} = \frac{12}{15}$ **c** $\frac{3}{7} = \frac{9}{21}$ **d** $\frac{2}{3} = \frac{12}{18}$
e $\frac{5}{7} = \frac{20}{28}$ **f** $\frac{3}{4} = \frac{75}{100}$ **g** $\frac{3}{8} = \frac{15}{40}$ **h** $\frac{6}{11} = \frac{18}{33}$

3 **a** $\frac{6}{5}$ **b** $\frac{5}{3}$ **c** $\frac{15}{7}$ **d** $\frac{11}{3}$
e $\frac{31}{7}$ **f** $\frac{32}{3}$

4 **a** $3\frac{1}{2}$ **b** $1\frac{1}{6}$ **c** $2\frac{3}{7}$ **d** $2\frac{1}{2}$
e $2\frac{2}{3}$ **f** $2\frac{1}{2}$

Practice 2B

1 **a** $\frac{4}{7}, \frac{11}{14}, \frac{23}{28}$ **b** $\frac{1}{4}, \frac{2}{5}, \frac{1}{2}, \frac{7}{10}$ **c** $\frac{4}{15}, \frac{3}{10}, \frac{7}{20}, \frac{2}{5}$

2 **a** $\frac{2}{5}, \frac{7}{20}, \frac{3}{10}$ **b** $\frac{13}{18}, \frac{2}{3}, \frac{7}{12}, \frac{5}{9}$ **c** $\frac{10}{12}, \frac{3}{4}, \frac{2}{3}, \frac{13}{20}$

Practice 2C

1 **a** $\frac{2}{3}$ **b** $1\frac{1}{5}$ **c** $\frac{3}{8}$ **d** $\frac{7}{10}$
e $\frac{11}{14}$ **f** $1\frac{1}{15}$ **g** $1\frac{7}{12}$ **h** $1\frac{7}{40}$
i $1\frac{11}{36}$ **j** $\frac{53}{56}$

2 **a** $\frac{1}{2}$ **b** $\frac{1}{4}$ **c** $\frac{1}{10}$ **d** $\frac{5}{12}$
e $\frac{1}{8}$ **f** $\frac{2}{15}$ **g** $\frac{7}{12}$ **h** $\frac{27}{40}$
i $\frac{1}{10}$ **j** $\frac{8}{21}$

3 **a** $2\frac{1}{10}$ **b** $3\frac{3}{8}$ **c** $2\frac{3}{4}$ **d** $5\frac{5}{8}$
e $5\frac{1}{6}$ **f** $7\frac{5}{12}$ **g** $8\frac{1}{3}$ **h** $4\frac{7}{24}$

4 **a** $1\frac{1}{6}$ **b** $1\frac{1}{4}$ **c** $2\frac{5}{12}$ **d** $1\frac{2}{3}$
e $2\frac{13}{20}$ **f** $2\frac{19}{24}$ **g** $\frac{11}{24}$ **h** $4\frac{1}{2}$

5 $2\frac{1}{4}$ kg **6** $4\frac{1}{8}$ yds **7** $1\frac{1}{8}$ pts

8 $1\frac{7}{8}$ kg **9** $\frac{11}{12}$ hr

Practice 2D

1 **a** $\frac{1}{5}$ **b** $\frac{5}{8}$ **c** $\frac{3}{10}$
d $\frac{1}{2}$ **e** $\frac{2}{9}$

2 **a** 1 **b** $6\frac{3}{5}$ **c** 4 **d** $7\frac{13}{20}$
e $3\frac{3}{32}$ **f** $11\frac{1}{4}$ **g** $12\frac{3}{4}$ **h** $15\frac{1}{3}$

3 **a** 2 **b** 9 **c** $10\frac{5}{6}$
d $16\frac{4}{5}$ **e** $11\frac{3}{8}$

4 £10.80p

5 Algebra marks: 9
Statistics marks: 6
Number marks: 9

6 $\frac{3}{10}$ kg

Practice 2E

1 **a** $2\frac{1}{2}$ **b** $\frac{5}{6}$ **c** $\frac{9}{14}$
d $\frac{6}{7}$ **e** $\frac{5}{6}$

2 **a** $1\frac{1}{5}$ **b** $3\frac{3}{7}$ **c** $1\frac{1}{10}$ **d** $4\frac{4}{11}$
e $\frac{23}{30}$ **f** $\frac{1}{16}$ **g** $\frac{1}{12}$ **h** 4

3 $6\frac{4}{13}$ (6 pieces of wood each $1\frac{5}{8}$ yards long and 1 piece $\frac{1}{2}$ yard long)

4 $13\frac{1}{3}$ (or 14 days as the 14th day is needed for the last $\frac{3}{4}$ mile)

Practice 3A

1 **a** 37.8 **b** 6.8 **c** 132.7 **d** 8.0
e 0.4 **f** 0.6

2 **a** 1.487 **b** 3.966 **c** 23.287 **d** 0.040
e 0.003 **f** 0.190

3 **a** **i** 2.35 **ii** 2.3 **iii** 2.348
b **i** 23.89 **ii** 23.9 **iii** 23.892
c **i** 1.31 **ii** 1.3 **iii** 1.310
d **i** 21.04 **ii** 21.0 **iii** 21.036
e **i** 6.20 **ii** 6.2 **iii** 6.200
f **i** 0.94 **ii** 0.9 **iii** 0.939

4 £3.33 per hour (2 dp is appropriate as it represents pence)

5 1.714 (3 dp is appropriate as it represents millimetres) or 1.71

6 **a** It is very difficult to measure less than $\frac{1}{10}$th of a cm (i.e. mm) and that is only one decimal place if the measurements are given in cm
b 29.6 cm

Practice 3B

1 **a** **i** 234 **ii** 23.4 **iii** 2340
b **i** 361.4 **ii** 36.14 **iii** 3614
c **i** 2302.1 **ii** 230.21 **iii** 23 021
d **i** 4120 **ii** 412 **iii** 41 200
e **i** 2.13 **ii** 0.213 **iii** 21.3

2 **a** **i** 2.34 **ii** 0.0234 **iii** 0.234
b **i** 3.792 **ii** 0.03792 **iii** 0.3792
c **i** 78.31 **ii** 0.7831 **iii** 7.831
d **i** 0.263 **ii** 0.00263 **iii** 0.0263
e **i** 0.201 **ii** 0.00201 **iii** 0.0201

3 **a** 8.06 **b** 15.58 **c** 31.04 **d** 0.0378
e 122.13 **f** 7.815 **g** 1.23 **h** 2.3088

4 **a** 85 **b** 119 **c** 49 **d** 0.59
e 2.3 **f** 27 **g** 6.9 **h** 580

5 130 kg

6 0.076 m long; 0.039 m high or 7.6 cm long, 3.9 cm high

7 34 stamps

8 1.017 kg

9 162.62 km

10 6 glasses

11 **a** 15.038 **b** 150 380 **c** 1.5038 **d** 1.5038

12 **a** 2.812 **b** 281.2 **c** 0.2812

Practice 3C

1 **a** $\frac{1}{2}$ **b** $\frac{3}{4}$ **c** $\frac{7}{10}$ **d** $\frac{9}{10}$
e $\frac{17}{20}$ **f** $\frac{9}{25}$ **g** $\frac{12}{25}$ **h** $\frac{1}{8}$
i $\frac{5}{8}$ **j** $\frac{1}{50}$ **k** $\frac{1}{40}$ **l** $\frac{1}{125}$

Practice 3D

1 **a** 0.5 **b** 0.75 **c** 0.4 **d** 0.3
e $0.2\dot{6}$ **f** 0.28 **g** $0.458\dot{3}$ **h** $0.\dot{6}$
i $0.\dot{3}\dot{6}$ **j** 0.41 **k** 0.136 **l** $0.8\dot{3}$

2 **a** 0.5, 0.49, 0.45, 0.4, 0.04
b $\frac{7}{10}$, $\frac{13}{20}$, 0.64, $\frac{3}{5}$, 0.06
c $\frac{15}{16}$, $\frac{7}{8}$, 0.84, 0.8, $\frac{3}{5}$
d $\frac{1}{2}$, $\frac{1}{9}$, 0.44, 0.404, 0.4

Practice 3E

1 **a** terminating **b** recurring
c terminating **d** recurring
e terminating **f** terminating
g recurring **h** terminating
i recurring **j** recurring

2 For example: $\frac{1}{2}$, $\frac{3}{4}$, $\frac{1}{5}$

3 For example: $\frac{1}{3}$, $\frac{1}{6}$, $\frac{1}{9}$

Practice 3F

1 **a** $\frac{1}{8}$ **b** $\frac{1}{5}$ **c** $3\frac{1}{3}$ **d** 4
e $3\frac{1}{13}$ **f** 9 **g** 20 **h** $\frac{8}{7}$
i 3 **j** $1\frac{1}{2}$

Practice 4A

1 **a** $\frac{1}{4}$ **b** $\frac{3}{10}$ **c** $\frac{13}{20}$ **d** $\frac{6}{25}$
e $\frac{8}{25}$ **f** $\frac{7}{40}$ **g** $\frac{17}{40}$ **h** $\frac{1}{3}$

2 **a** 0.5 **b** 0.2 **c** 0.45 **d** 0.28
e 0.48 **f** 0.375 **g** 0.025 **h** $0.\dot{6}$

Practice 4B

1 **a** 75% **b** 20% **c** 85% **d** 42%
e 88% **f** $32\frac{1}{2}$% **g** $82\frac{1}{2}$% **h** $33\frac{1}{3}$%

2 **a** 50% **b** 30% **c** 85% **d** 64%
e 82% **f** $42\frac{1}{2}$% **g** $92\frac{1}{2}$% **h** $66\frac{2}{3}$%

3 70%

4 55%

5 $\frac{7}{8}$; 0.8; $\frac{3}{4}$; 70%

6 $\frac{1}{4}$; 0.299; 30%; $\frac{1}{3}$; 0.35; $\frac{2}{5}$

Practice 4C

1 **a** £15 **b** £1.13 **c** 18 kg **d** 8 kg
e 125 g **f** £6.75 **g** 10.8 m **h** £1.92
i 200 m **j** £3.15

2 360

3 91 employees

4 £10.80

Practice 4D

1 £72 500 **2** £35.70 **3** £564

4 £923.55 **5** £0.18 or 18p

Practice 4E

1 **a** 20% **b** 5% **c** 17.5% **d** $32\frac{1}{2}$% **e** 15% **f** 8%
2 75%
3 45%

Practice 4F

1 20%
2 30%
3 35%
4 34%
5 25%

Practice 4G

1 Adult: £320
Junior: £170
Family: £640
2 Video player: £382.98
CD player: £102.13
Games player: £195.74
3 £1120
4 £18.40
5 £1100
6 £40

Practice 4H

1 £205.92
2 £122.85
3 £171.96
4 £231.53
5 £624.32
6 £288.26
7 £14 000

Practice 5A

1 **a** 4^2 **b** 8^3 **c** $3^2 \times 5^3$ **d** $8^4 \times 2^3$ **e** $9^3 \times 7^2$ **f** $3^4 \times 2^2$ **g** $9^3 \times 4^2 \times 3$
2 **a** 216 **b** 49 **c** 8 **d** 200 **e** 144 **f** 8100
3 1125
4 **a** 2592 **b** 35 721 **c** 2048 **d** 884 736
5 4.486 089 611
6 46.416 376 42

Practice 5B

1 **a** 3^7 **b** 2^7 **c** 7^6 **d** 5^8 **e** 10^4 **f** 6^2 **g** 5^5 **h** 7^5 **i** 2^5 **j** 6 **k** 9^4 **l** 4^7 **m** 2^8 **n** 5^6

Practice 5C

1 **a** $2^3 \times 3$ **b** $2^4 \times 3$ **c** $2^5 \times 3$ **d** $2^4 \times 3^2$ **e** $2^5 \times 3 \times 5$
2 **a** 2×5^2 **b** $2^2 \times 5^4$

Practice 5D

1 **a** 4.8×10^3 **b** 3.7×10^5 **c** 2.3×10^4 **d** 3.4×10^{-4} **e** 2.5×10^{-3} **f** 3×10^{-2} **g** 3.471×10^2 **h** 2.378×10^1 **i** 4.2×10^5 **j** 2×10^4 **k** 3.7×10^{-1} **l** 5×10^{-4}
2 **a** 4000 **b** 26 000 **c** 340 000 **d** 0.002 **e** 0.000052 **f** 0.62
3 **a** 50 100 **b** 9×10^{-4}

Practice 5E

1 **a** 6×10^9 **b** 5.1×10^{10} **c** 6×10^1 **d** 2×10 or 2×10^1 **e** 3×10^{-7} **f** 3.5×10^{-1} **g** 5.6×10^{10} **h** 1.25×10^4 **i** 8×10^{-3} **j** 7.5×10^{-8}
2 2.4×10^9
3 **a** 1.6×10^8 **b** 28 000
4 7.5×10^{-3}
5 **a** 1.9375×10^{-2} **b** 1.794×10^{10} **c** 2.4762×10^1 **d** 1.764×10^0
6 **a** 8.4×10^7 **b** 2.1×10^{-5}
7 8.01×10^{10}
8 3×10^{15} seconds
9 2.51×10^{-17} g
10 3.99×10^{13} km

Practice 6A

1 **a** 2 : 1 **b** 3 : 2 **c** 6 : 3 : 1 **d** 4 : 3 : 1 **e** 6 : 1 **f** 5 : 1 **g** 4 : 1 **h** 5 : 1 **i** 6 : 1 **j** 4 : 1 **k** 8 : 4 : 1 **l** 40 : 10 : 3
2 **a** $x = 1$ **b** $x = 2$ **c** $x = 1$ **d** $x = 5$ **e** $x = 12$ **f** $x = 5$
3 91
4 18 cm
5 100 men
6 £5.70 per hour
7 14 cm

Practice 6B

1 **a** 20; 25 **b** 25; 35 **c** 10; 8; 2 **d** £18; £12 **e** £70; £30; £20 **f** 42 cm; 30 cm; 18 cm
2 Tracy receives £4000
Wayne receives £3200
3 350 g plain flour
150 g wholewheat flour
4 20 sweets
5 0.5 m; 0.3 m; 0.2 m
6 £14.40
7 **a** £735 **b** £196
8 1 : 5
9 3 : 4

Practice 6C

1 **a** £1.50 **b** £19.50
2 **a** £2.94 **b** £38.22
3 £4.80
4 24 ℓ
5 £1.47
6 €533.33
7 150 g of butter
120 g of sugar
3 eggs
135 g of flour
45 ml of milk
8 1000 g tuna
1000 g mushroom soup
250 g grated cheddar cheese
10 spring onions
625 g breadcrumbs

Practice 7A

1 **a** $x + 5y$ **b** $6p + 2q$ **c** $6c + 11d$ **d** $a + 3b$ **e** $-4ef + 9gh$ **f** $7r - 11s$ **g** $7t^2$ **h** $5x^2y^2$ **i** $4c^2d + cd^3$ **j** $a^2b + 8ab^3$

Practice 7B

1 **a** $2a + 6$ **b** $6p + 9$ **c** $-4x + 12$ **d** $-8x - 12$ **e** $35x + 28$ **f** $-x - 3$ **g** $12x + 15y$ **h** $30x - 42$ **i** $6w - 4y - 10$ **j** $27b + 36c - 18d$
2 **a** $11x + 2$ **b** $7a - 8$ **c** $12m + 13$ **d** $6x + 2$ **e** $7x + 10$ **f** $-x - 3$ **g** $3x + 19$ **h** $26a + 11b$ **i** $-2p + 8q$ **j** $10a + 24$

Practice 7C

a **i** 16; 19 **ii** $3n + 1$
b **i** 6; 7 **ii** $n + 1$
c **i** 20; 24 **ii** $4n$
d **i** 23; 27 **ii** $4n - 1$
e **i** 37; 43 **ii** $6n + 1$
f **i** 16; 19 **ii** $3n - 2$
g **i** −9; −15 **ii** $-6n + 27$
h **i** −5.5; −8 **ii** $-2.5n + 7$

Practice 7D

1 **a**

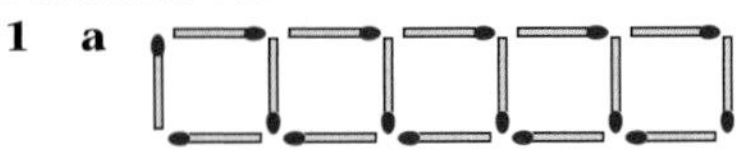

b $3n + 1$
c 115 matchsticks
d 79 matchsticks

2 **a** 1, 3, 5, 7 **b** $2n - 1$

3 **a**

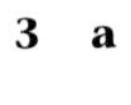

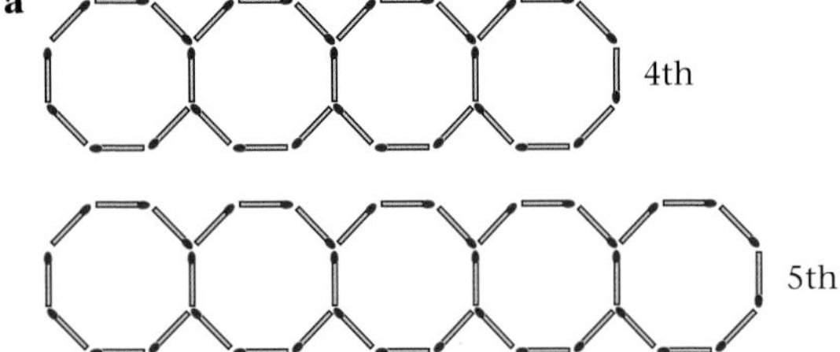

b $7n + 1$ **c** 351 matchsticks

4 **a**

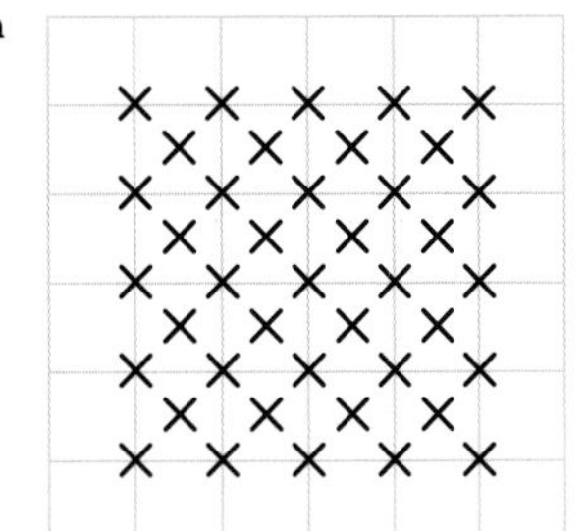

Pattern number 5

b

Pattern number	1	2	3	4	5
Number of crosses	1	5	13	25	41

c 113 crosses

Practice 7E

a b^2 **b** d^7 **c** e^5 **d** y^6
e x^{-3} **f** $3c$ **g** $6p^5$ **h** $9x^4$
i x^{-5} **j** $3b^3$ **k** $\frac{1}{2}d$ **l** $\frac{9}{2}y^{-1}$
m $16x^7$ **n** a^6 **o** $50e^5$ **p** $64x^6y^3$

Practice 7F

a $a^2 + 3a + 2$ **b** $b^2 + b - 6$
c $x^2 + x - 6$ **d** $2x^2 + 7x + 3$
e $3x^2 - 7x - 6$ **f** $6d^2 - 7d + 2$
g $15t^2 + 4t - 4$ **h** $x^2 - 4$
i $2a^2 + 7a - 15$ **j** $6e^2 - 13e + 5$
k $6p^2 - 13p - 8$ **l** $4x^2 - 14x + 6$
m $3d^2 - 19d + 20$ **n** $3t^2 + 10t - 8$
o $9p^2 - 1$ **p** $x^2 + 8x + 16$
q $4a^2 + 20a + 25$ **r** $9y^2 - 6y + 1$
s $25p^2 + 20p + 4$ **t** $9x^2 - 24x + 16$

Practice 8A

1 21 **2** 14 **3** 14 **4** 57
5 −10 **6** 31 **7** −9 **8** 32
9 16 **10** 25 **11** 19 **12** 12
13 −2 **14** $136\frac{1}{8}$ **15** 2.37 **16** 2.22

Practice 8B

1 **a** 15 **b** −40 **c** 48.9
2 **a** 15 **b** 20.7 **c** 16.82
3 **a** 64.4 °F **b** 212 °F **c** 23 °F
4 £52.80
5 $v = 26.58$
6 $K = 1258.58$

Practice 8C

1 $A = \frac{1}{2}b \times h$ **2** $T = p \times n$
3 $C = 27n$ **4** $C = 12x + 45y$
5 $S = 2v + u^2$ **6** $C = 15b + 8t$
7 $A = 3 + K$ or $A = 59 - K$; or $K + (K + 3) = 59$ or $2K + 3 = 59$

Practice 8D

1 $2(a + 4)$ **2** $7(c - 4)$
3 $3(4x + 5)$ **4** $4(2y - 3)$
5 $3(a - 3b)$ **6** $7(s + 3t)$
7 $3a(a + 2)$ **8** $2x(2x - 5)$
9 $6m(2m + 1)$ **10** $7x(2x - 3)$
11 $x(3x - y)$ **12** $2a(2b - 5)$
13 $rs(r - 2)$ **14** $9(d - 4e)$
15 $3x(5x - 7)$ **16** $7a(3 - 5a)$
17 $8g(1 - 3g)$ **18** $2x^2(4x - 1)$
19 $11a(a + 3)$ **20** $4b^2(3a - 1)$
21 $3t(2at - 5)$ **22** $cd(c - 1)$
23 $2a(b - 3a)$ **24** $4xy(2x - 5y)$

Practice 9A

1 $x = 4$ **2** $x = 14$
3 $a = -1$ **4** $p = 4$
5 $t = 2$ **6** $w = 2\frac{1}{3}$
7 $y = 3$ **8** $x = -2$
9 $b = 4$ **10** $d = 3\frac{1}{2}$
11 $e = -9$ **12** $m = 4$
13 $x = 2$ **14** $x = 3$
15 $x = 2$ **16** $x = 4$
17 $x = 6$ **18** $x = 3\frac{1}{2}$
19 $x = -2$ **20** $x = -2$
21 $y = -2\frac{1}{2}$ **22** $x = -1\frac{1}{2}$
23 $a = 9$ **24** $p = -2$

Practice 9B

1 $x = 2$ **2** $x = 5$
3 $x = 1$ **4** $x = 3$
5 $a = 2$ **6** $x = -5\frac{1}{2}$
7 $a = 1$ **8** $x = 2$
9 $x = 3\frac{1}{6}$ **10** $x = 3\frac{1}{2}$
11 $a = 4$ **12** $y = 20$

Practice 9C

1 $x = -8$ **2** $x = 3\frac{1}{2}$
3 $y = 3$ **4** $x = 4$
5 $a = 4\frac{1}{2}$ **6** $p = 4$
7 $m = 15$ **8** $y = -1$
9 $x = -2\frac{2}{3}$ **10** $y = -3$

Practice 9D

1 **a** $x = 3$ $y = 3$ **b** $x = 2$ $y = 1$ **c** $x = 2$ $y = 3$
2 $a = 2$ $b = -1$
3 **a** $x = \frac{3}{4}$ $y = -\frac{1}{2}$ **b** $x = 1\frac{1}{2}$ $y = -2$ **c** $x = 7$ $y = \frac{1}{2}$

Practice 10A

1

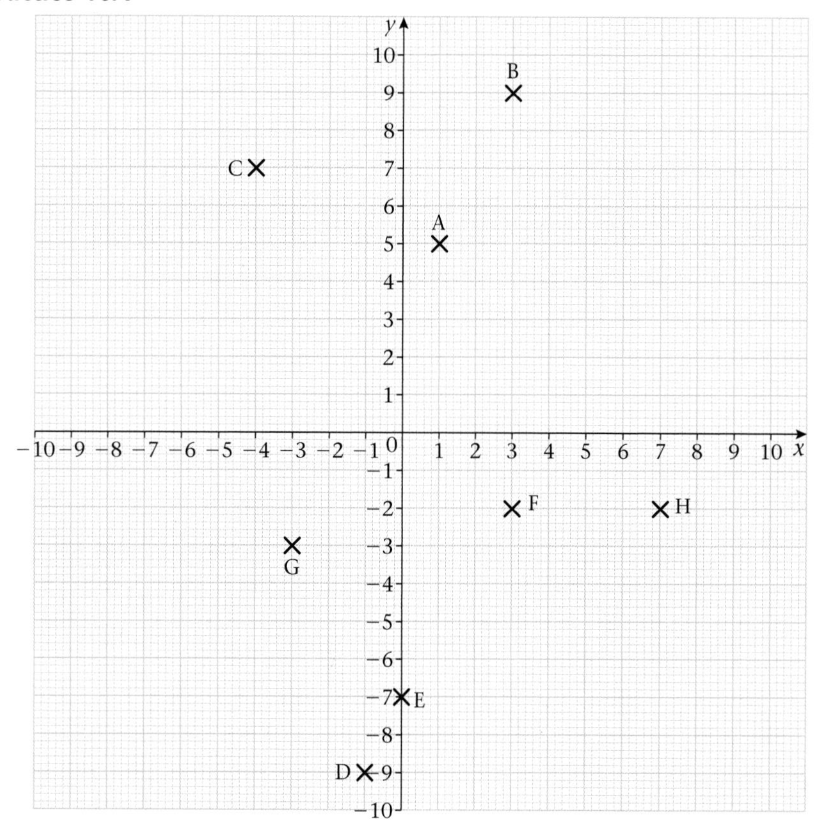

2 **a** (2, 7) **b** (1, 0)
c $(3\frac{1}{2}, -4\frac{1}{2})$ **d** (5, −2)
e $(-\frac{1}{2}, 2\frac{1}{2})$ **f** (−2, 0)
g $(-3\frac{1}{2}, 2)$ **h** (−1, 1)

Practice 10B

1 **a**

x	−3	−2	−1	0	1	2	3
y	**−3**	−1	**1**	**3**	**5**	**7**	**9**

b

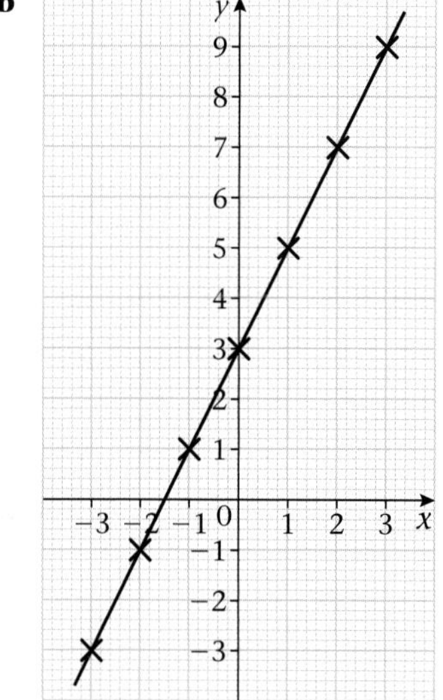

c **i** $y = 6$
ii $x = -1\frac{3}{4}$

2 a

x	−3	−2	−1	0	1	2	3
$y = 3x - 1$	−10	**−7**	−4	**−1**	**2**	5	**8**

b

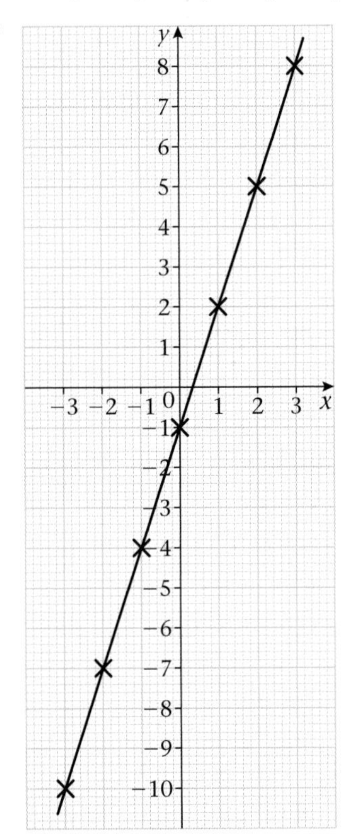

c $x = 2.5$

3 a $y = 2x + 3$ and $y = 4x + 3$

b $y = 2x + 3$ and $y = 2x - 7$

4 a $y = x + 1$

b $y = -2x - 1$

5 a 3, 2

b 5, −3

c $-\frac{1}{2}$, 3

d −4, −3

e −2,5

f 3, $-1\frac{1}{2}$

Practice 10C

1 a

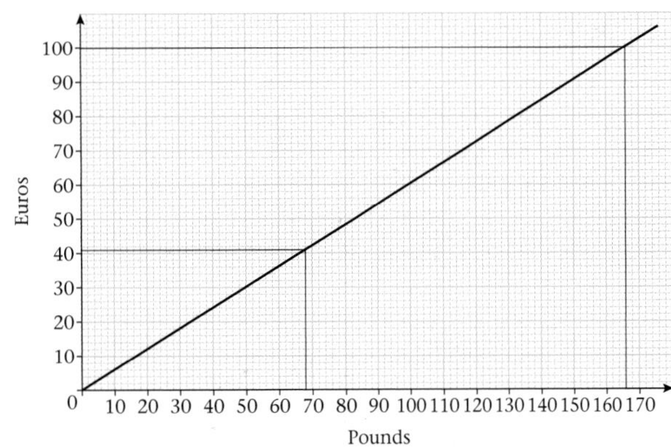

b i 40.8 Euros

ii £166.67

2 a 270 km

b 180 km/h

c

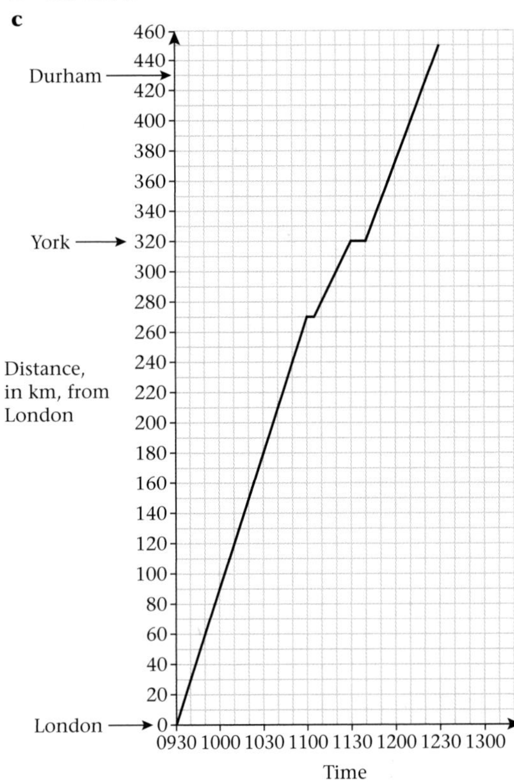

3 a 12 km/h

b He stopped at the lake

c

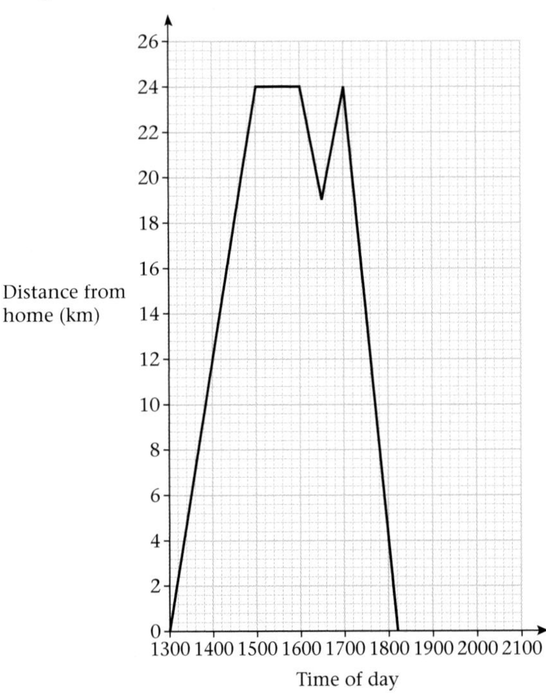

4 a 13h06

b Between 10h00 and 11h00

c 50 km/h

d He is cycling at 10 km/h

5 a i 1300

ii 20 km/h

b

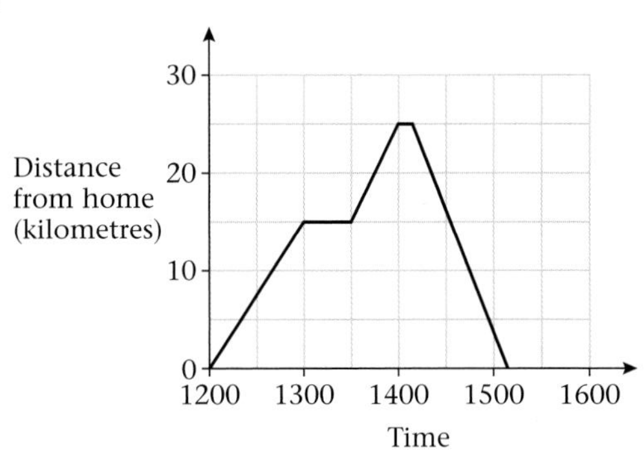

Practice 10D

1 a

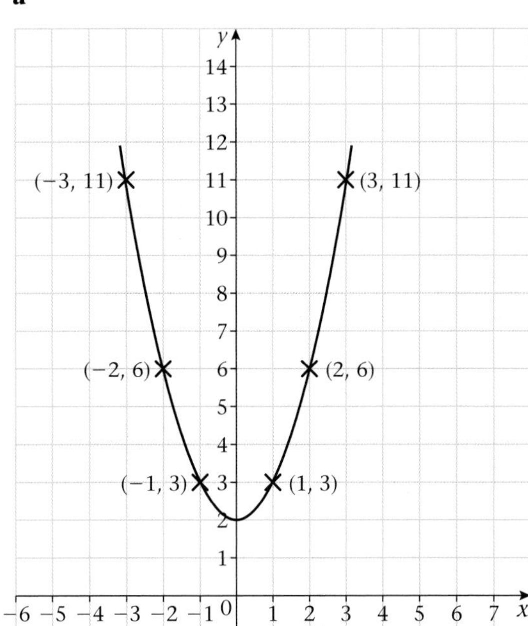

b

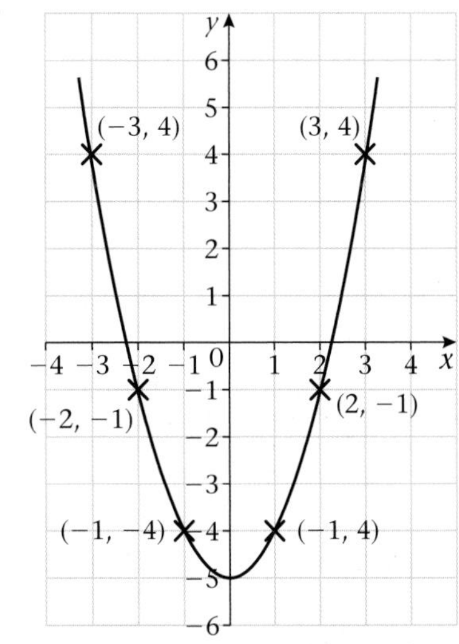

c

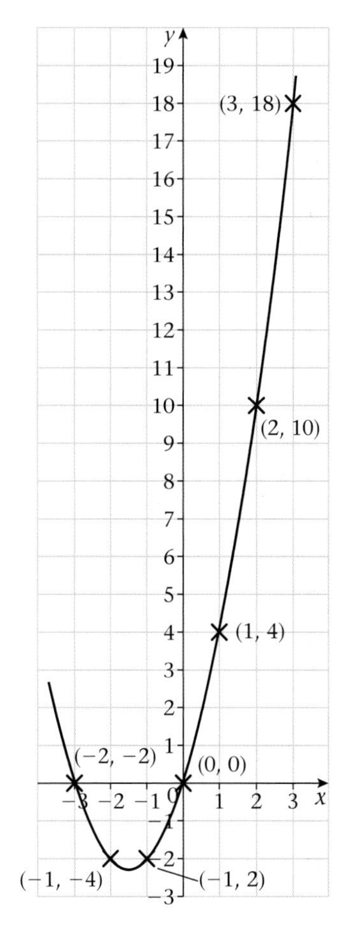

d

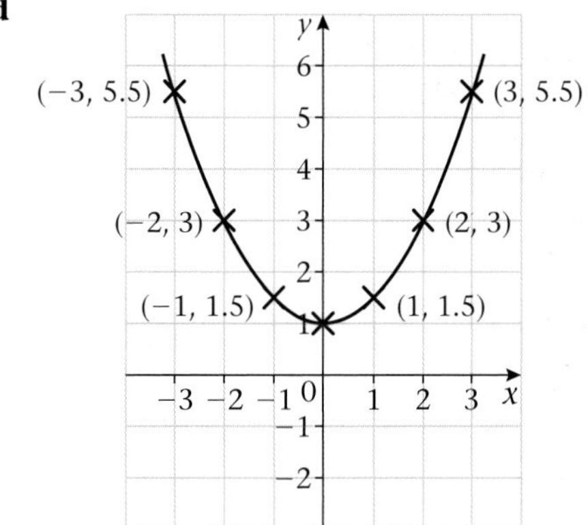

e

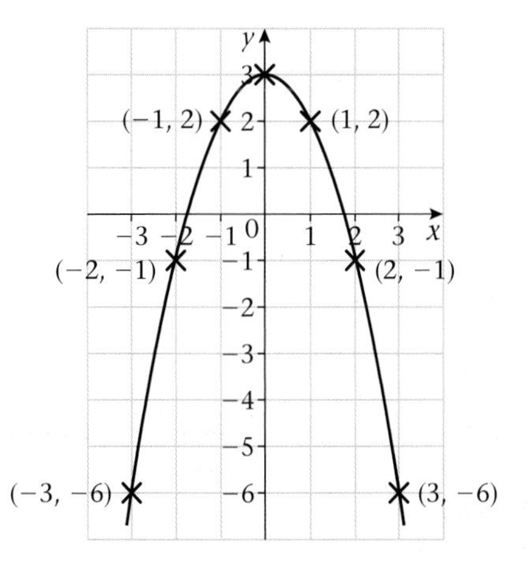

2 **a**

x	−3	−2	−1	0	1	2	3
y	18	**8**	**2**	**0**	2	8	**18**

b

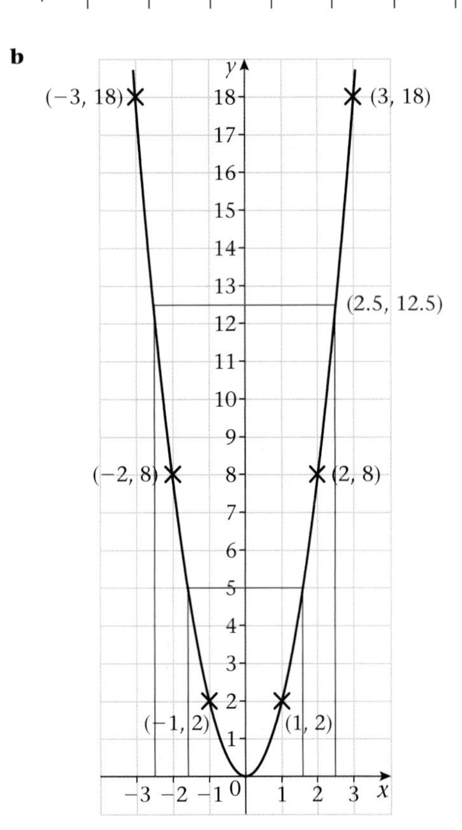

c **i** $y = 12.5$

ii $x = -2.4$ or 2.4

d (0, 0)

e $x = -1.6$ or 1.6

3 **a**

x	−4	−3	−2	−1	0	1
y	**5**	**1**	**−1**	−1	**1**	**5**

b

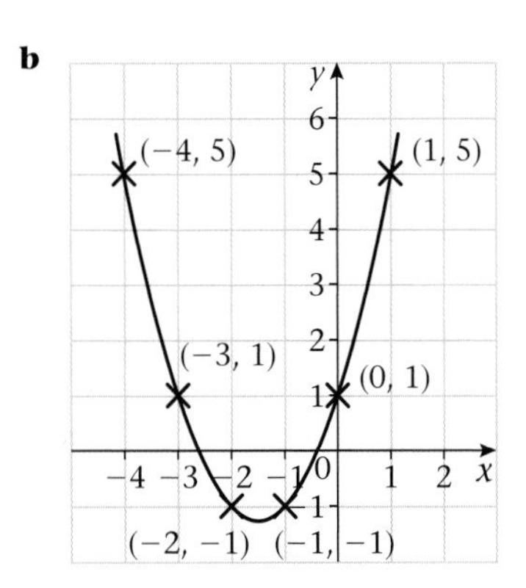

c $x = -2, 6$ or $-0, 4$

4 **a**

x	−1	0	1	2	3	4	5
y	6	1	−2	−3	−2	1	6

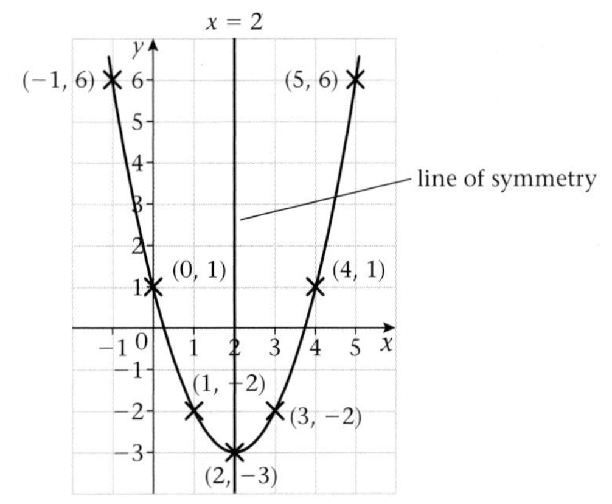

i minimum value: $y = -3$

ii $x = 2$ is the line of symmetry

iii $y = 0$ when $x = 0.3$ or 3.7

$y = -2$ when $x = 1$ or 3

b

x	−4	−3	−2	−1	0	1	2
y	29	12	1	−4	−3	4	17

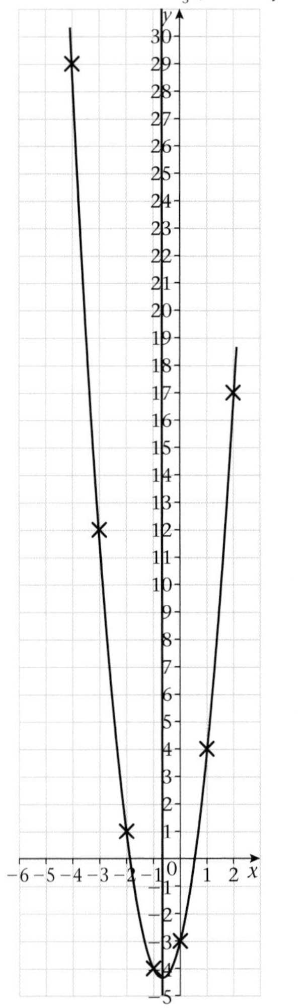

i minimum value: $y = -4\frac{1}{3}$

ii line of symmetry $x = -\frac{2}{3}$

iii $y = 0$ when $x = -1.9$ or 0.5

$y = -2$ when $x = -1.5$ or 0.2

c

x	-4	-3	-2	-1	0	1	2	3	4
y	-12	-5	0	3	4	3	0	-5	-12

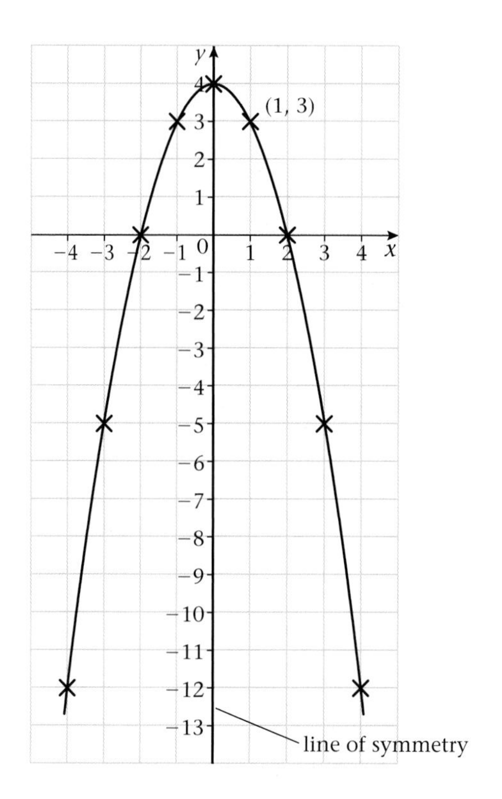

i maximum value $y = 4$

ii line of symmetry $x = 0$ (y-axis)

iii $y = 0$ when $x = -2$ or 2
$y = -2$ when $x = -2.4$ or 2.4

d

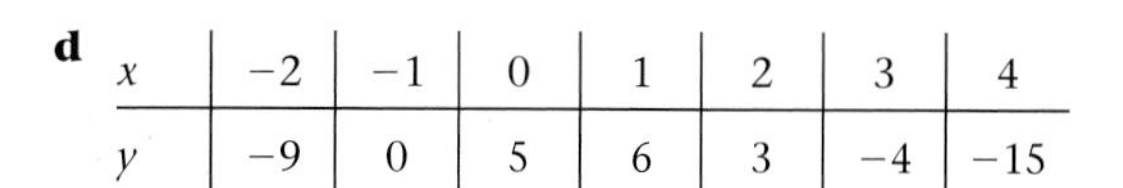

x	-2	-1	0	1	2	3	4
y	-9	0	5	6	3	-4	-15

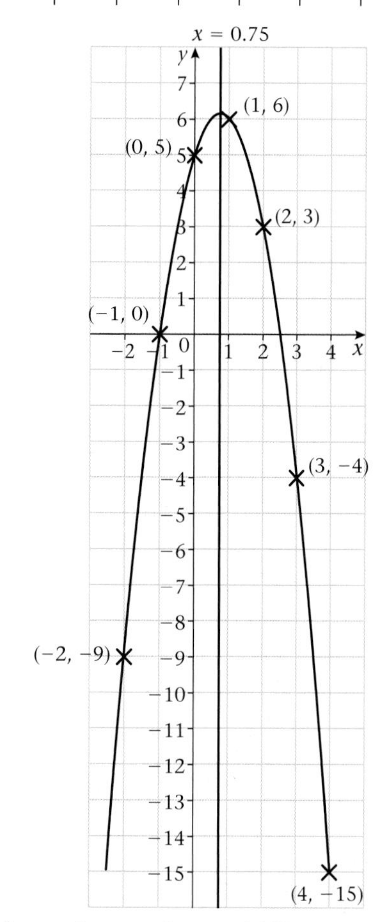

i maximum value: $y = 6.125$

ii line of symmetry $x = 0.75$

iii $y = 0$ when $x = -1$ and 2.5
$y = -2$ when $x = 1.2$ or $x = 2.4$

Practice 11A

1	a	$x(x + 3)$	b	$2x(x + 5)$
	c	$x(5x + 1)$	d	$2x(3x - 1)$
	e	$3x(2x + 5)$	f	$2x(2 - 3x)$
2	a	$(x + 1)(x + 4)$	b	$(x + 3)(x + 3)$
	c	$(x - 3)(x + 4)$	d	$(x + 3)(x - 5)$
	e	$(x + 6)(x - 1)$	f	$(x - 6)(x + 3)$
3	a	$(x - 3)(x - 1)$	b	$(x - 4)(x - 1)$
	c	$(x - 4)(x - 3)$	d	$(x - 2)(x - 1)$
4	a	$(x - 3)(x + 3)$	b	$(x - 5)(x + 5)$
	c	$(x - 2)(x + 2)$	d	$(x - 6)(x + 6)$
5	a	$3(x - 3)(x + 3)$	b	$2(x - 2)(x + 2)$
	c	$5(x - 2)(x + 2)$	d	$3(x - 5)(x + 5)$
6	a	$(x + 7)(x - 1)$	b	$(x - 4)(x - 2)$

Practice 11B

1 $x = -6$ or $x = 1$
2 $x = -1$
3 $x = -4$ or $x = 5$
4 $x = -5$ or $x = -2$
5 $x = -3$ or $x = 5$
6 $x = -3$ or $x = 4$
7 $x = -5$ or $x = -3$
8 $x = -5$ or $x = 3$
9 a $x = 2$ or $x = 4$
b $x = -7$ or $x = 1$

Practice 11C

1 $s = \frac{P}{4}$
2 $w = \frac{A}{\ell}$
3 $a = \frac{P - b}{2}$
4 $t = \frac{u - v}{a}$
5 $T = \frac{D}{V}$
6 $h = \frac{3V}{\pi r^2}$
7 $b = 2m - a - 2c$
8 $x = \frac{5}{2}y + \frac{35}{2}$
9 $x = \frac{y}{5} + 2$
10 $b = \frac{2A}{h} - a$
11 $t = \frac{v - u}{10}$
12 $x = \pm\sqrt{5y - 4}$
13 $u = \pm\sqrt{v^2 - 2as}$

Practice 12A

1 a, b

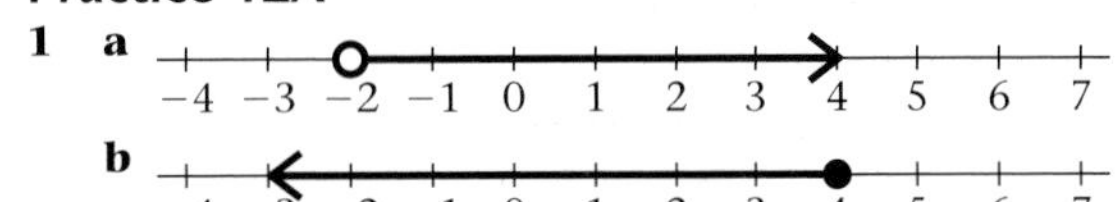

c

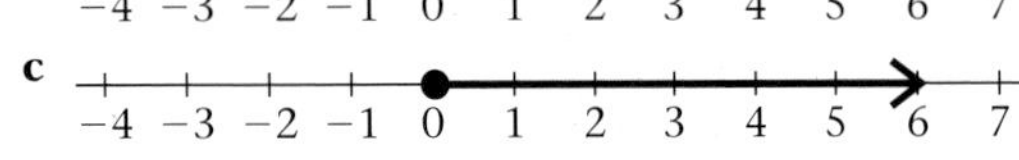

d

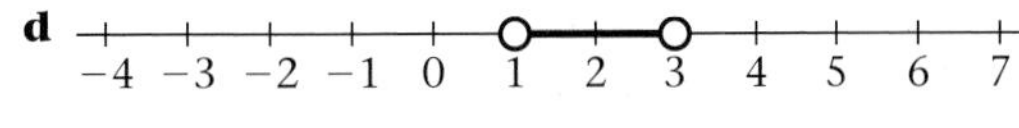

e, f, g, h

2 a $x \leqslant 2$
b $-3 < x < 1$
c $-1 \leqslant x < 2$
d $1 < x \leqslant 4$

Practice 12B

a $x \geqslant 2\frac{2}{3}$
b $x < -2$
c $x < \frac{1}{4}$
d $x > 4\frac{1}{2}$
e $x \leqslant 5$
f $x < 8$
g $x \geqslant 9$
h $x > -5\frac{1}{2}$
i $x \leqslant 3$
j $x \leqslant 2$
k $x \leqslant 2\frac{1}{2}$
l $y > -\frac{3}{5}$

Practice 12C

a

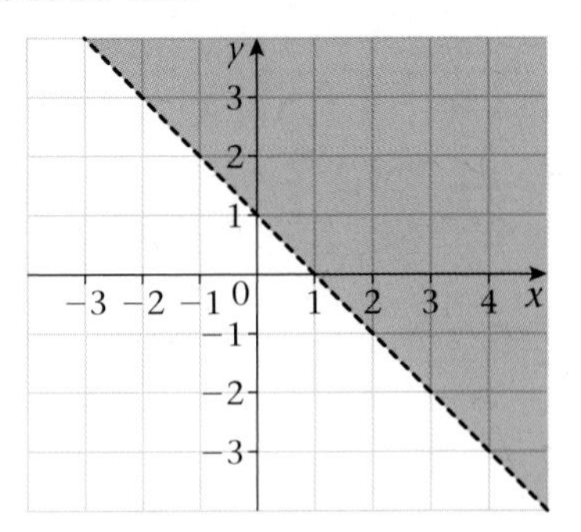

b

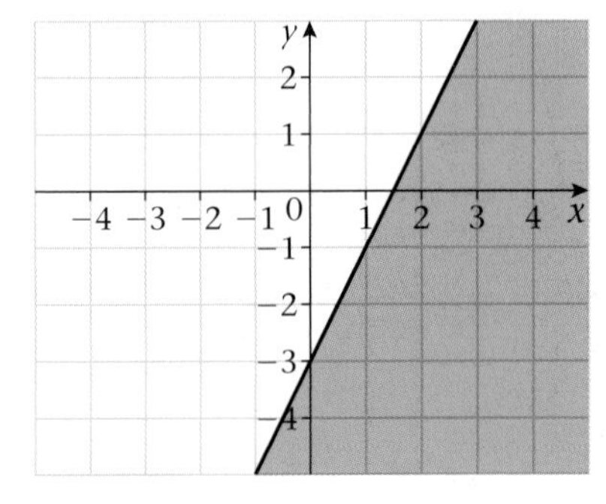

c

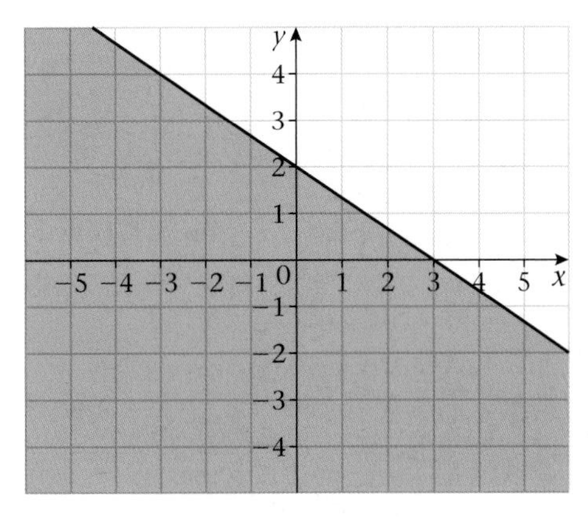

d

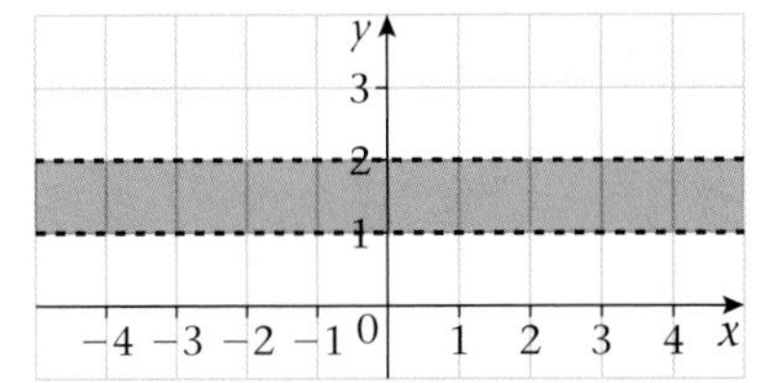

e

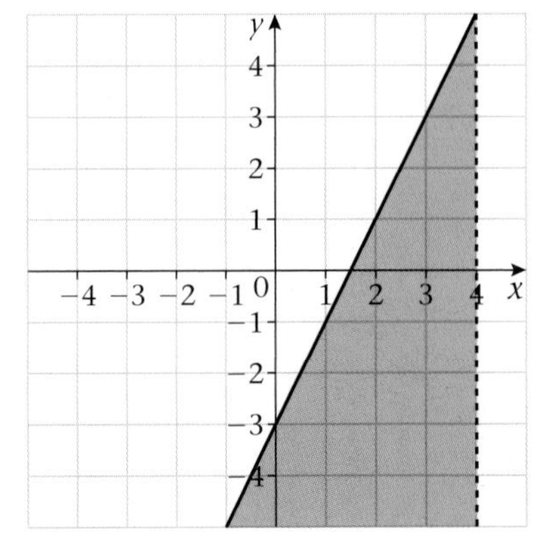

f

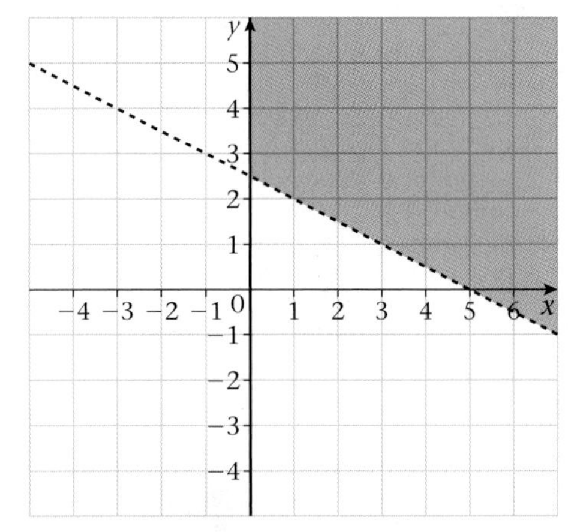

g

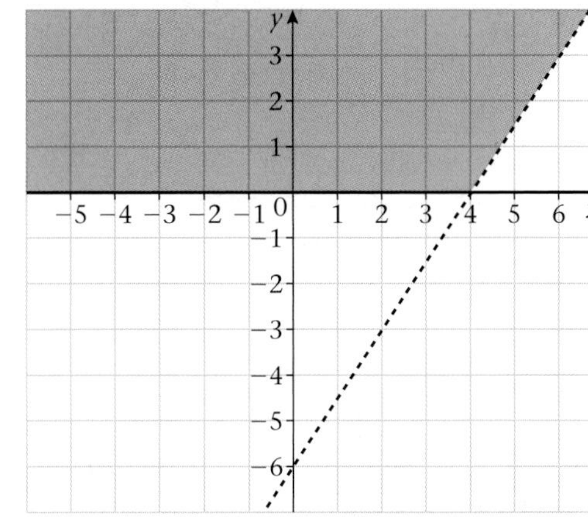

h

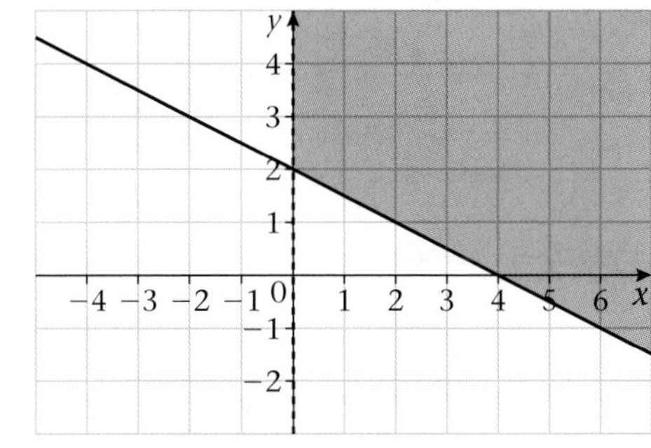

i

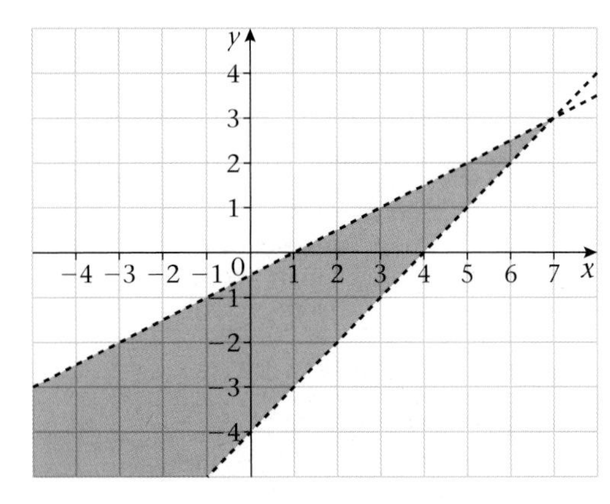

j

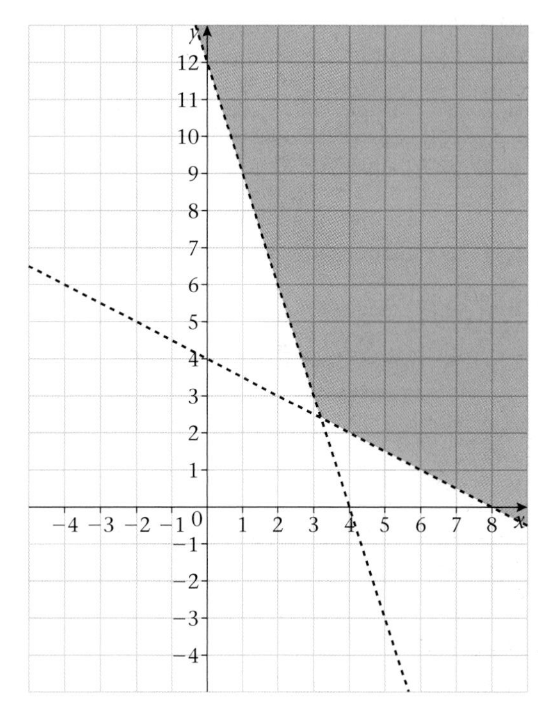

Practice 13A

1 $a = 107°$ **2** $b = 61°$
3 $c = 128°$ **4** $d = 47°$
5 $e = 48°$ **6** $f = 143°$ $g = 37°$
7 $k = 128°$, $j = 52°$, $l = 38°$ **8** $m = 45°$
9 $n = 62°$ **10** $p = 34°$
11 $q = 72°$ $r = 36°$ **12** $s = 67.5°$ $t = 22.5°$

Practice 13B

1 $a = 53°$ (alternate angles)
2 $b = 106°$ (alternate angles)
3 $p = 110°$ (corresponding angles)
$q = 30°$ (angles on a straight line; angles in a triangle add up to 180°)
$r = 30°$ (alternate angles)
4 $a = 68°$ (angles on a straight line)
$b = 68°$ (alternate angles)
$c = 70°$ (alternate angles)
$d = 42°$ (angles on a straight line)
$e = 42°$ (alternate angles)
$f = 42°$ (corresponding angles)
5 Angle MQR = 73° (corresponding angles)
$d = 73°$ (vertically opposite angles)
6 Angle XYH = 46° (alternate angles)
$e = 180° - 46°$
$= 134°$ (angles on a straight line)

Practice 13C

1 **a** $a = 34°$ (angles in a triangle add up to 180°)
b $b = 56°$ (base angles of an isosceles triangle are equal; angles in a triangle add up to 180°)
c $c = 64°$ (vertically opposite angles are equal; base angles of an isosceles triangle are equal; angles in a triangle add up to 180°)

2 **a** $d = 40°$ (angles on a straight line; base angles of an isosceles triangle are equal; angles in a triangle add up to 180°)

b $e = 5°$ (angles on a straight line; angles in a triangle add up to 180°)

c $f = 32°$ (angles on a straight line; base angles of an isosceles triangle are equal; angles in a triangle add up to 180°)

3 **a** $g = 80°$ (angles in a quadrilateral add up to 360°)

b $h = 135°$ (angles on a straight line; alternate angles)

c $j = 70°$ (base angles of an isosceles triangle are equal; alternate angles)

$k = 130°$ (angles on a straight line; angles in a quadrilateral add up to 360°)

4 **a** $m = 120°$ (angles in a pentagon add up to 540°)

5 **a** interior angle = 140°
exterior angle = 40°

b 30 sides

Practice 13D

1 Angle EBA = 25° (base angles of an isosceles triangle)
Angle BEA = 130° (angles in a triangle add up to 180°)
Angle DEB = 50° (angles on a straight line)
Angle DCB = Angle EBA (corresponding angles)
= 25°

2 Angle YXB = 110° (base angles of isosceles triangle are equal; angles in a triangle add up to 180°)
Angle YXC = 70° (angles on a straight line)
Angle CYX = 70° (base angles of isosceles triangle)
Angle YCX = 40° (angles in a triangle sum to 180°)
Angle AYB = 180° − 70° − 35°
= 75° (angles on a straight line)
Angle YAB = 75° (base angles of isosceles triangle)
Angle ABY = 30° (angles in a triangle sum to 180°)

3 Angle CQR = 65° (alternate angles)
Angle CR̂Q = 65° (base angles of an isosceles triangle)
Angle AĈB = 65° (corresponding angles)
Angle CAB = 65° (base angles of isosceles triangle)
Angle AP̂Q = 180 − 65° − 65° (angles of triangle APR add up to 180°)
= 50°

4 Angle ABC = 44° (corresponding angles)
Angle ACB = 124° (angles in triangle ABC add up to 180°)
Angle BCQ = 56° (angles on a straight line)
Angle CBQ = 56° (base angles of an isosceles triangle)
Angle BQP = 56° (alternate angles)

5 $(180° - a) + (180° - b) + (180° - C) + (180° - d) = 360°$ (angles in a quadrilateral add up to 360°)
$720° - (a + b + c + d) = 360°$
$720° - 360° = a + b + c + d$
$a + b + c + d = 360°$

Practice 13E

1 58° (angle in a semicircle is 90°, angle sum of a triangle is 180°)

Practice 13F

1 $d = 72°$ (angles on a straight line; opposite angles of a cyclic quadrilateral sum to 180°)
$e = 108°$ (opposite angles of a cyclic quadrilateral sum to 180°)
$f = 114°$ (opposite angles of a cyclic quadrilateral sum to 180°)

2 $g = 124°$ (angles on a straight line; opposite angles of a cyclic quadrilateral sum to 180°)
$h = 56°$ (corresponding angles; opposite angles of a cyclic quadrilateral sum to 180°)

3 $j = 50°$ (angle at centre = 2 × angle at circumference)
$k = 25°$ (angle at circumference = $\frac{1}{2}$ angle at centre)

Practice 13G

1 $m = 27°$ (angles in the same segment are equal)
$n = 51°$ (angles in a triangle add up to 180°; angles in the same segment are equal)

Practice 13H

1 $a = 47°$ $b = 124°$ $c = 74°$

2 $a = 61°$ $b = 76°$

Practice 13I

1 $x = 29°$, $y = 48°$

Practice 13J

1 $p = 31°$ (angle between tangent and radius at point of contact is 90°; angles in a triangle add up to 180°; angles on a straight line; base angles of an isosceles triangle are equal; angles in a triangle add up to 180°)

Practice 13K

1 $m = 44°$ (tangents from a point to a circle are equal, base angles of an isosceles triangle are equal).
$n = 22°$ (radius is perpendicular to tangent)

2 $p = 118°$, $q = 62°$

Practice 13L

1 $q = 16$ (angle in a semicircle is 90°; angles in a triangle add up to 180°)
$r = 16$ (angles in the same segment are equal)

2 Angle OTP = 37° (alternate angles)
Angle OPT = Angle OTP = 37° (base angles of an isosceles triangle are equal)
Angle PTY = 90° − 37° (angle between tangent and radius at point of contact is 90°)
PTY = 53°

3 Angle OTR = 90° (angle between tangent and radius at point of contact is 90°)
Angle TOR = 72° (angle in triangle TOR sum to 180°)
Construct AT and TP
Angle OTB = 54° (triangle OTB is isosceles)
Angle BAT = 36° (angle at circumference = $\frac{1}{2}$ angle at centre)
Angle ATP = 36° (alternate angles)
Angle ATO = 90° − 36° (angle between tangent and radius = 90°)
= 54°
Angle ATB = Angle ATO + Angle OTB = 54° + 54° = 108°
Angle AXB = 180° − 108° = 72° (opposite angles in cyclic quadrilateral sum to 180°)

4 Angle PAO = 90° = Angle OBP (angle between tangent and radius at point of contact is 90°)
Angle AOB = 105° (angles in a quadrilateral sum to 360°)
Angle AXB = 52.5° (angle at circumference = $\frac{1}{2}$ angle at centre)
Angle AYB = 127.5° (opposite angles of a cyclic quadrilateral sum to 180°)

Practice 13M

1 **a** 070°
b 089°
c 158°
d 241°
e 332°
f 252°
g 043°
h 229°

2

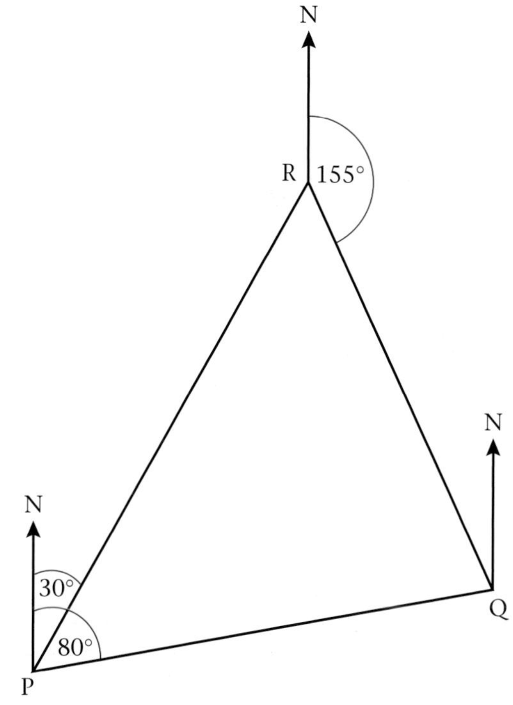

3 123°
4 295°
5 **a** 054°
b 344°
c 305°
d 164°

Practice 14A

1 **a**

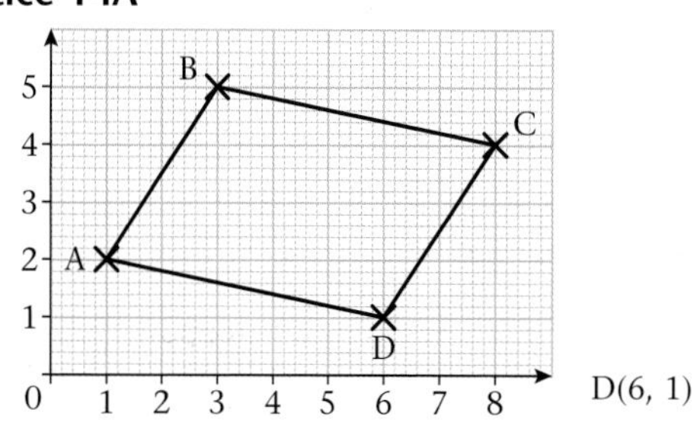

D(6, 1)

b

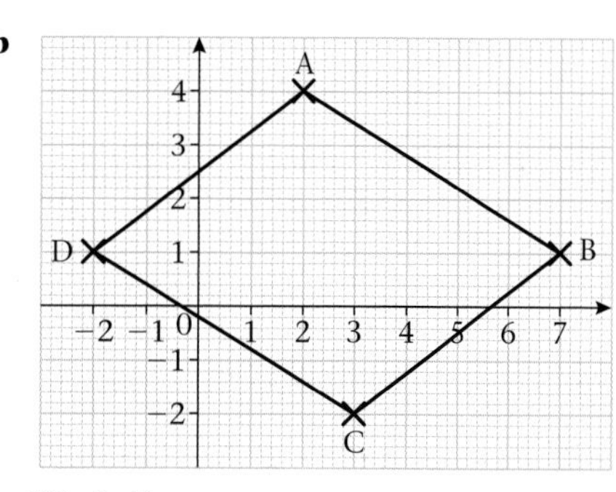

D(−2, 1)

c

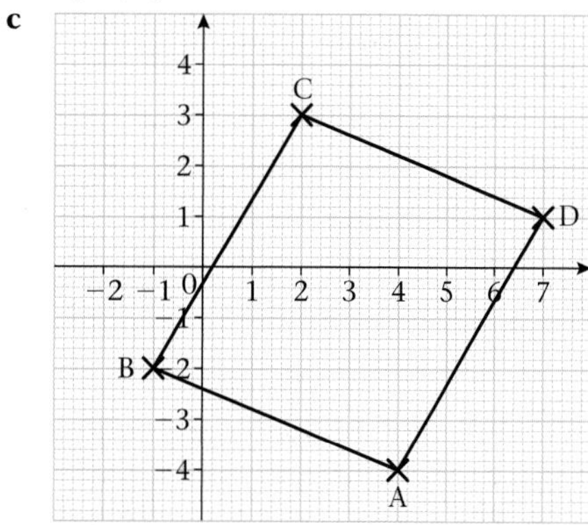

D(7, 1)

2 (−2, 8); (6, 2) or (8, 6)
3 (6, −3)
4 (5, 2) and (5, 4) (OR (1, 2) and (1, 4))
5 (5, 3) and (4, 6)
6 **a** For example (3, 3) **b** $y = 3$
7 **a** arrowhead **b** triangle
c kite **d** parallelogram
8 (4, 4)
9 $(-4\frac{1}{2}, -1)$
10 $x = 1$ and $y = -4$ LM = $2\sqrt{2}$
LN = $\sqrt{2}$
11 $x = -5$ and $y = -9$

Practice 14B

1

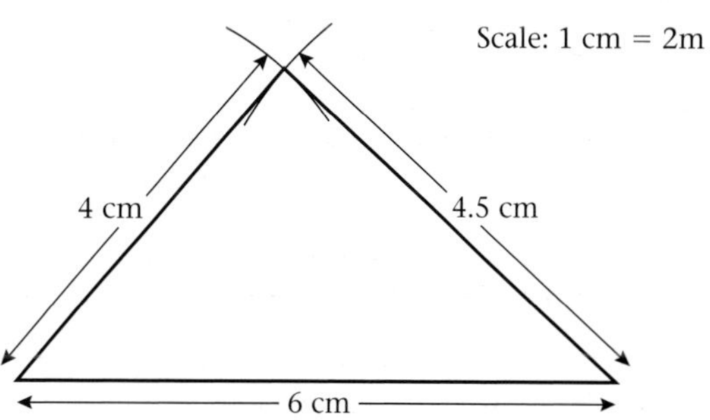

2

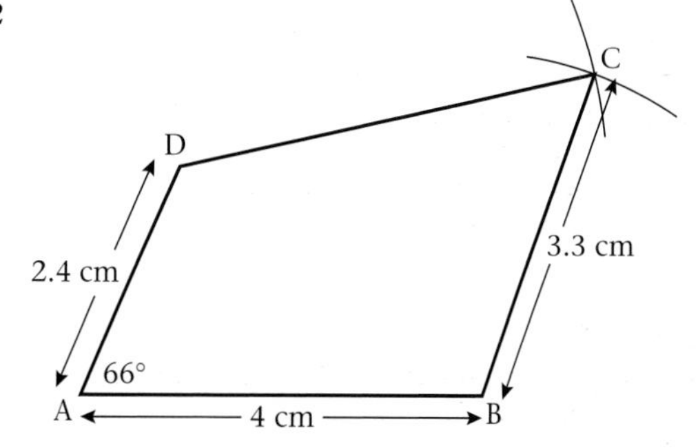

3

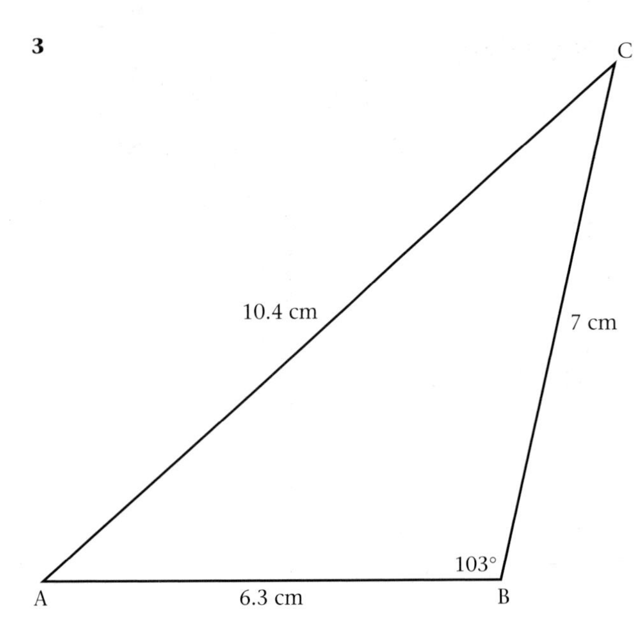

4

3 cm

2 cm

6 cm

5

6

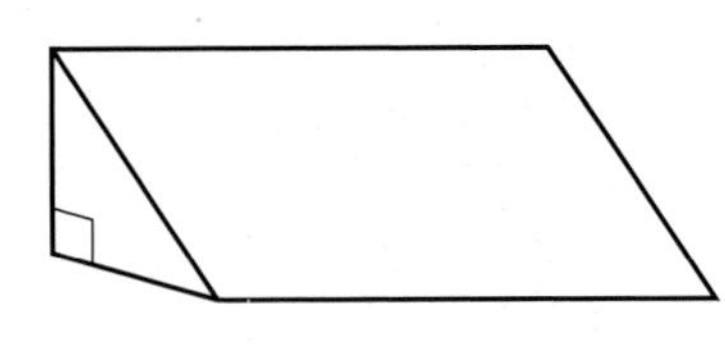

7 **a**, **b**, **c** and **e**

Practice 14C

1 **a**

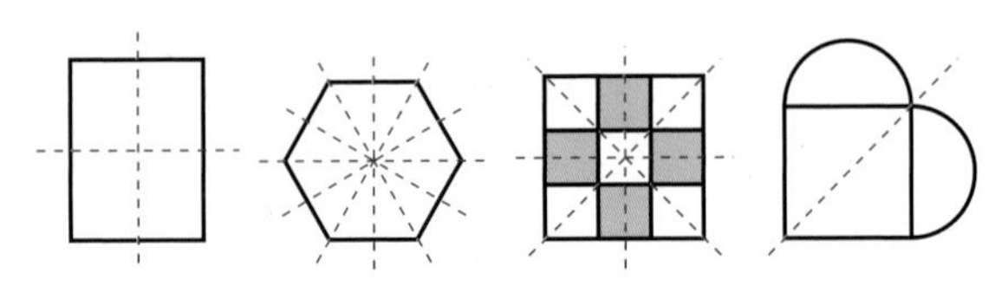

2 **a** 2
b no rotational symmetry
c 6
d 4

3

4 **a**

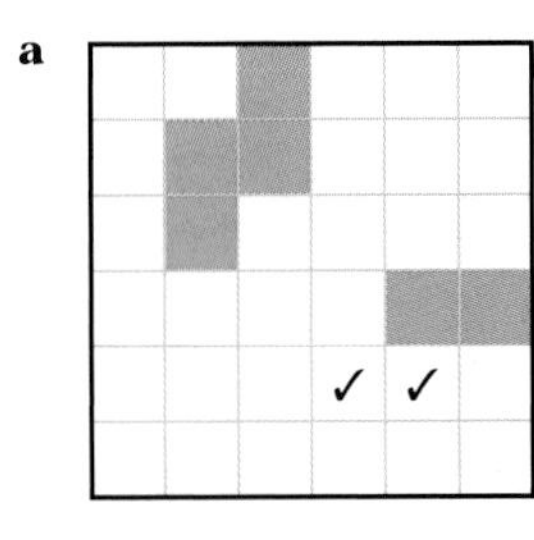

b

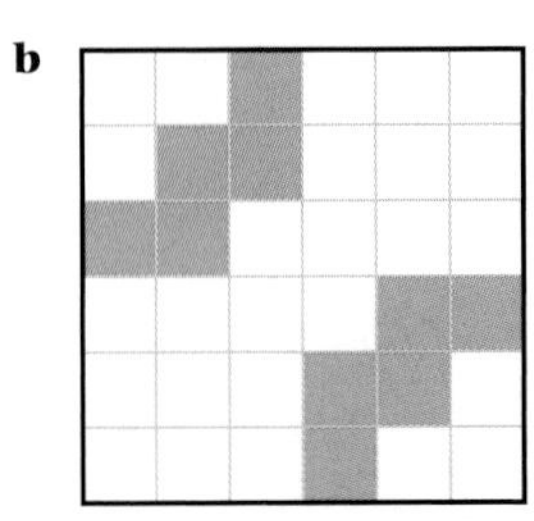

(4 squares shaded in)
order of rotational symmetry = 2

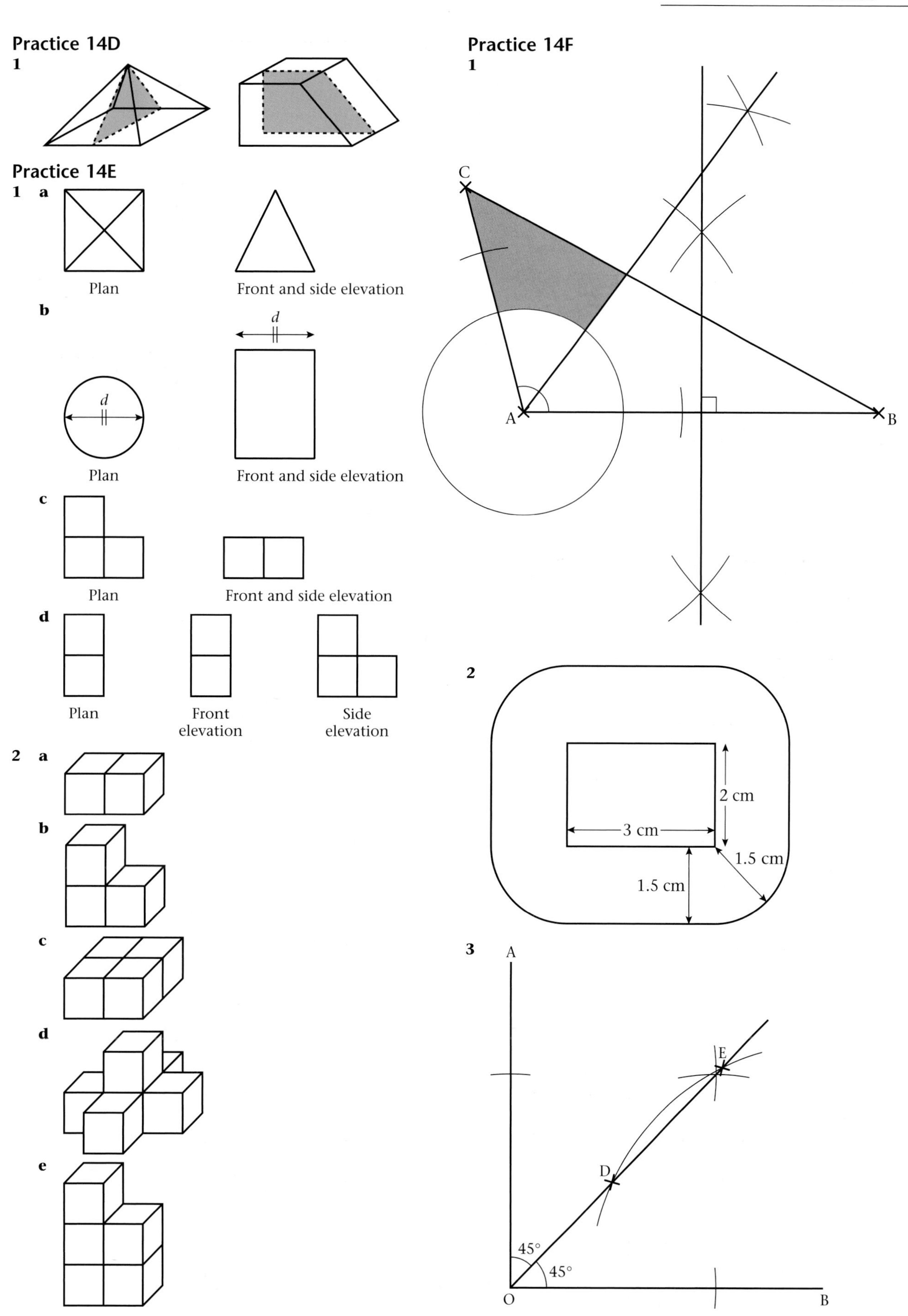

Practice 14D
1
Practice 14E
1 a
Plan
Front and side elevation
b
d
d
Plan
Front and side elevation
c
Plan
Front and side elevation
d
Plan
Front elevation
Side elevation
2 a
b
c
d
e
Practice 14F
1
C
A
B
2
2 cm
3 cm
1.5 cm
1.5 cm
3
A
E
D
45°
45°
O
B

4

A

10 cm

C

6 cm

10 cm

B

5

C

A B

C

6

A B

D C

Practice 15A

1 **a**

b

c

d

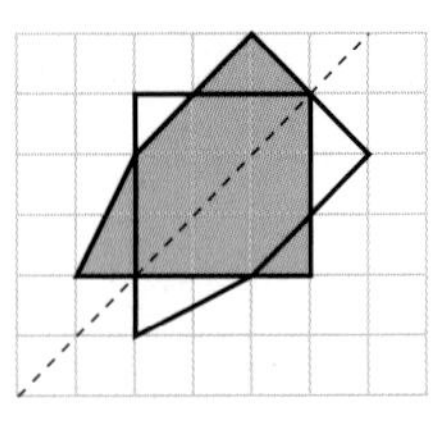

2 **a** reflect in line $x = 4$
b reflect in line $y = -\frac{1}{2}$
c reflect in line $y = -x + 5$

3

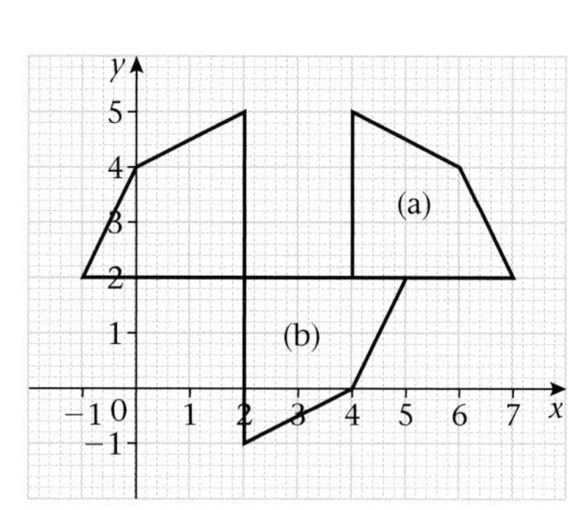

Practice 15B

1 **a** Quarter turn, clockwise with centre at (5, 1)
  b Quarter turn, anti-clockwise with centre at (0, 0)
  c Half turn with centre (2, 1)

2

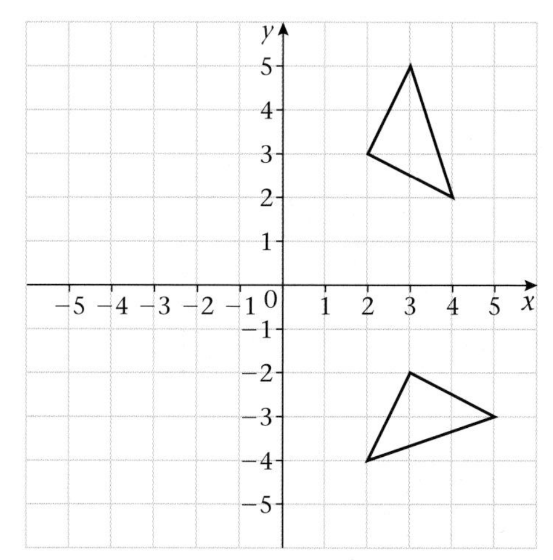

3 **a** Half turn, with centre (0, 0)
  b

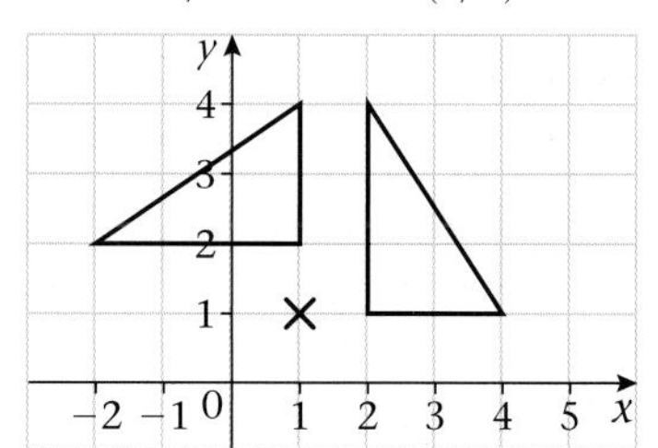

4

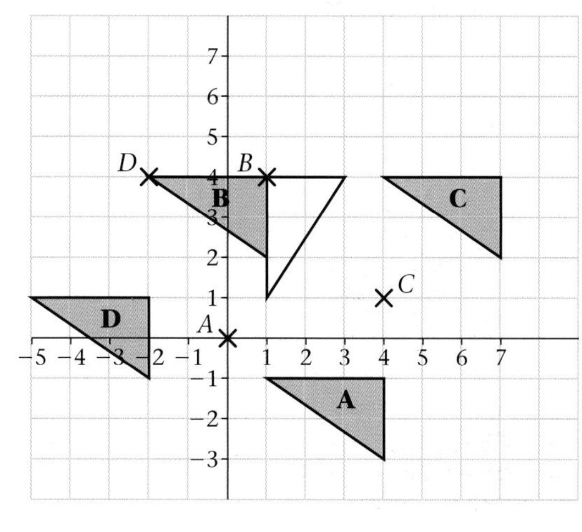

Practice 15C

1 $\begin{pmatrix} -3 \\ 1 \end{pmatrix}$

2

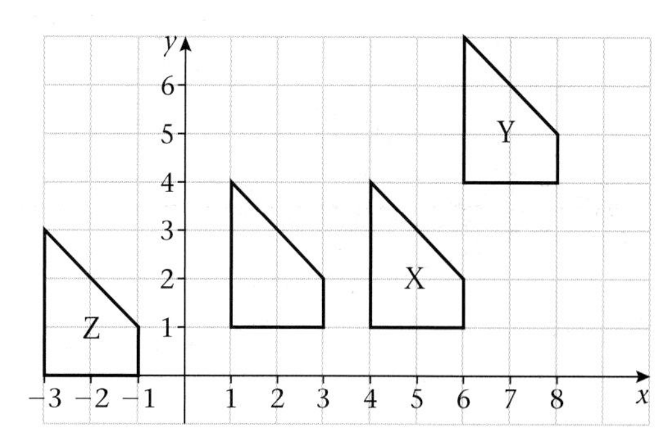

3

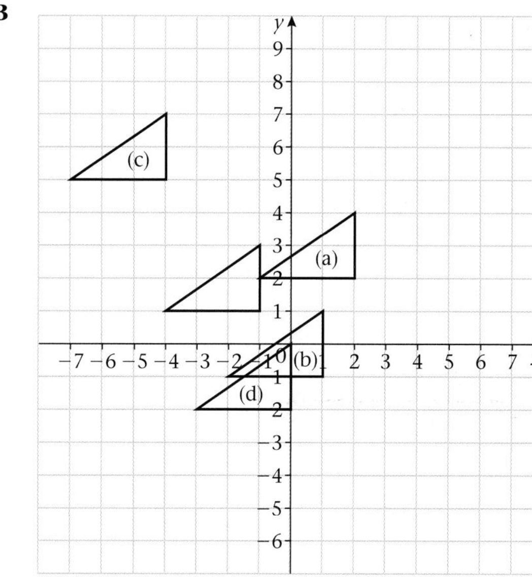

4 **a** $\begin{pmatrix} -7 \\ 0 \end{pmatrix}$

  b $\begin{pmatrix} 2 \\ -7 \end{pmatrix}$

  c $\begin{pmatrix} 2 \\ -4 \end{pmatrix}$

  d $\begin{pmatrix} -9 \\ 7 \end{pmatrix}$

Practice 15D

1

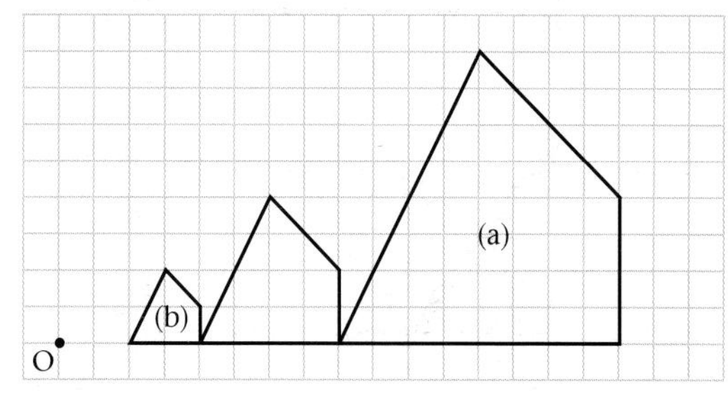

2 For **B** enlargement by a scale factor of $\frac{1}{3}$ with centre (3, 1)
For **C** enlargement by a scale factor of $\frac{4}{3}$ with centre (6, 1)

3

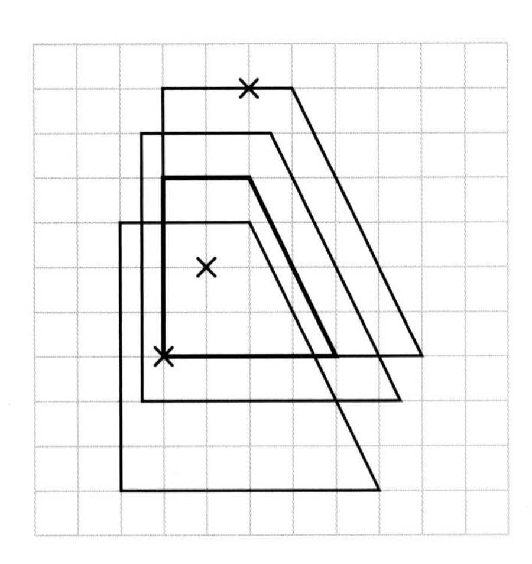

4

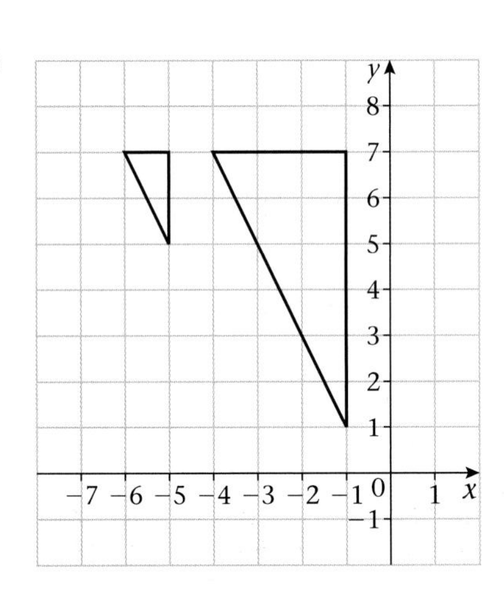

5 **a** 2

b

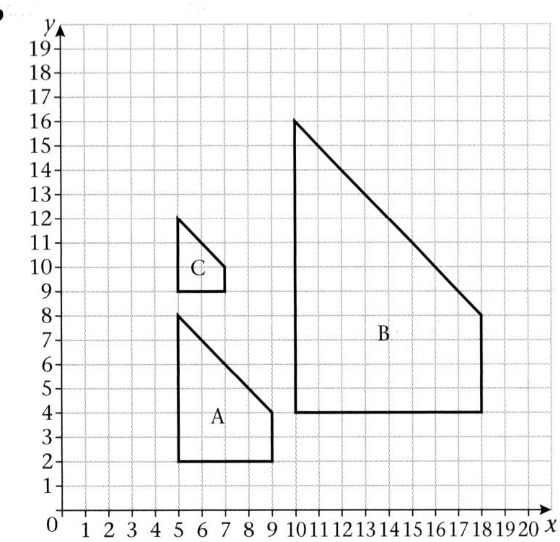

Practice 15E

1 **a** AD = 7.5 cm

b BC = 3 cm

2 CD = 8.25 cm

Practice 15F

1 **a** 4.2 km

b 7.4 cm

2 **a**

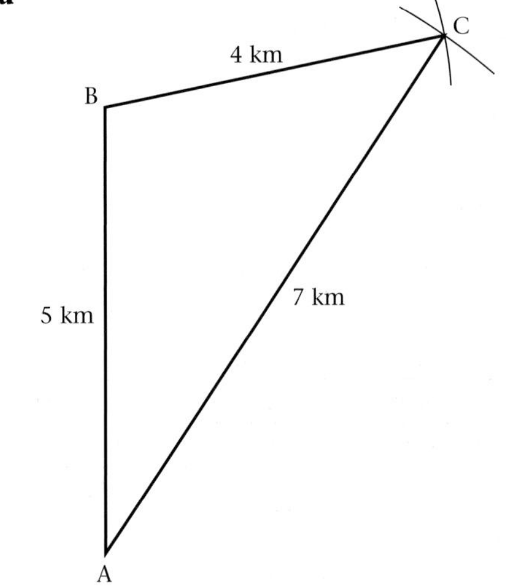

b 034°

c 214°

3

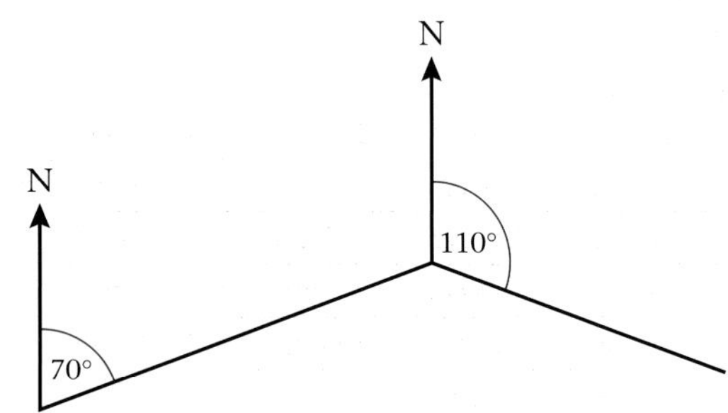

4

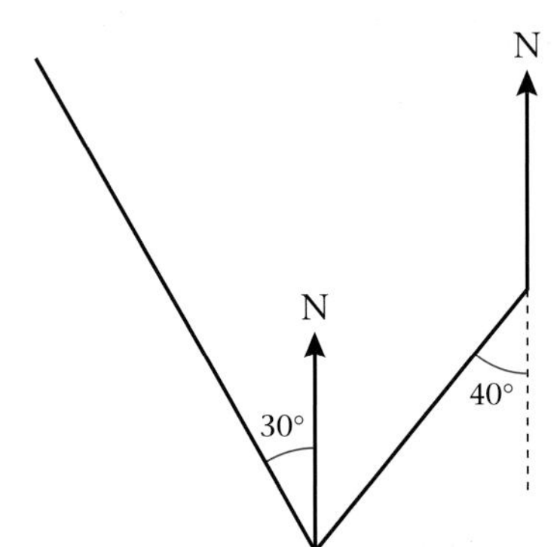

Practice 16A

1 **a** 35 cm^2 **b** 16 cm^2 **c** 0.1 m^2

d 30 cm^2 **e** 80 cm^2 **f** 4.5 m^2

2 **a** 46 m **b** 84 m^2

3 **a** **i** 30 cm **b** **i** 34 m

ii 46 cm^2 **ii** 65 m^2

4 **a** **i** Perimeter = 46 cm

ii Area = 126 cm^2

b **i** Perimeter = 54 cm

ii Area = 108 cm^2

5 Area = 120 cm^2

6 Area = 96 cm^2

7 8 m^2

Practice 16B

1 **a** 18.8 cm, 28.3 cm^2 **b** 37.2 mm, 113 mm^2
c 26.1 cm, 54.1 cm^2
2 **a** 26.6 m **b** 557 m^2
3 **a** 18.1 cm **b** 5.78 cm
4 15 313 cm^2,
5 87.5 cm^2, 54.8 cm
6 59.4 mm^2
7 94.3 cm, 224 cm^2

Practice 16C

1 408 cm^2
2

Length	Width	Height	Surface area
2 m	0.5 m	0.5 m	**4.5 m^2**
40 cm	25 cm	10 cm	**3300 cm^2**
50 cm	40 cm	5 mm	**4000 cm^2**
1.8 m	50 mm	40 mm	**0.328 m^2**
3 m	15 cm	8 mm	**0.9504 m^2**

3 1013 cm^2
4 **a** 7900 cm^2 **b** 3120 cm^2
5 3393 cm^2
6 2.91 m^2
7 radius = 7.96 cm
8 length = 4.074 m

Practice 16D

1

	Length	Width	Height	Volume
a	5 cm	6 cm	**3 cm**	90 cm^3
b	14 cm	**5.5 cm**	3 cm	231 cm^3
c	1.2 m	15 cm	**2 cm**	3600 cm^3
d	2 m	10 cm	35 mm	**7000 cm^3**
e	**5 cm**	34 cm	25 mm	425 cm^3
f	48 cm	**3.6 cm**	40 mm	691.2 cm^3

2 48 packets
3 16 cm
4 9077 cm^3
5 0.496 m^3
6 540 m^3

Practice 16E

1 7.07 m/s
2 5.04 m/s
3 166.95 km
4 9 min 28 sec
5 31.5 km/h
6 718.75 kg/m^3 OR 0.71875 g/cm^3
7 2300 kg
8 0.307 m^3
9 50 kg
10 **a** 700 cm^3 **b** 13.51 kg

Practice 16F

1 **a** 10.5 kg **b** 24 km
c 40.5 ℓ **d** 15 cm
e 125 g **f** 3.4 ℓ
g 120 cm **h** 164.8 km
i 21 kg **j** 6.25 cm
2 **a** 37.4 lbs **b** 15 miles
c 6 inches **d** 5.25 pints
e 3.5 ft or 42 in **f** 96 inches
g 0.44 lbs **h** 0.375 miles
i 1.313 pints **j** 1.87 lbs
3 49.5 kg
4 170 cm
5 77.2 km
6 5 ft 5 inches
7 £3.76 per gallon
8 35 miles
9 91 drinks
10 48 km/h

Practice 17A

1 $a = \sqrt{74}$ cm
2 $b = \sqrt{85}$ cm
3 $c = \sqrt{97}$ cm
4 $d = \sqrt{96}$ cm or $4\sqrt{6}$ cm
5 AC = 7 cm
6 QR = 3.4 m
7 186.7 km
8 BC = 4.4 m
BD = 7.0 m

Practice 17B

1 3.2 cm
2 **a** 7.0 cm **b** 41.8 cm^2
3 7.2 cm
4 16.2 cm
5 **a** PA = 10 m **b** 8 m
6 4.7 m
7 5.57 m
8 167 km

Practice 17C

1 **a** 13 **b** 27.6 **c** 13 **d** 17
2 **a** 2.5 **b** 16.2 **c** 5.7 **d** 12.5

Practice 17D

1 **a** a = 13.4 **b** b = 18.1 **c** c = 32.4 **d** 20.7
2 **a** 25.4° **b** 32.4° **c** 45.9° **d** 51.0°
3 p = 12.8
q = 11.2
r = 42.1
s = 31.3
4 AX = 6.8 cm
Angle C = 75.3°
5 61.9°

Practice 17E

1 21.9 km SOUTH
20.5 km EAST
2 16.1 km WEST
13.5 km SOUTH
3 47.1 km
4 **a** 321° **b** 141°
5 214.9 m
6 60.2 m

Practice 18A

	Small	Medium	Large	Total
Tea	**25**	40	**85**	150
Coffee	80	45	27	**152**
Chocolate	30	45	**23**	**98**
Total	135	**130**	135	400

Practice 18B

Type	Tally	Frequency
Saloon		
Hatchback		
MPV		
Van		
Bus		
etc.		

Practice 18C

Names can be randomly chosen from a hat OR starting from 1 assign each name a number. Use random numbers from the calculator to choose the 60 males.

Practice 18D

1 For example:

Film survey:
Please tick the boxes which apply to you.

Are you ☐ male ☐ female?

What is your age range?

☐ under 21 ☐ 21–40 ☐ 41–60 ☐ over 60

Which type of film do you enjoy most?

Romance ☐ Horror ☐
Comedy ☐ Fantasy ☐
etc.

2 a Question 1 is leading, question 2 is too vague and question 3 will elicit data that is difficult to use.

b For example:

1 The college needs a new restaurant.
agree ☐ disagree ☐

2 Do you use the restaurant
once a day ☐ two or three times a week ☐
once a week ☐ less than once a week ☐

3 The food at the restaurant is
☐ excellent ☐ good
☐ average ☐ poor

Practice 19A

1 a Saturday **b i** 30 **ii** 45

Practice 19B

1

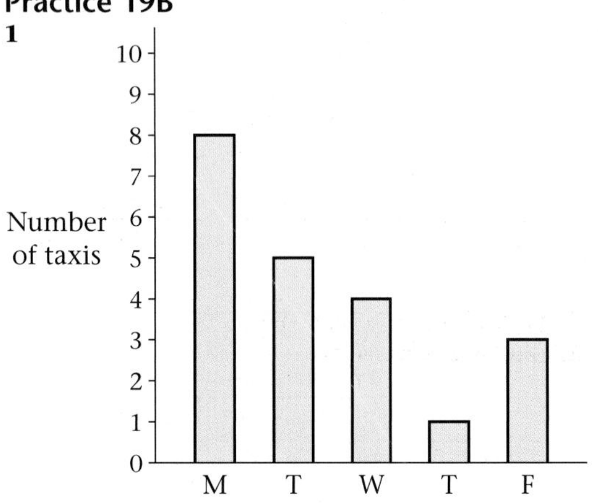

Practice 19C

1 a

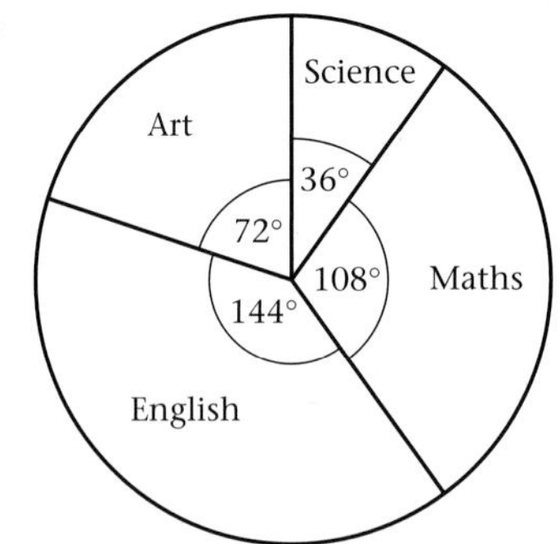

b

Subject	Frequency	Angle
Art	18	72°
English	36	144°
Science	9	36°
Mathematics	27	108°

Practice 19D

1 a 23°C
b 50°F
c December

Practice 19E

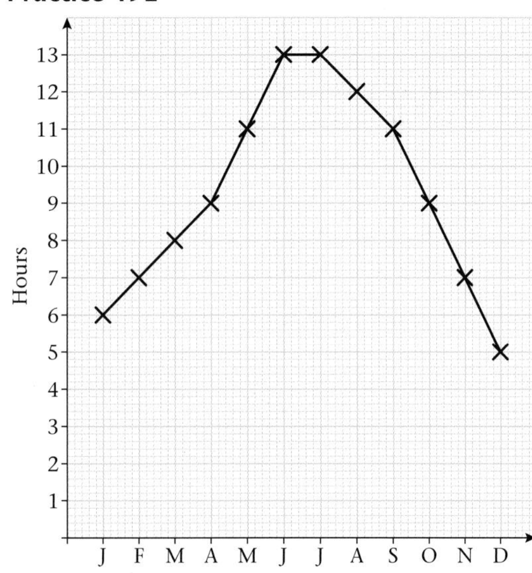

Practice 19F

1 a

3	4, 8, 9
4	0, 1, 2, 5, 8, 9
5	3, 3, 3, 4, 5, 6, 7, 7, 7, 8
6	2, 2, 4, 4, 5, 6
7	0, 5, 7
8	1, 4

b 50

2 a

0	3, 6, 7, 7, 8
1	0, 0, 1, 4, 4, 4, 5, 5, 6
2	
3	2, 6, 7, 9
4	0, 1, 1, 2, 2, 4
5	
6	5, 8
7	1, 7

b 74

Practice 20A

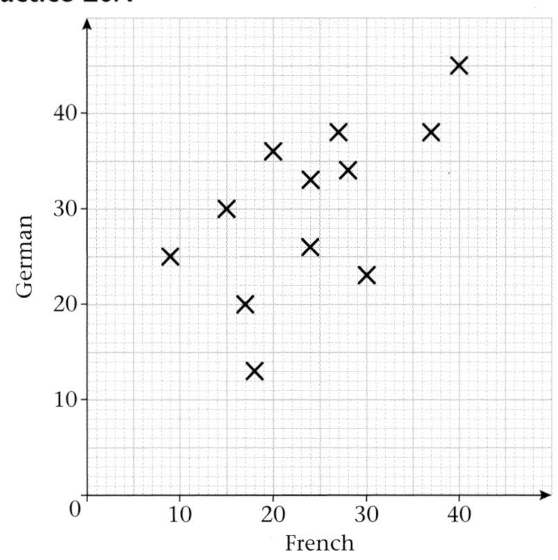

Practice 20B

1 Positive

2 **a** Negative

b None

c Positive

d None

3 **a** **i** No correlation

ii Length of hair

b **i** Positive

ii Shoe size *or* distance around neck *or* waist measurement

Practice 20C

1 **a**

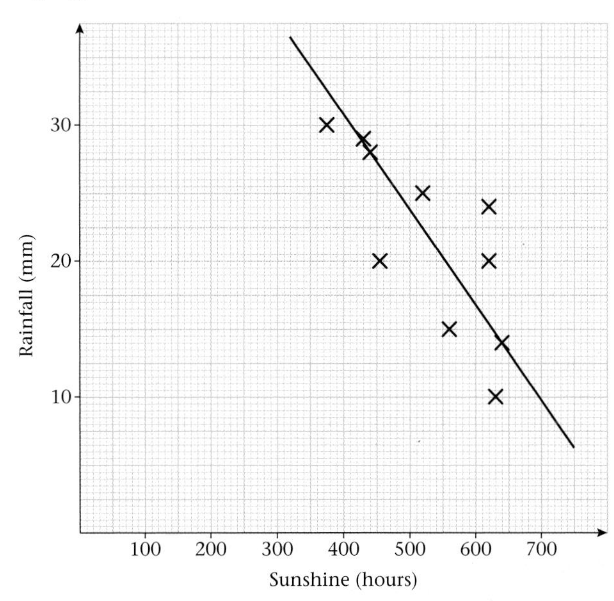

b The relationship between sunshine hours and rainfall is negative i.e. as number of sunshine hours increases, so the amount of rainfall decreases

c See graph above

d **i** 27 mm

ii 580 hours

2 **a**

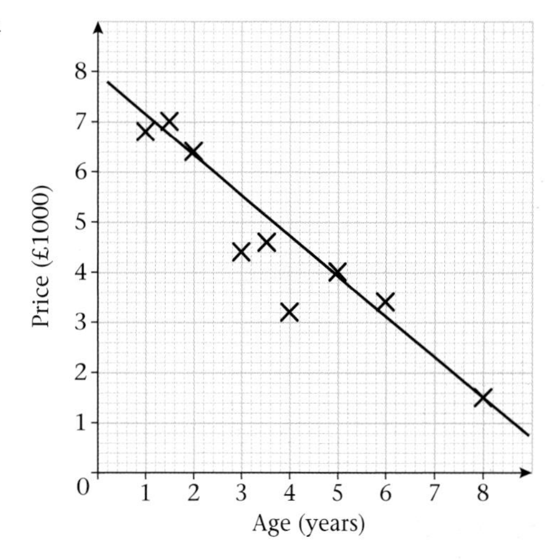

b As age of motorcycles increases the price of the motorcycle decreases (i.e. a negative correlation)

c See graph above

d **i** £2300

ii 3.6 years

3 **a**

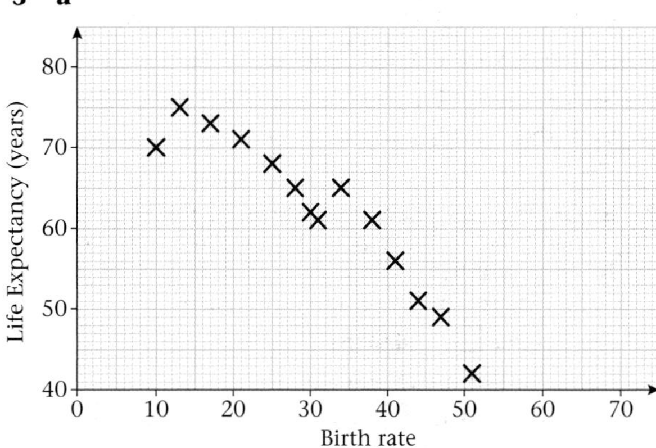

b As birth rate increases, life expectancy falls. There is negative correlation

c

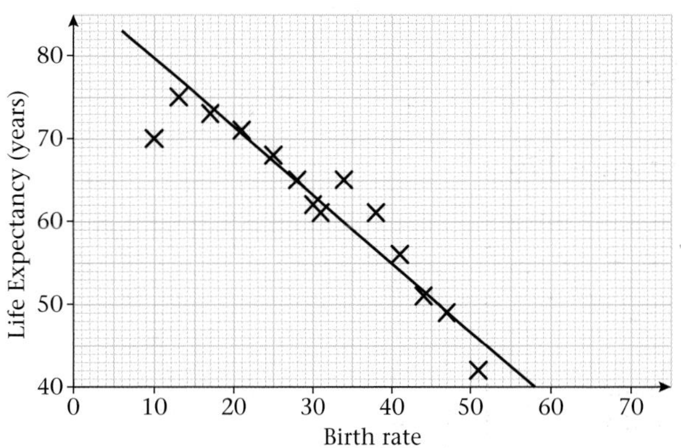

d 53 years (52–54 is acceptable)

e 27 (26–28 is acceptable)

4 **a**

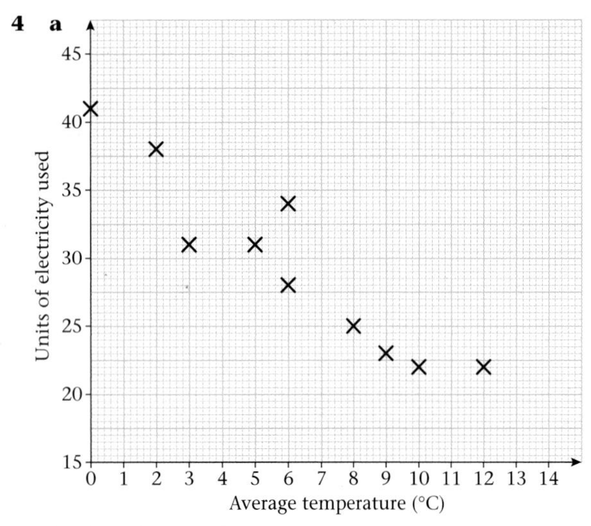

b Negative correlation

c

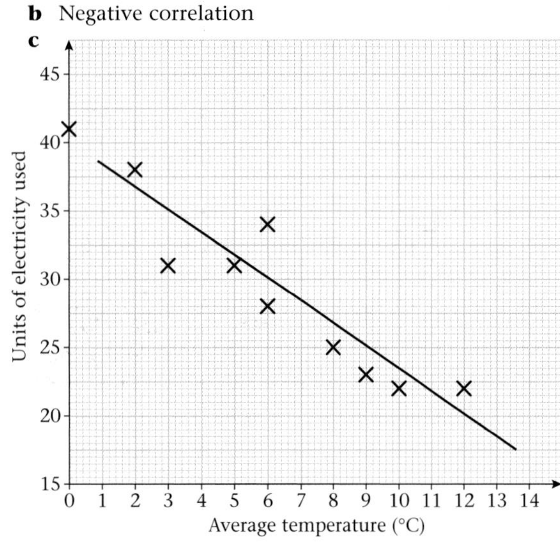

d **i** 3.2°C ± 10% **ii** 28 units ± 10%

Practice 20D

1 As size of floor space increases the house value increases (i.e. positive correlation)

2 **a**

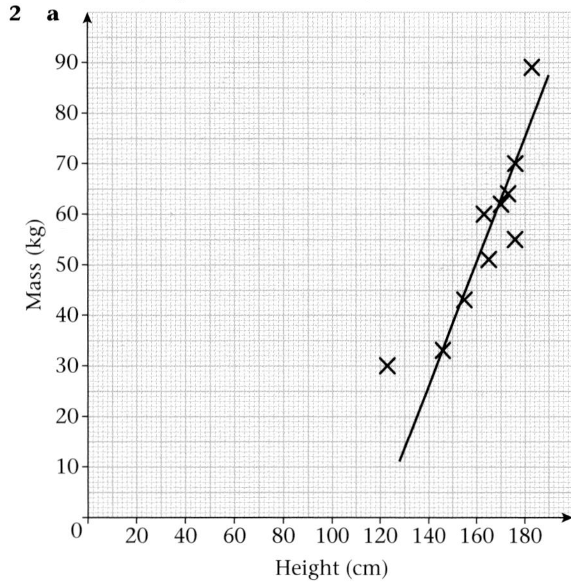

b See graph above

c As height increases, mass increases (i.e. a positive correlation)

d **i** 62 kg **ii** 172 cm

3 **a**

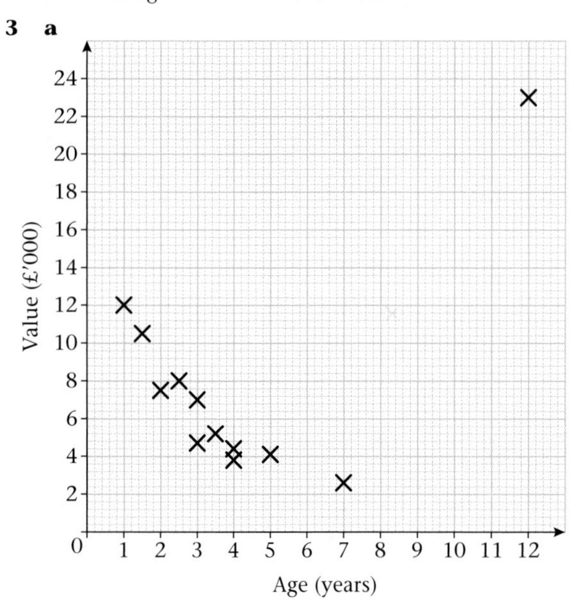

b 12, 23

c

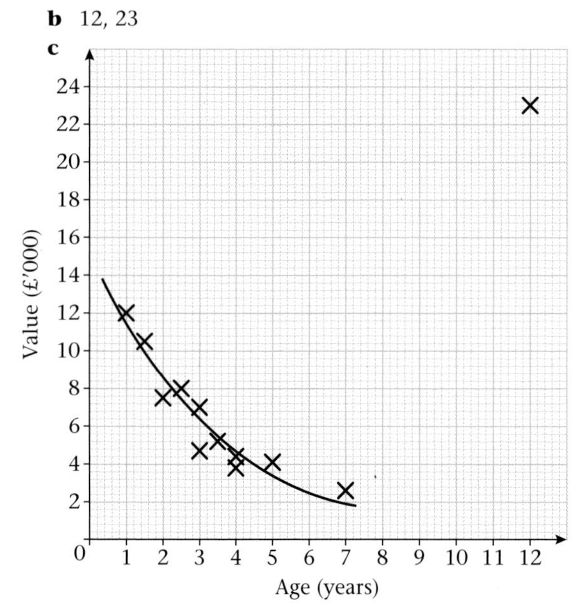

d As age increases, value decreases, but not a straight line (linear) link.

e **i** £2600 ± 10% **ii** 3.3 years ± 10%

Practice 21A

1 2.6

Practice 21B

1 1

Practice 21C

1 2

Practice 21D

1 8

2 **a** 2 **b** 5 **c** 2 **d** 2.2

Practice 21E

1 **a** 2.8 **b** 2 **c** 3

2 **a** 6 **b** 5.5 **c** 5.54

Practice 21F

1 **a** 271.83; 274.17; 234.5; 232.17; 232.5; 235.67; 223.33

b Mr Allis might have been on holiday

2 **a** 104; 106; 109; 112.5; 113; 114.25; 115.5; 117.25; 119

b

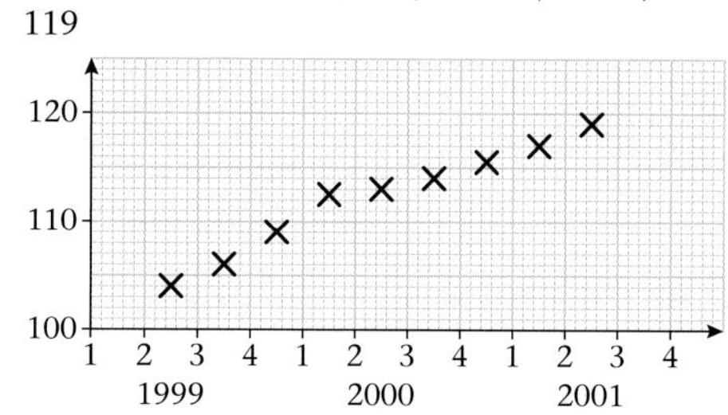

c There is an increase in the averages
The last quarter has the lowest cost
The third quarter has the highest cost

3 **a** 157.25; 158; 161.5; 163.25; 165.75; 167.5; 169.5; 171; 174; 177.5;183; 186; 189

b

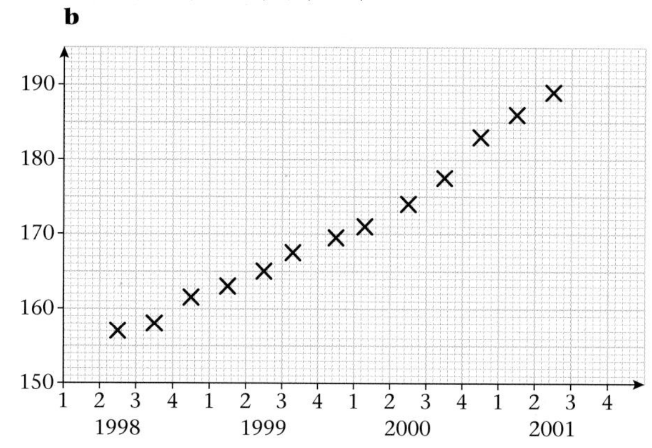

c There is a steady increase of average takings
The 1st quarter has the highest takings
The third quarter has the lowest takings

4 **a, c**

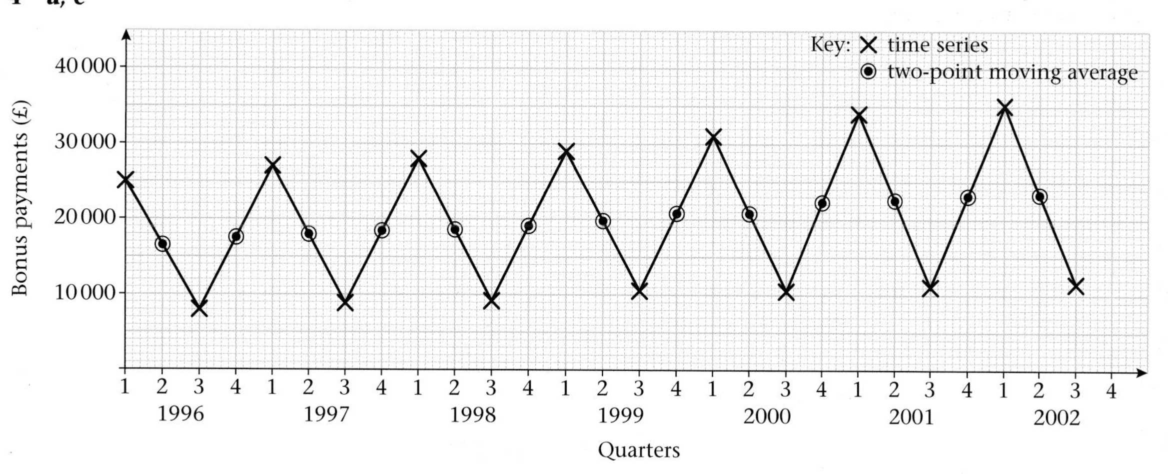

b 16 500, 17 600, 18 150, 18 600, 18 700, 19 250, 19 800, 20 700, 20 850, 22 100, 22 250, 23 100, 23 250

d The bonus is always higher in March than in September. There has been a steady increase in the bonus payments in both March and September over the years.

Practice 22A

1 **a**

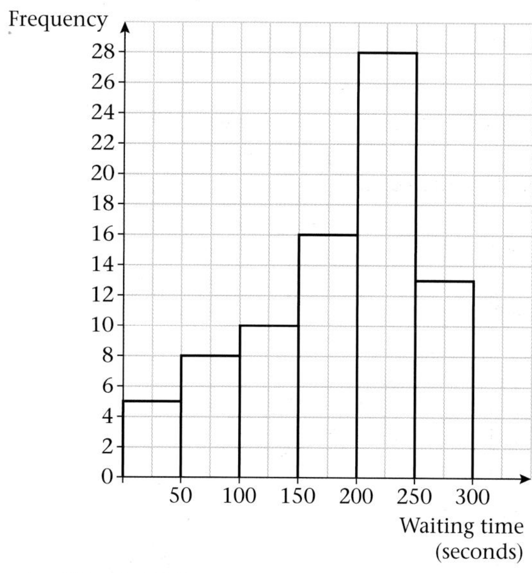

b $200 \leqslant t < 250$

Practice 22B

1 a

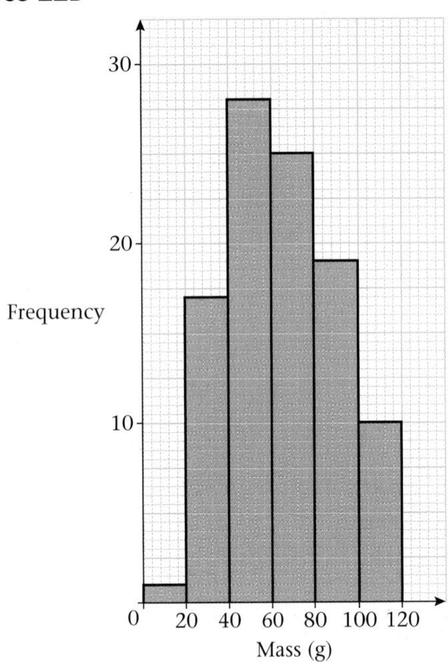

b

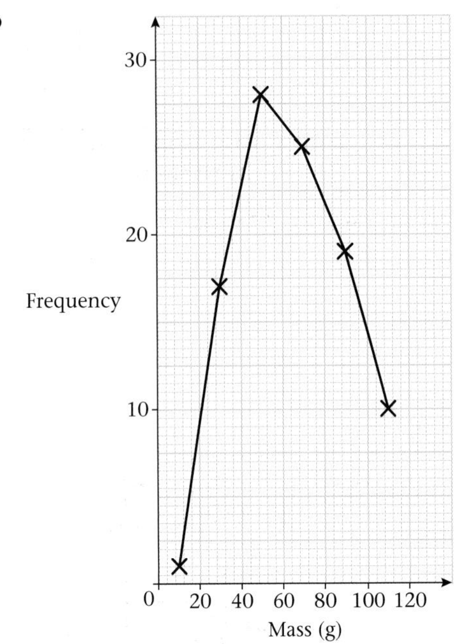

2

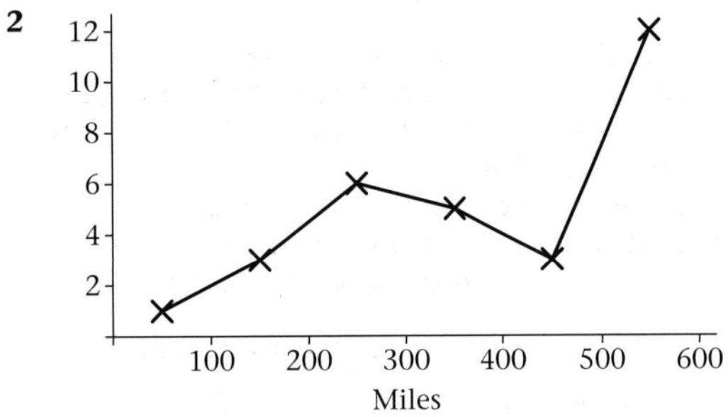

Practice 22C

1 There are more women than men at lower mass.
There are more men than women at higher mass.
Overall the men are heavier than the women.
The modal mass for men is 60–70 kg.
The modal mass for women is 50–60 kg. (Any 3)

Practice 22D

1 98.8 g

2 a $20 < a \leqslant 30$

b

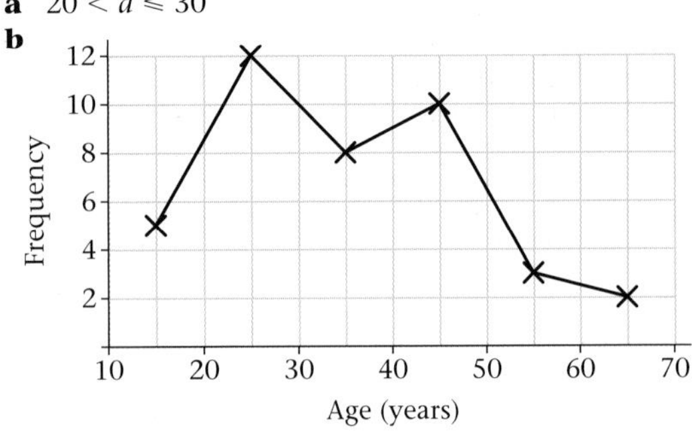

c mean age = 35

3 a

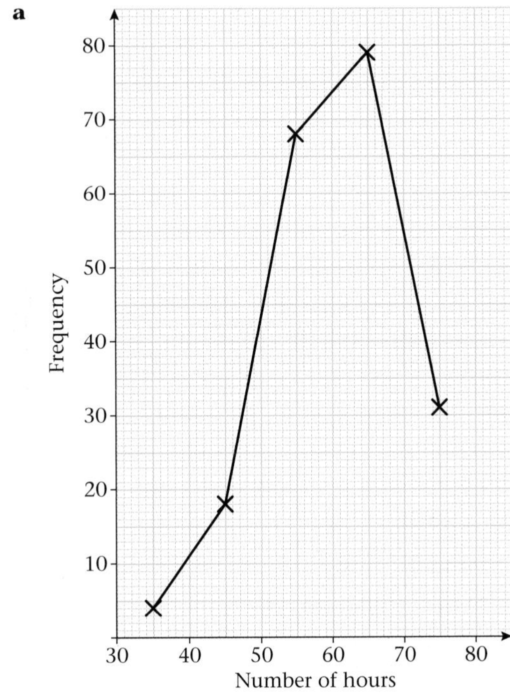

b The estimate for the mean number of hours = 60.75

c $60 < h \leqslant 70$

4 a Mean number of student absences = $13.8\dot{3}$

b

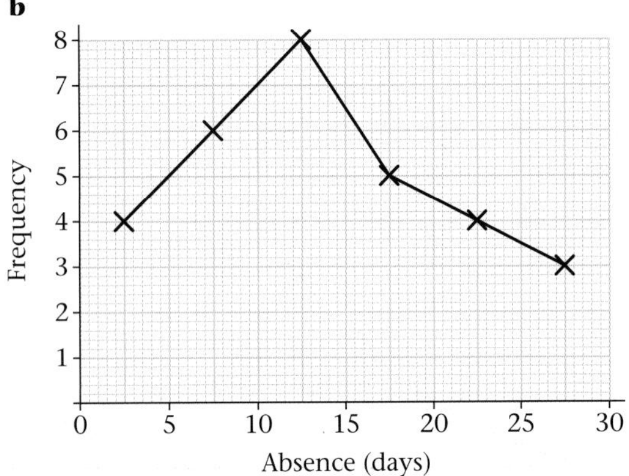

c $10 \leqslant d < 15$

Practice 23A

1 a

Duration d (seconds)	Cumulative frequency
$0 < d \leq 20$	5
$20 < d \leq 40$	15
$40 < d \leq 60$	33
$60 < d \leq 80$	45
$80 < d \leq 100$	54
$100 < d \leq 120$	57
$120 < d \leq 140$	60

b

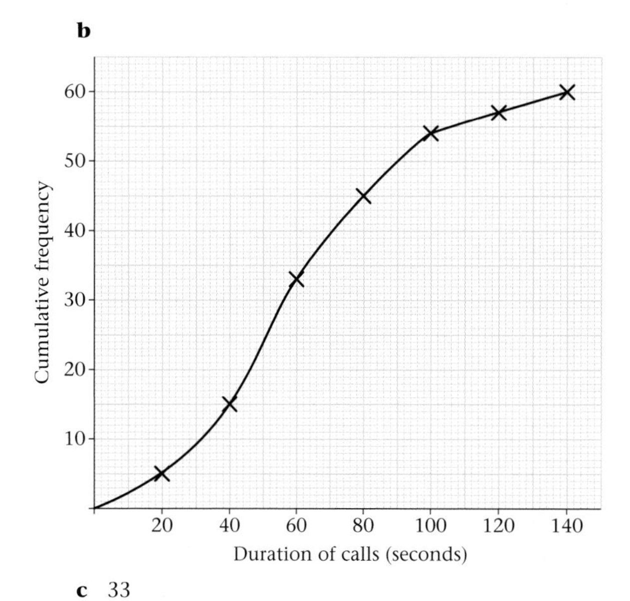

c 33

c **i** median time: 207 seconds

ii upper quartile: 238 seconds
lower quartile: 147 seconds
interquartile range: 91 seconds

iii 99 people

2 a 7.9

b

Age (t years)	Cumulative frequency
$0 < t \leq 5$	41
$5 < t \leq 10$	67
$10 < t \leq 15$	87
$15 < t \leq 20$	97
$20 < t \leq 25$	100

c

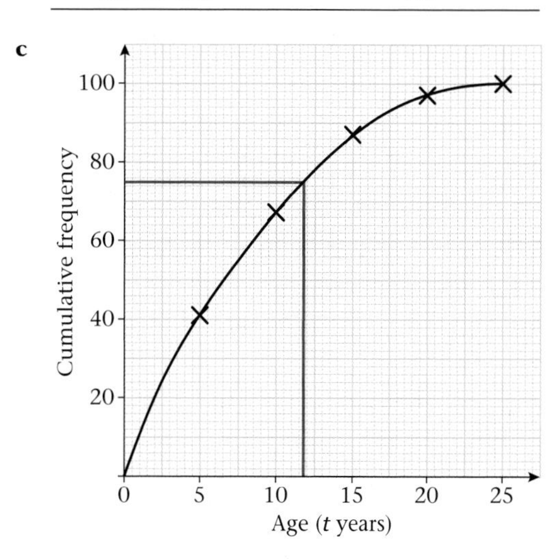

d upper quartile = 11.75 years

Practice 23B

1 a

Waiting time (t seconds)	Cumulative frequency
$0 < t \leq 50$	6
$50 < t \leq 100$	16
$100 < t \leq 150$	31
$150 < t \leq 200$	55
$200 < t \leq 250$	100
$250 < t \leq 300$	120

b

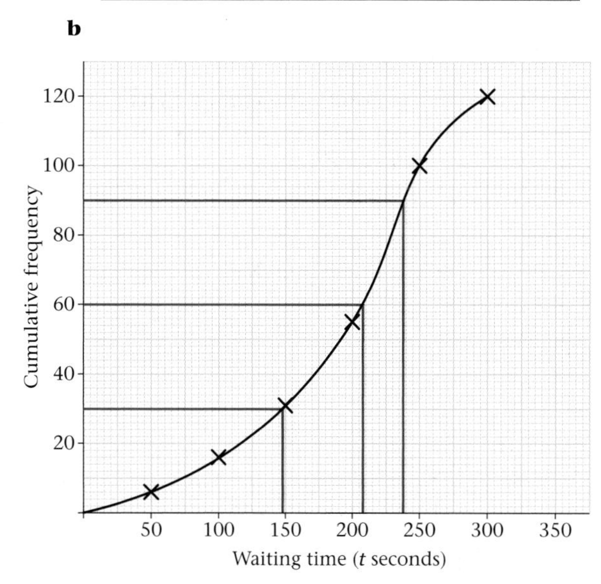

Practice 23C

1 a

0 147 207 238 300

b

0 120 190 210 300

c The median, lower quartile and upper quartile before the changes are all higher than those for the waiting times after the changes, by 17, 27, 28 seconds respectively.
The interquartile range before the changes is 91.
The interquartile range after the changes is 90.
The interquartile ranges differ by 1 second.

2 a

Mass (m kg)	Cumulative frequency
$60 < m \leq 65$	2
$65 < m \leq 70$	6
$70 < m \leq 75$	18
$75 < m \leq 80$	48
$80 < m \leq 85$	76
$85 < m \leq 90$	94
$90 < m \leq 95$	100

b

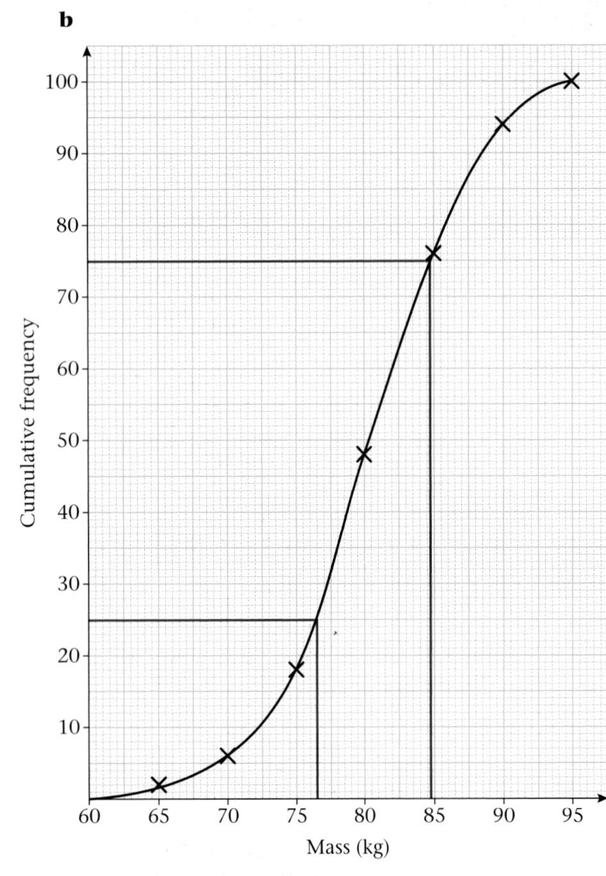

c $80 < m \leqslant 85$

d **i** 8.2 **ii** 12

e

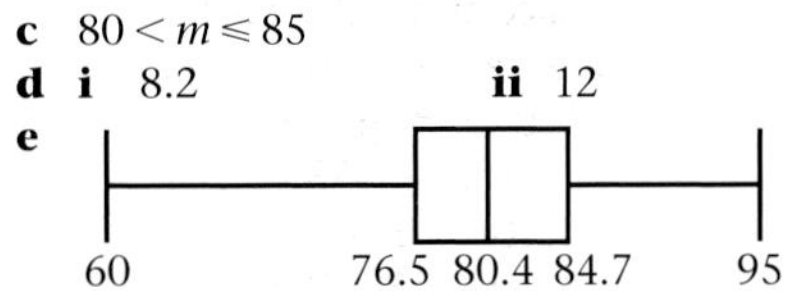

3 **a** 70 marks

b 90 students passed

Practice 23D

1 B and P A and Q D and R C and S E and T

Practice 24A

1

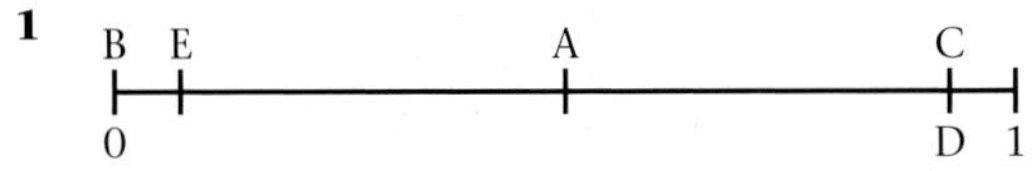

Practice 24B

1 **a** $\frac{1}{2}$ **b** $\frac{1}{6}$

c 0 **d** $\frac{1}{2}$

2 **a** $\frac{4}{12} = \frac{1}{3}$ **b** $\frac{6}{12} = \frac{1}{2}$ **c** $\frac{10}{12} = \frac{5}{6}$

d $\frac{10}{12} = \frac{5}{6}$ **e** 0

Practice 24C

1 **a** $\frac{3}{6}$ or $\frac{1}{2}$ **b** $\frac{4}{6}$ or $\frac{2}{3}$ **c** $\frac{3}{6}$ or $\frac{1}{2}$

Practice 24D

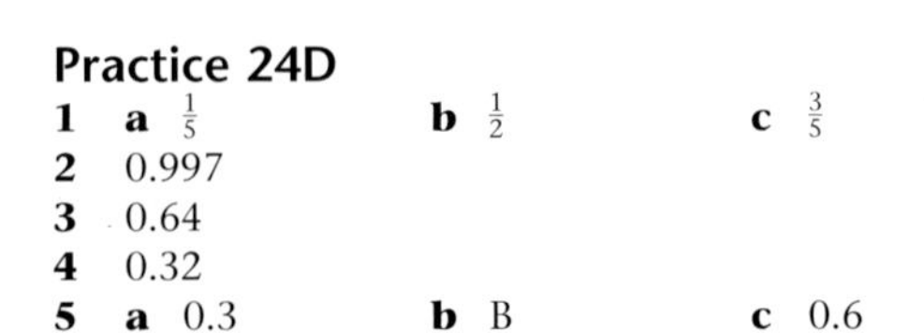

1 **a** $\frac{1}{5}$ **b** $\frac{1}{2}$ **c** $\frac{3}{5}$

2 0.997

3 0.64

4 0.32

5 **a** 0.3 **b** B **c** 0.6

6 0.15

Practice 24E

1 $\frac{12}{195} = 0.06$

2 1596

3 **a** 0.436 **b** 360

5 45

Practice 24F

1 **a** (1, B); (1, G); (1, R); (2, B); (2, G); 2, R); (3, B); (3, G); (3, R)

b $\frac{1}{9}$

2 **a** (H, 1); (H, 2); (H, 3); (H, 4); (H, 5); (H, 6); (T, 1); (T, 2); (T, 3); (T, 4); (T, 5); (T, 6)

b **i** $\frac{1}{12}$ **ii** $\frac{3}{12} = \frac{1}{4}$ **iii** $\frac{2}{12} = \frac{1}{6}$ **iv** $\frac{3}{12} = \frac{1}{4}$

3 $\frac{3}{12} = \frac{1}{4}$

Practice 24G

1 200

2 $\frac{31}{64}$ is approximately equal to 0.5

This indicates that the referee is NOT biased

3 Yes. The spinner is in favour of stopping on red (its probability = 0.51 as opposed to $0.\dot{3}$, if it was unbiased)

4 **a** If the spinner were unbiased the probability of stopping on each letter would be equal namely 0.2, which would mean out of 400 times each should have a frequency of 80. This is not the case since B's frequency is much smaller and D's is much larger, so the spinner is biased

b 0.28

Index